Darwin's Footprint

CEU Press Studies in the History of Medicine

Volume VI

Series Editor: Marius Turda

PUBLISHED IN THE SERIES:

Svetla Baloutzova
Demography and Nation
Social Legislation and Population Policy in Bulgaria, 1918–1944

Christian Promitzer · Sevasti Trubeta · Marius Turda, eds.
Health, Hygiene and Eugenics in Southeastern Europe to 1945

Francesco Cassata
Building the New Man
Eugenics, Racial Science and Genetics in Twentieth-Century Italy

Rachel E. Boaz
In Search of "Aryan Blood"
Serology in Interwar and National Socialist Germany

Richard Cleminson
Catholicism, Race and Empire
Eugenics in Portugal, 1900–1950

Darwin's Footprint

Cultural Perspectives on Evolution in Greece (1880–1930s)

Maria Zarimis

CEU PRESS

Central European University Press
Budapest—New York

Published in 2015 by
Central European University Press
An imprint of the Central European University Limited Liability Company
Nádor utca 11, H-1051 Budapest, Hungary
Tel: +36-1-327-3138 or 327-3000 · *Fax*: +36-1-327-3183
E-mail: ceupress@ceu.hu
Website: www.ceupress.com

224 West 57th Street, New York NY 10019, USA
Tel: +1-212-547-6932 · *Fax*: +1-646-557-2416
E-mail: meszarosa@ceu.hu

ISBN 978-963-386-077-9

ISSN 2079-1119

Library of Congress Cataloging-in-Publication Data

Zarimis, Maria, author.
Darwin's footprint : cultural perspectives on evolution in Greece (1880-1930s) / Maria Zarimis.
pages cm -- (CEU Press Studies in the History of Medicine ; Volume VI)
Includes bibliographical references and index.
ISBN 978-9633860779 (hardbound : alk. paper)
1. Evolution (Biology) in literature. 2. Greek literature, Modern--19th century--History and criticism. 3. Greek literature, Modern--20th century--History and criticism. 4. Xenopoulos, Gregorios, 1867-1951--Criticism and interpretation. 5. Natural selection--Philosophy. 6. Literature and science. I. Title.

PA5230.E86Z37 2014
889'.0936--dc23

2014037994

Printed in Hungary by Generál Nyomda Kft., Szeged

CONTENTS

LIST OF ILLUSTRATIONS

PREFACE

Evolutionism, and in particular Darwinism, play a major role in how we view human nature and how humanity relates to the natural world. The issues raised by Darwin and his followers are still important today and continue to be the subject of debate. My professional interest as a biomedical scientist and scholar of modern Greek would direct me in the first decade of 2000 to complete a PhD on the subject at hand—a preliminary work which provided the seed for this book, now reconfigured, much expanded, and revised.

Early conversations with academics had shown that the influence of Darwinism in various forms of Greek literature were generally viewed in terms of the loose idea of social Darwinism—that is, the struggle to survive in the name of progress, resulting in the elimination of those who were economically or physically weak. My investigations, however, would lead to a wider range of issues over which Darwinian evolutionism had challenged intellectual thinking in Greece on the traditional worldview. These issues I found were in line with what has already been well documented to be the case for other European countries and North America.

The idea that evolutionism was developed in different ways, each with its own implications, became apparent to me in the form of various themes found in the Greek works. These included themes on nature versus nurture, class, gender, race, politics, religion, national identity, language, and eugenics. In *Darwin's Footprint* I have endeavored to recontextualize the work of Greek writers, demonstrating that through these themes Darwinism did play a formative role in the Greek intellectual world in the late nineteenth century to the early decades of the twentieth century.

ACKNOWLEDGEMENTS

I am most grateful to Marius Turda for believing in my work and for initiating the idea to publish this book. I sincerely thank him for reading the various versions of my manuscript and for his suggested revisions. His support and guidance throughout the entire process have been crucial to the completion of this work. Further to this, I would very much like to thank the CEU Publishers for accepting the book for publication. I am greatly appreciative of the editor at CEU Press, Szabolcs László, for his ongoing support, advice, and for his patience during the process from manuscript to book. In addition, I would like to recognize the contribution of the copy editor at CEU Press, Adela Hîncu who, with much toil, has managed to transform my basic writing to a higher level. I am most grateful to the reviewers of the first submitted version of the manuscript, whose positive comments and insights have spurred me on to complete this work.

I would like to acknowledge the financial support which I received at the University of New South Wales in the very early stages of my work as a PhD research student; this was in the form of the Australian Post-Graduate Award (APA) Scholarship, and later in the form of a grant for Early Career Researchers. I am very grateful to the library of the University of New South Wales for having purchased the complete digitized collection of the magazine *Η Διάπλασις των Παίδων* (1879–1948), a collection which was critical to my research. I thank the excellent interlibrary loan service of this library, which made my research investigations so much easier.

I wish to thank the following individuals and organizations for their support, inspiration, and encouragement, and for having been there at various stages of the research associated with this book: Peter Bien, Vicky Doulaveras, Antonis Dracopoulos, John Gascoigne, Roslyn Haynes, Costas Krimbas, Antonis Liakos, Martyn Lyons, David Oldroyd, Olaf Reinhardt, Alfred Vincent, and Anna-Maria Vlassi-Galerou. Thank you also to the helpful staff at the Kostis Palamas Institute, Athens; to Varvara Tsaka at the Nikos Kazantzakis Museum, Crete; to Kazantzakis Publications, Athens, and its staff member Nikos Mathioudakis.

A very special thanks goes to Eleni Molfessi, Head Librarian of the Institute of Historical Research at the National Hellenic Research Foundation, Athens. Her tireless ongoing support and assistance were instrumental in the completion of this research.

Finally, I sincerely acknowledge the unending support of my husband Tom, who has always stood by me through this venture. To my children, Michael, Liana, and Nicola, I am thankful for your understanding and patience while I was writing this book. I dedicate this work to the loving memory of my parents, who passed on to me the passion to enquire.

NOTE ON TRANSLITERATIONS

The titles of publications in Greek and other languages, when first mentioned in the main text, are followed by the English translation. In the bibliography and footnotes, these titles have been kept in their original language; in the bibliography, they are followed by an English translation.

In the bibliography, where a non-English title of a book or article is followed by an English translation, the translation is capitalized sentence-style. Elsewhere, where a non-English book or article is mentioned that has not been published in English, then the translated English title of the book or article is also capitalized sentence-style; and it is italicized as well. An example of this is the Greek novel *Rich and poor*; this is used rather than *Rich and Poor*. For all books and articles published in English, the titles are always capitalized headline-style and italicized. An example is *The Descent of Man*.

Names of authors and publishers have been mainly transliterated, except where they have been translated, e.g., the name George rather than Giorgios.

With the transliterations from Greek to English I have used a phonetic principle, although it is difficult to be absolutely consistent. Some proper nouns, such as names of individuals and places, have existing English versions which I have used, and which do not conform to the phonetic principle. An example of this is Petros Charis rather than Petros Haris (*Πέτρος Χάρης*).

ΦΙΛΟΣΟΦΙΚΗ ΚΑΙ ΚΟΙΝΩΝΙΟΛΟΓΙΚΗ
ΒΙΒΛΙΟΘΗΚΗ ΦΕΞΗ

CH. DARWIN

ΠΕΡΙ ΤΗΣ

ΓΕΝΕΣΕΩΣ

ΤΩΝ ΕΙΔΩΝ

ΜΕΤΑΦΡΑΣΙΣ
Ν. ΚΑΖΑΝΤΖΑΚΗ

ΕΝ ΑΘΗΝΑΙΣ
ΕΚΔΟΤΙΚΟΣ ΟΙΚΟΣ ΓΕΩΡΓΙΟΥ Δ. ΦΕΞΗ
1915

First Greek translation of *On the Origin of Species*, trans. Nikos Kazantzakis, 1915.
Courtesy of the Nikos Kazantzakis Museum, Myrtia, Greece.

INTRODUCTION

[Darwinism] unifies in a stroke these two completely disparate worlds, until then, of the meaningless mechanical physical sciences, astronomy, physics and chemistry on the one side, and the world of meaning, culture, art and of course the world of biology. One stroke shows how to unify all the sciences.
DANIEL DENNETT[1]

The main objective of this book is to unveil how Darwinism formed a part of the Greek intellectual and cultural life during the late nineteenth and early twentieth centuries. What will unfold are the informative, intriguing, and often controversial details drawn from the writings of prominent literary figures, historians of science, scientists, politicians, feminists, and other intellectuals from various academic disciplines. Most observations are derived from the literary world, focusing particularly on the important writer Grigorios Xenopoulos (1867–1951) and placing him and the other Greek intellectuals within the context of the international scene.

Internationally, the ideas and concepts that arose from Darwinism and other associated evolutionary theories exhibited a multitude of themes; some of which contradicted each other depending on the perspective taken. Apart from Darwin's biological theories, I will be paying particular attention to how his theories influenced various post-Darwinian[2] movements such as eugenics; how his theories played a role in formulating new thoughts about religion, class, gender, race, and sexuality; and how these theories ultimately contributed to the intellectual thinking in Greek society between 1880 and the 1930s.

1 Daniel Dennett quoted in Bragg, *On Giants' Shoulders*, 160.

2 I define post-Darwinian thought as those theories, concepts, and ideas which were strongly influenced in their formation by any of Darwin's theories. It is a term commonly used by historians of science.

No other evolutionary theories have been absorbed so readily and pervasively as the theories of the British naturalist Charles Darwin (1809–1882). I have specifically chosen to include both "noncreative" and creative writing in this book because what is evident with the scientific and literary writing of the nineteenth and early twentieth centuries is that there was—and indeed often still is—a fine line between science and the imagination. Charles Darwin's *Origin of Species* is a clear example of a factual scientific work which is also highly valued as a creative work of literature.

Literary responses to Darwinism have been abundant and everchanging since its beginnings with *The Origin of Species* (1859), *The Descent of Man* (1871), and *The Expression of the Emotions in Man and Animals* (1872).[3] Greece, however, is an exception to this general pattern found throughout the Western world. Even a cursory reading of any number of books or periodicals in the literary field of modern Greek scholarship shows that a detailed discussion of the topic of Darwinism in modern Greek culture is surprisingly lacking.[4]

Standard histories of modern Greek literature omit the influence of Darwinian thought as an entity on its own. Although they do discuss naturalism—in which Darwinism played a formative role together with many other works in the sciences, such as Claude Bernard's *Introduction to the Study of Medicine* (1865)—this role is never brought to light. In fact, Darwin's name is never mentioned directly in connection with Greek literature. I found this to be the case for the following key histories: Roderick Beaton's *Introduction to Modern Greek Literature* (1994), P. D. Mastrodimitris' *Introduction to modern Greek literature* (*Εισαγωγή στη νεοελληνική φιλολογία*) (1974), Linos Politis' *A History of Modern Greek Literature* (1973), Mario Vitti's *History of modern Greek literature* (*Ιστορία της νεοελληνικής λογοτεχνίας*) (1978), K. Dimaras' *History of modern Greek literature* (*Ιστορία της νεοελληνικής λογοτεχνίας*) (1948), and Ilias Voutieridis' *Short*

3 All references in this book to Darwin, *On the Origin of Species* (hereafter *OS*) are to the following edition, which contains the text of the 1859 first edition, as well as "An Historical Sketch" and the Glossary, added by Darwin later: John Burrow, ed., *The Origin of Species by Means of Natural Selection, or the Preservation of Favoured Races in the Struggle for Life* (Harmondsworth: Penguin Group, 1968). All references to Darwin, *The Descent of Man* (hereafter *DM*) are to the facsimile photoreproduction of the 1871 first edition: Charles Darwin, *The Descent of Man, and Selection in Relation to Sex*, 2 vols. (Princeton NJ: Princeton University Press, 1981). All references to Darwin, *The Expression of the Emotions in Man and Animals* (hereafter *EE*) are to the following edition: Charles Darwin, *The Expression of the Emotions in Man and Animals* (Chicago: University of Chicago Press, 1969). Other works by Darwin have influenced the literary mind, though to a much lesser extent, such as *The Variation of Animals and Plants* (1868).

4 Politi, "Δαρβινικό κείμενο και *Η φόνισσα*," 157–158.

history of modern Greek literature (*Σύντομη ιστορία της νεοελληνικής λογοτεχνίας*) (1933).[5] It should also be noted that numerous foreign literary influences were recorded by Greek literary historians; for example, Voutieridis comments on Freudianism, Nietzscheism, and other "isms," but not Darwinism.[6]

Notably though, historian Gianis Kordatos (1891–1961) in his *Great history of Greece* (*Μεγάλη ιστορία της Ελλάδας*) does include Darwinism as one of the theories which had a cultural influence in shaping intellectual debates in Greece around 1880:

> So within the intellectual field many things were brewing, and ideas were expressed which fought religious mania and monastic charlatanism. Every now and then learned individuals would come from European universities and they would openly or secretly fight to free the minds of the youth from every kind of superstition, spreading the new pedagogical, philosophical, and sociological ideas. The university opened the road scientifically preparing the youth, and from the university chair could be heard the past momentous scientific truths which, as would be expected, scandalized the upper clergy, the musty minds and the reactionaries. ... Darwinism and the theories of Galileo, Copernicus, and the other European learned and naturalists started to waken the minds of the youth.[7]

Historian Efthymios Nicolaidis notes that concerning the presence of Darwinism in the histories of science in Greece, it was only in the decade of the 1980s that a professional community of historians of science was formed.[8] To add to this, these historians rarely dealt with Greek topics.[9] Any Greek content that existed in the historiographies of science at the time was focused

5 See Beaton, *Introduction to Modern Greek Literature*; Mastrodimitris, *Εισαγωγή στη νεοελληνική φιλολογία*; Vitti, *Ιστορία της νεοελληνικής λογοτεχνίας*; Voutieridis, *Σύντομη ιστορία της νεοελληνικής λογοτεχνίας*. Reference to Darwin's 1859 *OS* occurs in Dimaras, *Ιστορία της νεοελληνικής λογοτεχνίας*, 648, in a chronological table under the heading Γενική Παιδεία (General Culture). The other headings are Γενική Ιστορία (General History), Ελληνική Ιστορία (Greek History), and Ελληνική Παιδεία (Greek Culture). These timelines cover the world and Greek history and culture of the period 1453–1940. A similar chronological table is found in L. Politis, *History of Modern Greek Literature*, 291, which also notes Darwin's *OS*. The table dates range from the ninth century AD to 1971.

6 Voutieridis, *Σύντομη ιστορία*, 231–255.

7 Kordatos, *Μεγάλη ιστορία της Ελλάδας*, 494. Unless otherwise noted, all translations from Greek to English are my own.

8 Nicolaidis, "Ιστοριογραφία των επιστημών," 536.

9 Ibid. According to Nicolaidis, this was due to the trend for those working in the natural sciences to focus on the international aspects rather than the local, and also due to the current Anglo-Saxon historiographic school, which Nicolaidis maintains is moving away from the study of primary sources and is oriented towards sociological approaches.

on the Ottoman period and to a lesser degree on the ancient and Byzantine periods.[10] Specifically related to the history of Darwinism, evolutionary biologist, geneticist, and historian of science Costas Krimbas became interested in the early 1980s in writing the Greek history of the theory of evolution.[11] His approach to the reception of Darwinism in Greece has provided a contextual framework for this book. Since the 1990s, historians of science have been collating primary sources on the history of evolutionary science, but apart from Krimbas' work, information on the history of Darwinism in Greece remains scarce.[12]

Even though this book deals mainly with Darwinian and post-Darwinian thought, I will also be referring to modern evolutionary ideas in general. Evolutionary theories were not new in Darwin's time. Prior to Darwin's theories, they had been the subject of speculation for half a century, but his publication of the *OS* and the *DM* marked a major change in the way in which humanity viewed itself with respect to the natural world and to the cosmos. Darwinism placed humanity in nature. From the nineteenth century to the present, "man's place in nature"[13] has been the subject of an ongoing debate. The way we interpret Darwinism plays a major role in the way we decide to place nature in our culture.

Darwin's influence then permeated many other fields, including the social sciences (such as criminology, psychology, and political science), philosophy, and the arts (such as literature and the visual arts). In the social sciences and in literature Darwinism and evolutionism were used to interpret "man's place in nature" and aspects of class, gender, and race, which were major issues in the late nineteenth and early twentieth centuries.[14]

10 Ibid.

11 Ibid.

12 See the bibliographic references to the articles by Mihail Stefanidis on the history of the natural sciences in Karas, *Ιστορία των επιστημών*, 56.

13 The phrase "man's place in nature" is a concept which was derived from the nineteenth-century debate of the same name and was adopted by academics as a theme in the discourse on the implications of Darwinian evolutionary thought. It dealt with the Darwinian application of the theory of common descent to humanity, which deprived "man" of his former unique position. English biologist and Darwinian advocate Thomas Huxley (1825–1895) used it as the title of his book *Man's Place in Nature*. When referring to this phrase, I have chosen not to change it to "humanity's place in nature," as this loses the background associated with the original phrase.

14 For further information on the general impact of Darwinism in the late nineteenth and early twentieth centuries see Oldroyd, *Darwinian Impacts*; Oldroyd and Langham, *Wider Domain of Evolutionary Thought*; Greene, *Darwin and the Modern World View*; Bratchell, *Impact of Darwinism*; Degler, *In Search of Human Nature*; Greene, *Science, Ideology, and World View*.

In Europe and America controversial literary treatments of Darwinian issues proliferated in the nineteenth century and continued in various forms into the twentieth. In Greece, too, the impact of Darwinism was significant enough to exert considerable influence on modern Greek literature and culture. Indeed, modern Greek prose fiction and other genres of the late nineteenth to early twentieth centuries show abundant evidence of Darwinian and other evolutionary influence. This will be examined further in chapter 2, in the work of several modern Greek writers.

Further to my argument concerning the lack of scholarship on Darwinism in Greece, I note in particular that there is no significant attempt to integrate the work of Grigorios Xenopoulos within its Darwinian framework. To this effect, this book addresses a significant gap in modern Greek literary scholarship by examining Xenopoulos' responses to Darwinian and post-Darwinian theories in a selection of his prose fiction, essays, and letters, in the form of themes, issues, imagery, and motifs, from 1900 to around 1930. In particular, I endeavor to demonstrate that Xenopoulos—probably more so than any of his contemporaries—interweaves predominantly Darwinian and other evolutionary theories with his study of society. He shares this technique with several Greek, other European, and American writers, in naturalist and then realist literature. Like many intellectuals of his time who were well-read in the sciences and Darwinism, he responds to issues of religion and "man's place in nature" in his writings. I attempt to place Xenopoulos in the context of these other writers by also selectively examining some of their work. In this book I view the research on Xenopoulos, along with the Darwinian exploration of the work of other Greek writers, such as Emmanuel Roidis, Kostis Palamas, Kalliroi Parren, and Alexandros Papadiamantis, as a stepping stone towards a more unified and comprehensive study of the influence of Darwinism and evolutionism in modern Greek literature—a study which is long overdue.

Charles Darwin's 2009 double anniversary (of his birth and *OS* publication) along with major ongoing world developments in genetics have accelerated scholarship by academia and the public arena on Darwinian-related matters, such as its teaching in schools. It is timely, then, to consider Darwin's impact on Greek literature and culture.

There are several reasons why I have chosen Xenopoulos as the main writer through which to explore the influence of Darwinism in Greece.

While he had turned to literature later in his life, his university background was science-oriented, studying at the Physical and Mathematical Sciences School at the University of Athens. The academics and students of this department had played an integral part in supporting Darwinism during the controversy of the 1880s, as will be dealt with later.

Xenopoulos documented that he had read most of Darwin's work. According to Darwinian literary commentator Gillian Beer, there is a difference between the writer known to have read Darwin himself and the writer who may have read about him from secondary sources. To illustrate this, Beer argues that due to the enormous influence of Darwin and those influenced by him (post-Darwinism), "everyone found themselves living in a Darwinian world in which old assumptions had ceased to *be* assumptions, could be at best beliefs, or myths, or, at worst, detritus of the past." In her explanation, Beer uses as an analogy our post-Freudian world, where we are affected by Freud's ideas "even if we have not read a word of Freud, even—to take the case to its extreme—if we have no Freudian terms in either our active or passive vocabulary." She continues to argue that Freud's ideas "have been so far institutionalized that even those who query his views, or distrust them, find themselves unable to create a world cleansed of the Freudian." In a similar manner, Darwinian ubiquity gave way to a plethora of writers who were affected by Darwin's theories and, as Beer suggests, it would allow one to find certain Darwinian elements "in almost anyone [she] chose." Although she does not consider this "an improper enterprise," she maintains that it is "insufficient ... because it does not take account of the [writer's] *act of reading* and reaction." She asserts that "ideas pass more rapidly into the state of assumptions when they are *unread*," whereas reading produces questions. So in her study, limiting her choice to the writers who are known to have read Darwin (Beer also indicates Lyell, Spencer, and Huxley for her analysis), she tracks "the difficult flux of excitement, rebuttal, disconfirmation, pursuit, forgetfulness, and analogy-making, which together make up something of the process of assimilation."[15]

Xenopoulos, then, was not one of those who imbibed Darwin from secondary sources, rather he felt that it was crucial to read Darwin thoroughly

15 Beer, *Darwin's Plots*, 3–4 (italics in the original). See also Zarimis, "Darwinian Reception and Selected Responses," 656.

at the primary source. He wanted to avoid the anecdotal misinformation circulating at the time and to be able to comment responsibly on Darwin's theories. In chapter 3 of this book an excerpt from the "Athenian Letter" entitled "Things are serious" will be discussed, where Xenopoulos wrote about this issue.

Xenopoulos was writing over a long period of post-Darwinian influence. As a prolific writer and an intellectual, he published over eighty novels between 1891 and 1945. In addition to published numerous short stories, plays, and essays, he wrote in a children's magazine and was also a literary critic. I have chosen to examine selected work from 1900 to around 1930 because, as will be seen in chapter 3, Xenopoulos was actively writing about Darwinism, science and religion, and other associated issues in the magazine during those years. This work will be examined together with a series of novels and essays which he wrote in that time.

As will be seen in the following chapters, non-Greek Darwinian literary scholarship appears to find less to say about creative works published after 1930. It seems that by about 1930 Darwinian ideas were not as popular in literature as before, and that general interest in the biological perspective on man and society dissipated with the horrors perpetrated by Nazi Germany during World War II. This is despite the fact that in science by 1940 the new "synthetic theory" combining Mendel's inheritance theory and Darwinian theory reinforced the idea that natural selection was the mechanism for evolution. Later scientific discoveries, such as Watson and Crick's construction of a model of the complex DNA molecule and later the mapping of the human gene, brought to light the major role Darwinian thought plays in the nature versus nurture debate,[16] and also confirmed the evolutionists' claim that humans share a common ancestry with animals.

Xenopoulos' novels that I have chosen here clearly reveal a post-Darwinian or other evolutionary lexicon dealing with Darwinian concepts.[17] In Xenopoulos' *Rich and poor* (*Πλούσιοι και φτωχοί*) (1919), on a number

16 According to neo-Darwinism, characteristics which individuals acquire during their life cannot be passed on to their offspring; in other words, nurtured qualities cannot be inherited as would be expected under Lamarckian conditions. Equally, genetics has and continues to determine whether a particular human quality is genetic or nurtured. Hence genetics and Darwinism play a critical role in the nature-nurture debate.

17 Note that only two of his works mentioned in this study have published English translations: *The Three-Sided Woman* (1992) and *The Downfall* (1992).

of occasions he mentions natural selection and Darwin's theory of evolution, which he associates with the social classes and socialism. A eugenic approach is revealed in his dark novel *The night of degeneration* (*Η νύχτα του εκφυλισμού*) (1926), with disturbing reference to natural selection and sexual selection. His novel *Tereza Varma-Dacosta* (*Τερέζα Βάρμα-Δακόστα*) (1926) exhibits clear motifs of transformation when the protagonist, Tereza, appears to "mutate." Xenopoulos' representation of Tereza appears to be of the New Woman and has Darwinian undertones. His *Three-Sided Woman* (*Τρίμορφη γυναίκα*) (1922) presents a different version of the degenerate New Woman. Both latter novels reflect his interest in the Women's Movement in Greece. I examine also the "Athenian Letters" ("Αθηναϊκαί Επιστολαί") that he wrote in the popular children's magazine *Children's Guidance* (*Η Διάπλασις των Παίδων*) (1896–1947). A cluster of "Letters" dealing with Darwinism within the period from around 1900 to the 1930s reveals his use of Darwinian influences and motifs. I will be expanding on the fact that these "Letters" played a significant role in educating the general Greek society at a time when there was no other source accessible to them. Also, some of these "Letters" play a crucial role alongside the analysis of his novels by way of themes in common.

It is important to note at this point that Xenopoulos' views on Darwinism or any other ideological concepts cannot, of course, be deduced from his prose fiction. However, in his autobiographical novel *My life as a novel* (*Η ζωή μου σαν μυθιστόρημα*), first published in 1939 in the later part of his writing career, he admits to being more open, particularly regarding his early "philosophical crisis" and his views on the sexual instinct. This novel gives a more accurate reflection of his views. In contrast, his semi-autobiographical *Rich and poor* and *Kosmakis* (*Ο Κοσμάκης*) (1923) need to be carefully considered before making any judgment on his overall ideology, even though he states in *My life as a novel* that his own philosophical crisis was the same as that of his protagonist in *Kosmakis 1: The first awakening.*[18] As will be shown in chapter 3, it seems that towards the end of his life Xenopoulos was more open in expressing his views.

He was extremely aware in his "Athenian Letters" of the image and views that he wanted to project. As a writer who lived from his profession, he avoided offending his readers or the establishment, and attempted to

18 Xenopoulos, *Η ζωή μου*, 136.

maintain a "moderate" view on issues such as politics and religion. For this reason, generally speaking, he manipulated the account of his own views, approaches, and experiences in his works. Therefore it is difficult to say that the view expounded in a piece of writing was what he believed at the time and, as will be demonstrated throughout this book, each work needs to be assessed bearing this in mind.

Finally, Xenopoulos has tended not to be taken seriously by literary historians and commentators, as is often the case with very prolific writers. Beaton notes that like Papadiamantis, he was a professional writer whose work was criticized for "sloppiness and repetition."[19] He also states that in "Xenopoulos' case longstanding popularity has stood in the way of critical reevaluation."[20] Beaton maintains that with the reissuing since 1984 of his novels in paperback, there may be a revival of literary commentary.[21] However, as this study will show, some of Xenopoulos' better known novels have not been reevaluated, and a significant number of his novels have only had minimal reviews; some originally published in serial form, such as *The night of degeneration* and *From the kitchen to the harem* (*Απ' την κουζίνα στο χαρέμι*) (1923), were only published in book form by Vlassis Brothers in 1994.

In general with Xenopoulos' work, there has tended to be a lack of study, rather than a lack of work to study. The broad spectrum of his work, from novels, plays, literary criticism, to his children's magazine, spanned over fifty years. Not only did he contribute to the more serious type of literature, but also to semi-popular literature. This has created the perception that his work sits between these two extremities, with neither field of literary commentators taking on his work systematically and comprehensively.[22] He was able to channel his work to reach a large proportion of the literate Greek population in one way or another. He read widely, passing his knowledge to his readers. Moreover, through his work he contributed significantly to popular culture.

19 Beaton, *Introduction to Modern Greek Literature*, 102.
20 Ibid.
21 Ibid.
22 There are few exceptions, such as the work by Georgia Farinou-Malamatari and Konstantinos Malafantis, which will be referred to later in this book.

GENERAL RECEPTION OF DARWINISM IN GREECE

In this section I examine the reasons for the interest in Darwinism by only a restricted group of Greek writers, and also the lack of interest by many Greek commentators. The reception of Darwinism, known as the Darwinian revolution, was felt widely throughout the world.[23] This section can only attempt to achieve some understanding of what issues affected the response of Greek writers to Darwinism. In addition, it may also provide some understanding of why literary critical discourse in the area is not developed.[24]

Costas Krimbas has written extensively on Darwinism, its history, and relevant to this study, on the reception of Darwinism in Greece. His 2009 magnum opus *Darwinism and its history until our present* (*Δαρβινισμός και η ιστορία του έως τις μέρες μας*) is an international and multidisciplinary work which follows the history of Darwinism and associated evolutionary theories and scientific findings, being a major contribution to the scholarship in Greece.[25] In addition, his authoritative 1993 essay "Darwinism in Greece: The first steps" is frequently referred to by academics.[26] The essay highlights certain points which are relevant to this study; so I deem it necessary to note them here, making some extrapolations. Most significant is his statement that "Darwinism constituted one of the 'external' ideological movements which entered our country, in contrast with those which were born and formed to respond to deeper 'internal' needs; the 'internal' ideological currents, such as the views on the continuity of the Greek nation and the three stages of its history, the Great Idea, the language restoration, etc."[27] The implications of this are that for at least up to 1930 writers and commen-

23 For the reception of Darwinism see Ellegård, *Darwin and the General Reader*. For the comparative reception see Engels and Glick, *Reception of Charles Darwin in Europe*. This book examines seventeen European countries, including England, but does not include Greece. Further to this, for the reception in England, Scotland, Germany, France, the United States, Russia, the Netherlands, Spain, Mexico, and the Islamic world see Glick, *Comparative Reception of Darwinism*. For Italy see Pancaldi, *Darwin in Italy*. For the reception in Belfast, Edinburgh, Australia, New Zealand, Canada, and America see Numbers and Stenhouse, *Disseminating Darwinism*.

24 A brief account of the broader reception of Darwinism is found in Zarimis, "Darwinian Reception and Selected Responses."

25 Krimbas, *Δαρβινισμός και η ιστορία του*.

26 Krimbas, "Ο δαρβινισμός στην Ελλάδα," 81–108. This essay is a revised version and is the one used in this study. For earlier versions see *Τα Ιστορικά*, no. 2 (1984): 335–348; *Materia Medica Greca* 10, no. 5 (1982): 465–471. Further on Darwinism and also on its history in Greece see Krimbas, *Δαρβινικά* and Krimbas, *Εκτείνοντας τον δαρβινισμό και άλλα δοκίμια*.

27 Krimbas, "Ο δαρβινισμός στην Ελλάδα," 82.

tators tended to draw their intellectual inspiration from the forementioned national issues. Needless to say, there were those Greeks who wrote about the broader political and philosophical issues. Further to Krimbas' observation, this study will investigate those who did appropriate a Darwinian and post-Darwinian approach of various sorts in their discourse on these "internal" matters. Other such associated matters include Greece's general economic and social status.

Darwinism as an entity on its own (separate from the naturalist school) did not change the general course of Greek culture in the period under consideration. My investigations show that it was writers such as Xenopoulos, who had studied in the sciences or medicine, or who had read Darwin's works for themselves, who were keen to appropriate Darwinian and post-Darwinian thought to their work for whatever purpose.

Interestingly, other imported intellectual trends that Krimbas mentions, such as Nietzscheism and Bergsonianism, were absorbed more readily into the Greek literary world. The reason for this may be that these two trends had a strong metaphysical basis, unlike Darwinism, which was based on the natural sciences. Greece's general lack of concern with Darwinian thought was perhaps associated with a more general lack of interest in the natural sciences.[28]

Krimbas records that in the late nineteenth century, Darwinian ideas, like many ideas and theories, were received in Greece chiefly from Germany and more specifically through Ernst Haeckel (1834–1919).[29] Haeckel propounded his theory of "ontogeny recapitulates phylogeny," but also viewed Darwin's theory of evolution as including Lamarckian elements.[30] Many Greek intellectuals, including George Skliros, Alexandros Delmouzos, and George Papanicolaou (of pap-test fame) were students of his and followed his lectures.[31] In Greece Darwin's theories were also received through France, and many of the educated Greeks would have read the German or French translations. Krimbas points out that Xenopoulos, when discussing the first Greek performance of Ibsen's *Ghosts* (*Οι βρυκόλακες*) in 1894,

28 Krimbas details problems associated with the lack of scientific interest and research in Greece. See Krimbas, "Η έρευνα στην Ελλάδα: 1977," 225–239.

29 Krimbas, "Ο δαρβινισμός στην Ελλάδα," 83.

30 Krimbas, *Δαρβινισμός και η ιστορία του*, 128.

31 Ibid.

stated regarding Greece: "in terms of literature we constitute a province of France."[32]

Those who played a crucial role in the reception of Darwinian ideas were either German- or French-educated, or of German background. These included French-educated botanist and poet Theodoros Orfanidis (1817–1886), medically trained Spyridon Miliarakis (1852–1919), who also trained in Germany as a botanist, and the German Theodor von Heldreich (1822–1902), the Hellenized botanist and poet who wrote to Darwin in French. In 1878, in one such letter Heldreich wrote on the reception of Darwinism in Greece at the time, disclosing that Darwin's fervent supporters were still quite rare in Greece.[33] He also revealed: "It is not free from some danger and it requires some courage of one's convictions to acknowledge and accept the principles of your work in this country, where we still find ourselves under the rule of dogmatism. Minds must gradually be prepared with caution; nevertheless the Truth will triumph even here, and we must hope that day will not be too far away."[34]

In 1895, Petros Apostolidis (1866–1937), doctor and literary writer (pseudonym Pavlos Nirvanas), wrote an article "Science and the Arts" ("Επιστήμη και Τέχνη"). Apart from the general negative reception, he noted the response of many Greeks in the literary world:

> I still remember the terrible scandal at our university which the exposition of the Darwinian theory generated a few years ago. The professor of medicine, who was at risk by introducing the law of evolution to the Greek students, was not only in danger of being excommunicated by the Holy Synod, like the other innocent author of *Pope Joan*, but to his surprise, it is said, in those days he found the doors of most of his customers closed to him. Don't think however that this social flag-waving was supported by religious fanatics, monks, and little old ladies. The sound of the rebel sermon had not reached them. The delicate world of the day-dreamers and the hypersensitive constituted the crusade's lead—those who saw the poetry of life extinguished by the breath of the scientific geniuses.[35]

32 See further Xenopoulos, "Οι Βρυκόλακες," 360.

33 Krimbas, "Ο δαρβινισμός στην Ελλάδα," 90. This article contains Heldreich's other letters and Darwin's replies.

34 Ibid.

35 Apostolidis, "Επιστήμη και Τέχνη," 25.

In the cited passage Apostolidis is referring to Emmanuel Roidis (1836–1904), who wrote the satirical masterpiece *Pope Joan* (*Η Πάπισσα Ιωάννα*), published in 1866; the novel was condemned by the Greek Orthodox Church for its irreverent themes on the Church. The pro-Darwinian Apostolidis was very well aware of, and probably associated with the Darwinian controversy—that is, the disputes between the academics of philosophy, medicine, and theology at the University of Athens during the 1880s. In this passage it is most likely that the professor of medicine he is referring to is Ioannis Zochios, who will be discussed later. Apostolidis also reveals that Darwinism had not been received yet by the broader public and that it was Greek intellectuals in the literary world who rejected Darwin's theory. This last point will be further discussed later in this chapter.

Before World War I, initiatives to establish new universities apart from the one in Athens were postponed most likely due to Greece's preoccupation to execute the "Great Idea."[36] Nonetheless, Krimbas reveals that by 1879 Darwinism in Greece had already had an impact not just on general biology but also more specifically in the branches of zoology, phytology, anatomy, and embryology, as well as in anthropology, ethnology, philosophy, and psychology.[37] Darwin's theories were somewhat delayed in coming to Greece from German and French sources, and it was not until 1915, over half a century after the original English publication, that the *OS*, translated by Nikos Kazantzakis, was able to be read in Greek, albeit in "katharevousa," a formal learned form of the language.[38] Kazantzakis translated the *OS* from the English edition, most likely from the sixth and final one.[39] Greece was one of the last Western countries to have the book in translation.[40] Prior to this, those who wanted to read about Darwin and who could not read French or German were obliged to rely on the translations of essays, excerpts from

36 Nicolaidis, *Science and Eastern Orthodoxy*, 175.

37 Krimbas, "Ο δαρβινισμός στην Ελλάδα," 92.

38 See Darwin, *Περί της γενέσεως των ειδών*, trans. Nikos Kazantzakis. A second translation, authored by Andreas Pangalos, was published in 1956, with a second edition circa 1974. In 1997, a translation by the Biology Department of the University of Patras was also published, with a prologue by Prof. Stamatis Alahiotis.

39 For the circumstances surrounding the source of Kazantzakis' translation see Prevelakis, *Τετρακόσια γράμματα*, 520. Also, further on the reasoning for the sixth edition see Zarimis, "Darwinian Reception and Selected Responses," 650.

40 Krimbas notes that Germany and France had it in 1862; Holland, Italy, and Russia in 1869; Denmark, Hungary, and Poland in 1873; Spain, Serbia, and Japan in 1896; China in 1903; it was translated into Czech and Lithuanian languages in 1914. Krimbas, "Ο δαρβινισμός στην Ελλάδα," 101.

biographies of Darwin, and summaries of his theories published in Greek newspapers and periodicals.[41] The first complete Greek translations of the *DM* were published in 1976 and 1977, over a century after the original English publication in 1871.[42] Based on these observations, Krimbas argues that "these show the limited interest in Greece in the natural sciences and more generally in research on nature, the numerically limited Greek middle class, and the almost nonexistent scientific community."[43] As will be seen in chapter 3, Xenopoulos discussed the lack of interest of the sciences in Greece and showed his great interest in nature in his "Athenian Letters" in the *Children's Guidance* magazine. Similar comments could be made regarding a general lack of interest by Greek creative writers and also Greek literary critics. In his final paragraph and footnote, Krimbas maintains that Darwinism and other scientific ideas had been received passively by the Greek scientific world, and that in the case of Darwinism Greek botanists accepted and mimicked the Western prototype without challenging these ideas or producing further original ideas.[44]

Due to the delay with which complete Greek translations of Darwin's key works the *OS* and the *DM* appeared, his work was communicated to a Greek readership by means of brief publications in periodicals, journals, and magazines. Information arriving on Darwinism was often disjointed. To add to this, there were no specialized scientific journals in Greece in the nineteenth century. According to George Vlahakis, such weekly periodicals as *Estia* (*Εστία*) (1876–1895) "were the only channels of dialogue on subjects which even in Europe were at the time at the peak of scientific interest, such as the theory of evolution."[45] *Estia* was liberally oriented and viewed Darwin's ideas favorably. (Note that after 1895 *Estia* became a daily newspaper and is still being published.) Miliarakis, for instance, wrote a number of articles on Darwin and his theories. In 1877, he also translated into Greek from a German version of "A Biographical Sketch of an Infant," which was

41 Several Greek translations of foreign articles were by Miliarakis and Heldreich.

42 Earlier translations from 1917 and 1927 existed, but were only sections of the *DM* and were little known. See Krimbas, "Ο δαρβινισμός στην Ελλάδα," 102. See further on Darwin's works and translations: Freeman, *Charles Darwin, a Companion*; Freeman, *Works of Charles Darwin*.

43 Krimbas, "Ο δαρβινισμός στην Ελλάδα," 102.

44 Ibid., 102–103. Krimbas' views on the passive reception are supported by George Vlahakis. See Vlahakis, "Science in Society in 19th Century Greece," 117.

45 Ibid., 123.

Darwin's famous biographical essay on one of his children.[46] It was the first of Darwin's works to be translated into Greek. Dimitrios Vikelas published in 1900 a partial translation of Darwin's *Voyage of the Beagle* in the series of "The society for the dissemination of useful books." The translation was of Darwin's chapter 10, "Tierra del Fuego."[47]

It is important to note that Darwin's *EE*, which complements his *DM*, has not as yet been translated into Greek. However, Spyridon Miliarakis wrote *The mental traits of animals* (*Αι ψυχικαί ιδιότητες των ζώων*), which was first published posthumously in 1926.[48] Miliarakis' book is not a translation of *EE*, but simply a study inspired by Darwin's book. In this ninety-page book Miliarakis draws attention to the expression of those emotions that humans have in common with animals, using similar headings to those used by Darwin, such as happiness, fear, melancholy, grief, and weeping (though the latter is not an emotion but the expression of an emotion). It is clear that he had read *EE* because many of the examples he uses are the same as those of Darwin. To the best of my knowledge, there has not been any commentary on Miliarakis' book, and commentary on *EE* in Greece is limited.[49]

While by the 1880s Greek botanists had accepted Darwin's works and many theologians had rejected his theories,[50] this book provides evidence of published polemic, predominantly pro-Darwinian, by Greek writers of the same period, albeit written under pseudonyms. Take, for example, an article published in March 1880 under the pseudonym "Caliban" by Vlasis Gavriilidis,[51] editor and contributor to the early Greek literary period-

46 Miliarakis, "Κάρολος Δάρβιν: βιογραφικόν σχεδίασμα μικρού τινος παιδίου," 817–824. For other articles in *Estia* on Darwin see Miliarakis, "Κάρολος Δάρβιν," 451–456; Miliarakis, "Ο Δάρβιν και η θεωρία του: μια εκατονταετηρίς," 5; Miliarakis, "Ο Δάρβιν και η θεωρία του: ο δαρβινισμός εν Ελλάδι," 5. See also the periodical *Parnassos* (*Παρνασσός*) in 1890 on Darwinism in Kalopothakis, "Η θεωρία της εξελίξεως," 365–378; and the periodical *I Fysis* (*Η Φύσις*) in 1904 in Vlisidis, "Η κατά Δαρβίνον θεωρία της εξελίξεως," 385–389.

47 Vikelas, *Η Γη του Πυρός*.

48 Miliarakis, *Αι ψυχικαί ιδιότητες των ζώων*.

49 From the Greek perspective, Fotini Stylianopoulou claims that after its initial popularity, it is only in the last thirty to forty years that Darwin's *EE* has been recognized for its study of expressions. See Stylianopoulou, "Δαρβίνος και συγκινήσεις," 26–30.

50 Note that prior to this, in 1876 at the University of Athens, theologian and assistant professor Spyridon Sougras launched a significant attack on Darwin's *DM* with his publication entitled *The most recent phase of materialism: Darwinism and its unfoundedness* (*Η νεοτάτη του υλισμού φάσις, ήτοι ο δαρουϊνισμός και το ανυπόστατον αυτού*).

51 Καλιβάν also cited as Κάλιμπαν.

ical *Don't Get Lost* (*Μη Χάνεσαι*) (1880–1883).[52] He writes about the professor of physiology and anatomy Ioannis Zochios (1840–1912) of Athens University, who in 1880 lectured on evolution to his students. It is said a lecture of his mocked a sermon that had been given by theologian and cleric Latas. Caliban satirizes the matter, showing that he was strongly in favor of Darwin and in agreement with Zochios, at the expense of Latas. Caliban's final message is: "We think that religion has nothing in common with science, and if Mr. Latas wants to refute Darwin, he ought to first study natural science, anthropology and biology, because scientific matters can only be discussed scientifically. The pulpit, from where it is necessary to teach exclusively pure Christian morality, has nothing in common with science."[53] Pressure by the Church to stop the lectures only went as far as the university recommending that in future Zochios should avoid making any antireligious comments in his lectures.[54]

In her PhD dissertation "The emergence of the theory of evolution of the species: data from the Greek domain," Anthi Sotiriadou confirms in a quantitative and qualitative study that from the eighteenth century until recent times Greeks have shown little interest in the studying or teaching of the natural sciences.[55] In addition, she notes that educational texts at the end of the nineteenth century only made general references to evolution, not naming Darwinism. She maintains that it was only in the 1930s that Darwinian evolution was referred to by name in high school textbooks.[56] Nicolaidis asserts that the new Greek state, formed after the Greek Revolution of 1821, preferred to endorse the pursuit of letters over that of science because "Greece's strength rested on the glorious past and hence it was more profitable to study this past than to develop science."[57] He claims many within the

52 The poet Giorgos Stratigis (1860–1938) refers to Gavriilidis as "charming despite borrowing his pseudonym from the name of the Shakespearean man-beast." For this see Valetas, *Η γενιά του '80*, 5. Note also Caliban was the half-man half-beast in Shakespeare's play *The Tempest* (1612). Robert Browning in his poem "Caliban upon Setebos" (1864) depicts Caliban in evolutionary terms, as an intermediate somewhere between humans and the lower primates.

53 Caliban [Vlasis Gavriilidis], "Δαρβίν-Λάτας," 5.

54 For a detailed description of the occurrences associated with Zochios see Krimbas, *Θραύσματα κατόπτρου*, 100–101. Krimbas does not make any reference to the article found in *Μη Χάνεσαι*. Also see Nicolaidis, *Science and Eastern Orthodoxy*, 183–184.

55 Sotiriadou, "Η εμφάνιση της θεωρίας της εξέλιξης," 90–95.

56 Ibid., 201–202. Sotiradou also points out that it was only in 1934 that a university department associated exclusively with the natural sciences was established at the University of Athens and in 1943 at the Aristotle University of Thessaloniki. Ibid., 95.

57 Nicolaidis, *Science and Eastern Orthodoxy*, 174.

Orthodox Church then feared the West, and that they considered the teaching of science as synonymous with the teaching of Western culture; hence science could distance people from Orthodoxy.[58] This view would be exacerbated with the advent of Darwinism, which was notoriously associated to materialism. Generally speaking, Western Churches have come to terms with the Darwinian evolution of humanity by excluding an atheistic component.[59]

Further to this, the investigations carried out by Lucia Prinou and associates, about the teaching of the Darwinian theory of evolution in secondary schools in Greece from 1930 until very recently, highlighted the lack of interest in it. They reveal that in this period regarding Greek education "it is ascertained that evolution remains a peripheral unit in the biology lesson, a topic which is addressed fragmentarily."[60] Due to the media publicity surrounding the 2009 double anniversary of Darwin and promotion of his theory, it is understood that the Greek Ministry of Education and Religious Affairs[61] has reintroduced the chapter on evolution in the curriculum of the upper high school. This includes Darwin's original theory but not the current revised theory. So, generations of Greek youth until now have had little, if any, formal secondary education on the Darwinian theories which play a crucial role in modern thinking.

Two late nineteenth-century Athenian periodicals were in conflict over Darwin's theory of evolution. These were the religion-oriented *Anaplasis* (*Ανάπλασις*), first published in 1887, which tended to be anti-Darwinian, and *Prometheus* (*Προμηθεύς*) (1890–1892), which represented science and was pro-Darwinian. Sotiriadou's investigation of the two periodicals indicates: "The common characteristic of the two is the imitating, even to the point of copying the stance of the West, in its dealing with the theory."[62] The

58 Ibid.

59 These include the Anglican Church and the Roman Catholic Church.

60 Prinou, Halkia, and Skordoulis, "Θεωρία της εξέλιξης," 260. Lucia Prinou, Lia Halkia, and Constantine Skordoulis have written extensively on the education (and the lack thereof) of Darwinism in Greek schools. Several of their academic articles are accessible on the internet. See also Sariggelis, "Η θεωρία της εξέλιξης στην εκπαίδευση," a 2014 article by a biology teacher on the teaching of Darwin's theory in schools in Greece, which contains an outline of its history from the early 1900s to the present.

61 Nicolaidis notes an ideology which arose in society after Greece's independence—namely, that during the Ottoman occupation, the Greek nation owed its survival to the Greek Orthodox Church. This view would contribute to the establishment and ongoing existence of a government ministry which combines education and religion. Nicolaidis, *Science and Eastern Orthodoxy*, 179.

62 Sotiriadou, "Η εμφάνιση της θεωρίας της εξέλιξης," 201.

arguments that were dealt with included the convergence of Darwinian theory with religion and the counter-discourse—that is, their irreconcilability.[63] According to Sotiriadou, the Eastern Orthodox Church did not take an official stand on evolutionary theory, and its reaction in general was considered mild.[64] There were, however, various theologians and clerics who demanded that evolutionary theory not be taught at university, and also that the lecturers who taught it to be sacked; neither of these expectations would eventuate.[65] Since then many Orthodox Christians, theologians, and clerics have supported an inability to align biological evolution with Orthodoxy, whereas others have maintained some compatibility between the two.[66]

According to Roxani D. Argyropoulou, Darwinian and positivist approaches dealing with the natural sciences in society were not accepted readily by the establishment in Greece, whereas, she argues, the application of evolutionary theory in society was easily received by liberal and Marxist intellectuals, who used it as a validation for their ideologies.[67] At odds with this argument is Krimbas, who recently commented on how the interpretative power of Darwinism with respect to various phenomena, social and other, was initially received in Greece: "It is an effort that met with fierce opposition, mainly for political reasons. The retort came from Marxists who would have wanted all social phenomena to be reduced again to social and not to biological phenomena. It is an interesting conflict which still exists today with less intensity."[68]

The socialist George Skliros (1878–1919) is primarily referred to for his work *Our social question* (*Το κοινωνικό μας ζήτημα*) (1907),[69] a Marxist analysis of Greece's social structure. A heated intellectual debate surround-

63 Further on religion versus evolution in Greece see Makridis, "Εξελικτική και Ορθοδοξία στην Ελλάδα," 173–220. On the conflict between the two rival periodicals see Kritikos, "Science and Religion in Greece," 35–50.

64 Sotiriadou, "Η εμφάνιση της θεωρίας της εξέλιξης," 202.

65 Ibid.

66 The scholarship from a religious perspective is considerable. Most recently published is the very interesting book by Buxhoeveden and Woloschak, *Science and the Eastern Orthodox Church*. Also Woloschak, "Compatibility of the Principle of Biological Evolution with Easter Orthodoxy," 209–231.

67 Argyropoulou, *Η φιλοσοφική σκέψη*, 36, 39–43. She cites the following Darwinians who applied evolutionary approaches to society: Neoklis Kazasis, Pavlos Karolidis, Ion Dragoumis, Avrotelis Eleftheropoulos, Mihail Stefanidis, Dimitris Glinos, and George Skliros.

68 For this see Kouvelas, "Η θεωρία που προκαλεί ακόμη και σήμερα," in which Krimbas discusses the impact of Darwinism with colleagues.

69 See further on this Stavridi-Patrikiou, *Δημοτικισμός και κοινωνικό πρόβλημα*. Also, George Skliros is the pseudonym of George Konstantinidis.

ing this work continued for two years in the primary literary and political periodical *Noumas*. Up until his death, however, among other intellectual trends a crucial aspect of his writing would go on to utilize Darwinian language and patterns to discuss the issues of Greek society and its language. Petros Vlastos responded to Skliros' 1908 evolutionary ideas[70] with the following:

> Mr. Skliros stipulates that *Evolution* has ironclad laws and guides society from the predetermined to its free state. But how can the ironclad *Evolution* and a free society stand together? Maybe *Evolution* stops as soon as the society becomes free or maybe the term "freedom" is not precise enough? And if also free society is itself something predetermined and hardwired why should we bother ourselves with so many battles and kill ourselves in the *struggle for life*? Isn't it better that we fold our arms and let *the famous laws of evolution* do whatever they want? And maybe all the difference here is—because we believe that man is also a law of nature—that as much as nature acts on him so does man act on nature. We know that this is not yet proven scientifically but each philosophy requires, deep down, some belief. Even the strictest scientists rest on some belief, that is, the objective world.[71]

The word for evolution used in the original Greek text was an early form, "Ξετυλιξιά," which actually means "the unravelling" or "the unfolding." Prior to Darwinism the word "evolution" in English also had a different meaning biologically. It was the "unfolding of form in the development of the embryo, and in particular to the notion of preformation—that the adult organism is preformed in the fertilized egg."[72] The contemporary Greek word for evolution in biological terms is "εξέλιξη."

The observations in the cited passage suggest that the nature of Darwinian evolution—whether it was considered deterministic or not and was applied to individuals and society— was relevant in 1908. Intellectuals such as Vlastos speculated about a deterministic evolution where humans have no control and can only sit back and let nature take its course. These Darwinian ideas contributed to the age-old nature versus nurture debate, which

70 Skliros, "Διαλεκτικός Υλισμός," 2–8.

71 Vlastos, "Πραγματισμός," 11 (my italics). This essay is also in Stavridi-Patrikiou, *Δημοτικισμός και κοινωνικό πρόβλημα*, 134.

72 See Gregory, *Oxford Companion to the Mind*, 314. In chapter 2 I will note early terms such as "νόμοι της Ξετυλιξιάς" (laws of evolution) and "η θεωρία του Ξετυλιμού" (the theory of evolution).

questioned whether it was worth applying oneself in an attempt to improve the Greek nation. This mode of thinking is examined in the next chapter. Further to this, a decade later Xenopoulos would be writing his novel *Rich and poor,* examining these exact issues of Darwinian influence in socialism and capitalism in the Greek society.

CHAPTER I

EVOLUTIONARY THEORIES

> Language ... is not a simple array of words and forms, but a living organism which mirrors the organism of the modern Greek mind, its perceptiveness, its plasticity, [and] the magnitude and depth of our intellectual and mental world.
>
> ALEXANDROS DELMOUZOS[1]

It is vital at this point to investigate the main scientific ideas that may have influenced society and the literary world in the late nineteenth to early twentieth centuries. These include aspects of Darwin's theory, its philosophical implications, and other evolutionary and associated theories popular at the time. I have included the social implications of Herbert Spencer's theories even though they are not biologically based. I have only done so because Xenopoulos had been keen on Spencer's theories before these were eclipsed by neo-Darwinism, around 1900; also, because Xenopoulos does allude to them in some of the novels examined in this book. Darwin featured aspects of Spencer's thoughts in his key works (*OS*, *DM*, and *EE*); for instance, Darwin eventually included in his sixth edition of the *OS* Spencer's idea of the "survival of the fittest." Many Greek intellectuals often confused the concepts of Spencer with those of Darwin.

DARWINISM

Darwin was the founder of modern evolutionism, the science of evolutionary biology—that is, the study by which living organisms have developed following the origin of life. His *OS* provided the basic argument for evolu-

1 From an address in 1918 to the primary school teachers of Athens. Papanoutsos, *Α. Δελμούζος*, 153-154.

tion by proposing a mechanism of change in animals and plants—that is, natural selection. This is a process in the life of every generation by which individuals who are not sufficiently fit are eliminated from the population, whereas the individuals with certain heritable traits who survive the environment will pass on those traits to their offspring. Darwin defined it: "This preservation of favourable variations and the rejection of injurious variations, I call Natural Selection."[2] His announcement of it coincided with that of Alfred Russell Wallace (1823–1913), who propounded the same idea although they had worked independently of each other.

As will be seen in the forthcoming chapters, there are a number of tenets of Darwinian evolution that have created an impact on fields outside biology and, relevant here, that have been absorbed by the literary imagination, and intellectual thought in general. Evolutionist and twentieth-century commentator Ernst Mayr summarized the concepts that collectively constitute Darwin's theory of evolution via natural selection. These are: "the non-constancy of species," known as "the modern conception of evolution itself"; "the notion of branching evolution, implying the common descent of all species of living things on earth from a single unique origin"; the "gradual" nature of evolution, "with no breaks or discontinuities";[3] and "the mechanism of evolution [which is] natural selection."[4] These are the concepts from which many of the themes, motifs, and ideas in the literary sphere originate, as will be demonstrated in this book.

The concept of natural selection was new and it revolutionized thinking—not only in science but also in many other disciplines—in a way no other evolutionary theory had previously done. Historian of science Peter Bowler argues that some commentators have claimed that others authors (William Charles Wells, Patrick Matthew, and Edward Blyth) developed the selection

2 Darwin, *OS*, 131.

3 Philosopher Gottfried Wilhelm Leibniz (1646–1717) is attributed for the concepts of gradualness and continuity when in 1712 he states: "Everything goes by degrees in nature, and nothing leaps, and this rule controlling changes is part of my law of continuity." Quoted in Mayr, *Growth of Biological Thought*, 325. As noted by Mayr, Leibniz's thought rejected Platonism and by doing so this was an "indispensable prerequisite for modern evolutionary thinking [and] was one of the foundation stones of Darwin's explanation of evolution. However, Leibniz also believed in "the inner drive toward progress, if not perfection," an incorrect view later taken up by Herbert Spencer and others in applying the concept to evolution. Ibid.

4 Mayr, "Darwin's Influence on Modern Thought," 80. Here Mayr also notes that evolution prior to Darwin had always been considered linear by all proposers of such theories, such as Jean-Baptiste Lamarck, and even back to Aristotle's concept of *Scala Naturae*, known as the chain of being. Linear evolution was viewed as "a teleological march toward greater perfection."

theory before Darwin, so that it was not an original discovery by Darwin. Bowler highlights that Darwin's notebooks show that "he drew no inspiration from Matthew or any of the other alleged precursors."[5] Similarly, both Bowler and Mayr argue that although Thomas R. Malthus in his *Essay on the Principle of Population* (1797) is credited with the term "struggle for existence," it was very limited in its meaning. Darwin's natural selection was a development way beyond what Malthus saw in primitive tribes—that is, the geometric rate of increase of a population exceeding the arithmetic rate of food resources.[6] It is therefore important to emphasize here that, contrary to some beliefs, the influence of Darwin's theory of natural selection was of a revolutionary nature both in science and in terms of ideology, profoundly changing thinking and perceptions in society. Darwin's ideas were to go against the prevailing ideas of his time. This was because, according to his theory, the creation of life did not rely on a designer or creator, as evolution via natural selection functioned without one. For this reason Darwin delayed discussing the origin of the human species in the *OS*, in which he spoke of living things, animals, and plants, leaving humanity till the *DM*.[7] As will be seen in this study, natural selection is a theme that Greek intellectuals absorbed in their writings, including Xenopoulos in the novels to be discussed in chapters 4 and 6.

Darwin's theory eliminated the idea of a teleological force that was supposed to lead to a higher perfection. This belief, which originated with the ancient Greeks,[8] still continued in various non-Darwinian the-

5 Bowler, *Evolution* (2009), 158.

6 Bowler, *Evolution* (2009), 104–105; Mayr, *Growth of Biological Thought*, 478–479, 484, 491–493. Bowler's general opinion of "precursor-hunting," which means looking for forerunners to a particular discovery, is that it is misleading. He indicates it is claimed by scholars who select information from various naturalists preceding Darwin in an attempt to "fit," say, the selection theory and hence show that these past naturalists have anticipated Darwin's theory. Bowler maintains that in order to make any such comparisons between past theories and those of Darwin, "historians insist that a scientist's writings must be read as a whole in order to reconstruct the context within which he or she studied nature." Bowler, *Evolution* (2009), 20. My approach here, therefore, is not to examine past non-Darwinian theories as precursors to Darwinism.

7 Darwin "had lost his orthodox belief and come to the conclusion, which he retained to the end of his life, that questions of ultimate causes and purposes were an insoluble mystery." J. W. Burrow, "Editor's Introduction," in Darwin, *OS*, 24. Also, so as not to incense the Church, Darwin attempted to avoid discussion of the metaphysical in his *OS* and *DM*. He did not succeed with this because there are contradictory elements which make reference to a creator or which imply a creator in his theory of evolution. Darwin, *OS*, 458–460.

8 Prior to Darwin, as far back as Plato, species (*eide*, kinds, or types) were argued as being stable and invariable. However, the debate associated with their constancy began in the eighteenth century, and transformation and transmutation were terms utilized in science well before Darwin.

ories after Darwin's theory of natural selection. In fact, Darwinian evolution was often misinterpreted as leading to a greater perfection, particularly due to some of the contradictory or unclear comments found in his books.[9] This idea of higher perfection, relevant to Xenopoulos' work, is dealt with in chapters 2, 3, and 4; it is also discussed in chapter 2, regarding the works of Nikos Kazantzakis. Natural selection occurs firstly due to the variation of species, based on genetics, and this is due to chance (that is, it is random). However, it must be realized that the actual selection process, dependent on one's response to conditions, is not a random agent but a directional one.[10]

Although the most significant and critical Darwinian influence was Darwin's theory of natural selection, it was not his only idea that affected writers. In his *DM* Darwin not only placed natural selection in the context of humanity, but he also propounded the theory of sexual selection—that is, the selection of certain attributes in a sexual partner to promote the chance of the fittest offspring. "Sexual selection," Darwin argued, "depends on the success of certain individuals over others of the same sex in relation to the propagation of the species; whilst natural selection depends on the general conditions of life."[11] These concepts of natural and sexual selection are very relevant here, as Xenopoulos explicitly and implicitly fuses them into his works; therefore, they will be discussed in the chapters to come.

Perhaps the most devastating aspect of Darwin's theory was the affirmation in his *DM*, and later in his *EE*, of humanity's common descent from one progenitor, in alignment with animals. Of course, traditional religious explanations conceived of one set of progenitors, Adam and Eve. But with Darwinian theory, the issue was human descent from animals. Religion and philosophy had always placed humans above and distinctly separate from other living beings, whereas Darwinian evolution placed them close to the apes. On mental abilities Darwin maintained that "the difference in mind between man and the higher animals, great as it is, is certainly one of degree and not of kind."[12] As will be revealed in chapter 3 of this book, in his "Athenian Letters" Xenopoulos agrees with Darwin regarding the common

9 These comments by Darwin will be discussed later in this book.

10 Mayr, "Darwin's Influence on Modern Thought," 82.

11 Darwin, *DM*, 2:398.

12 Darwin, *DM*, 1:105.

descent of humanity. This concept is also taken up by other writers such as Emmanuel Roidis, discussed in chapter 2.

After the *DM*, Darwin in his *EE* further elaborates on the kinship of the human species with other animals, by demonstrating the similarities in the facial expressions of their emotions. Among scientists, criminologists, and scientific physiognomists there perpetuated a hereditarian discourse linking an individual's facial expressions with their inner being. Although Darwin's *EE* was not as well-known to literary writers as his earlier two works, it did influence some writers, including Xenopoulos, who wrote using physiognomy and facial expressions in a Darwinian manner. This theme will be developed in chapter 4 on Xenopoulos' novel *Rich and poor.*

The controversy in science raged for almost a whole century as to which was the valid evolutionary theory—that is, Darwinism versus the non-Darwinian theories such as Lamarckism. In science, Darwinian theory came into disfavor around 1900. Evolutionism was well accepted but the mechanism by which it occurred, natural selection, was in dispute.[13]

Lamarckism

The theory of the French naturalist Jean-Baptiste de Lamarck (1744–1829) found in his evolutionary treatise *Philosophie zoologique* (1809) resurfaced later in the century in the new form of "neo-Lamarckism." Only one aspect of his dated theory was adopted—"the inheritance of acquired characteristics" (also known as characters) or "use-inheritance." Characteristics or traits acquired during the life of an individual were supposed to be passed on to the next generation. Acquisition of these traits in one's life was achieved by willing it to happen, by the intentional use or disuse of physical or mental features causing structural or functional change. Hence a new behavior pattern is established due to new habits which are adaptive, and use-inheritance becomes the mechanism of change rather than natural selection. However, the issue of such acquired characteristics being inherited is very controversial. The popularity of neo-Lamarckism was due to its philosophical consequences. It allowed individuals to direct their own evo-

13 For the non-Darwinian theories which proliferated at the turn of the century, including the variations of orthogenesis, see Bowler, *Evolution* (2009), 224–273; Bowler, *Non-Darwinian Revolution*; and Bowler, *Eclipse of Darwinism.*

lution by will, or choice. Here humanity was seen as being the controller of its destiny and its environment. However, with the early twentieth century came neo-Lamarckism's complete collapse in science.[14] It should be noted, though, that writers in general did not always follow the trends of science. Playwright George Bernard Shaw (1856–1950) continued to write in favor of it into the third decade, in *Man and Superman* (1901) and *Back to Methuselah* (1921). Particularly, in the early twentieth century there was a trend by neo-Lamarckian eugenicists, using their ideology to provide a framework for the improvement of the human race.

Neo-darwinism

Although Lamarckism foregrounded environmental influence, it was refuted by the work of German biologist August Weismann (1834–1914). In his *Essays upon Heredity* (1885 in German, 1889 in English) Weismann proposed that living things contain a hereditary substance, the "germ plasm" (effectively, the gene), which unlike the perishable body of the individual, the "somatoplasm," it is transmitted from generation to generation.[15] The germ cell (eggs and sperm) contained the factors which determined the transmission of characteristics or traits from the parent to the next generation. He argued that the germ plasm was isolated completely from the main body or "soma" of the organism which carried it, and hence remained unchanged from one generation to the next. He further indicated that the germ-cell directed the development of the whole organism and could not be modified by the environment.

Not only was Weismann's theory an early forerunner to modern genetics, but it also became the hard-inheritance theory known as "neo-Darwinism." Important too is his reference to the "perpetual continuity" of the germ plasm. This refers to the theory that the germ plasm is passed down through each generation and is seen as having an immortal property.[16] This "immortality" is taken up by literary scholars in the late nineteenth and early twentieth centuries in the sense of "continuity," and relevant to the biological continuity of a race. In addition, it was taken up by writers who tended to align

14 Bowler, *Evolution* (2009), 315.

15 Weismann, *Essays upon Heredity*, 104.

16 The idea of the continuity of hereditary material is still the basis of our understanding of physical inheritance.

the evolution of literature and language with biological evolution. It will be mentioned later in this chapter and discussed in more detail in chapter 2.[17] The concept of continuity from an evolutionary perspective did pertain initially to Darwin's theory which was based on his tenet of "no breaks and discontinuities"; however, Darwin was not able to take it the step further to the germ plasm. It is worth pointing out that the general theory of continuity was noted as early as 1867 as "one of the grandest conceptions of man, and we have rather poets than men of science who work it out."[18]

The concept of variation was explained in 1865 by the Austrian monk Gregor Mendel (1822–1884), who presented the results obtained from his experiments crossing varieties of the garden pea. He was able to show that genetic variation followed simple statistical laws. His work was ignored for quite some time, in part because it was published in a little-known journal. It was only with the appreciation of Darwinian evolution that Mendel's work was realized. The rediscovery in 1900 of Mendel's laws of inheritance highlighted his theory that genetic variation produced a possible range of genetic combinations in the progeny; and that variation was continuous in generation after generation. Scientists realized that Mendel's research on heredity and variation worked well with Darwin's theory of selection. Hence, the environment selects those genetic combinations which favor the adaptation of an organism. With Weismann's theory as a forerunner to neo-Darwinism, the so-called new synthesis of Mendel on heredity and Darwin on selection, named the neo-Darwinian synthesis, emerged in the 1930s. This doctrine is still today basically the accepted scientific theory of evolution and inheritance.

New synthesis

A mechanism of non-Darwinian evolution that became popular in evolutionary biology at the turn of the twentieth century was orthogenesis. It was popularized by Theodor Eimer and was used predominantly by North American

17 More recently, Richard Dawkins utilizes the concept of Weismann's theory of the continuity of the germ plasm—more specifically, "perpetual continuity"—as a basis for his book *The Selfish Gene* (1976). He views "the gene as the fundamental unit of natural selection" and "the fundamental unit of self-interest," which has "potential for immortality." See Dawkins, "Immortal Coils," in *The Selfish Gene*, 21–45. Note also his later book *The Blind Watchmaker* (1986), which defends natural selection against intelligent design. Dawkins is now assuredly the most well-known defender of Darwinism, at least for the general public.

18 Hunt, "On the Doctrine of Continuity," 120.

scientists. Orthogenesis was evolution which occurred in a straight line. Here evolution was not adaptive—that is, it did not rely on a response to the environment to change, unlike the adaptive evolutions such as Darwinism, which was selective, and Lamarckism, which relied on the inheritance of acquired characteristics. Like other non-Darwinian theories, orthogenesis was teleological—directed towards fixed goals, preordained, regular, and linear.[19] It is worth noting that all evolutionary theories up until Darwin's theory were linear; this included Lamarckism. However, many scholars still viewed Darwinian evolution as linear, with this view filtering through to many of the literary writers at the end of the nineteenth century who wrote about it in this form. Greek writers were not excluded from this, as will be discussed later.

When evolutionists realized that "'evolution' was not synonymous with 'progress,'" the "devolution"—that is, retrogression or degeneration of humanity—was perceived as a possibility to be feared.[20] Although many theories of degeneration surfaced prior to Darwin, it was Darwin's work, particularly in England, which managed inadvertently to "validate" the possible biological degeneration of humanity via regression, atavism, and decline. In his *OS* Darwin maintained that with variations of characters (or characteristics) within a species, some characters could "revert to some of the characters of an early progenitor."[21] Also, he stated in his *DM*: "Whenever a structure is arrested in its development, but still continues growing until it closely resembles a corresponding structure in some lower and adult member of the same group, we may in one sense consider it as a case of reversion."[22] This was "reinforced" by Edwin Lankester in his *Degeneration. A Chapter in Darwinism* (1880).[23] "Retrogressive metamorphosis" (or devolution), he argued, occurred when a simple organism or lower form, such as a parasite, evolved from a complex organism; this he believed reflected on society, where the human species could also regress to a "parasite" or some lower form of life. The theme of retrogression to a lower form is taken up by Xenopoulos and will be examined predominantly in chapter 5.

19 Bowler, *Evolution* (2009), 247–250.

20 Trotter, *English Novel in History*, 111. Note that the term "devolution" was used biologically to refer to the evolution into more primitive forms. Evolution was often perceived by lay people as being progressive. However, Darwinian evolution is directionless, hence "devolution" is still evolution.

21 Darwin, *OS*, 195.

22 Darwin, *DM*, 1:122.

23 Lankester, *Degeneration*, 30.

Social implications of evolutionism

Darwinism was not just a biological theory; it also had major social implications. Social Darwinism is a loose term which has been used for any number of applications and mistaken applications of Darwin's theories of evolution.[24] In this regard it has been used in opposing political situations, such as militarism and pacifism, capitalism and socialism, and with reference to eugenics, laissez-faire economics, and racism. Some of these ideas will be expanded in the relevant chapters where they are associated with certain literary works and ideologies. The term social Darwinism should perhaps be avoided because it cannot cover the different interpretations and the frequent contradictions in ideologies.

Marxism is a classic case where Darwin's idea of struggle was supposedly applied. In 1861, referring to the *OS*, Karl Marx (1818–1883) stated in a letter to Ferdinand Lassalle in Berlin: "Darwin's book is very important and it suits me well that it supports the class struggle in history from the point of view of natural sciences."[25] Karl Marx and Frederick Engels (1820–1895) were well aware of the reality of the struggle for existence in society, which for them was an avoidable human condition. They believed that the conditions which determined the struggle for existence were sociohistorical and not biological. In chapter 4 there will be a discussion on the application of Darwinian ideas to socialism, related to Xenopoulos' novel *Rich and poor*, and also in chapter 3, related to his "Athenian Letters."

The English philosopher and critic Herbert Spencer (1820–1903) propounded a type of social Darwinism that encouraged the struggle for existence and merciless competition. It is better to refer to Spencer's Victorian version of social Darwinism as social Spencerism. He defined evolution as "an integration of matter and concomitant dissipation of motion; during which the matter passes from an indefinite, incoherent homogeneity to a definite, coherent heterogeneity; and during which the retained motion

24 Social evolutionism is a term which covers evolutionary theories in the social sciences. These evolutionary theories are not necessarily based on evolutionary biology and can be based on other ideas or disciplines. For instance, Herbert Spencer's "Synthetic Theory" had a metaphysical component. Auguste Comte's evolutionary concepts are also not based on evolutionary biology. See further Sanderson, *Social Evolutionism*.

25 Ryazanskaya, *Marx, Engels*, 115.

undergoes a parallel transformation."[26] This definition first appeared in his essay "Progress: Its Law and Cause" (1857), prior to Darwin's *OS*.[27] Spencer saw the evolution of biological species as a progression from a lower to a higher level of complexity. He thought that evolution was teleological and Lamarckian based, to a degree, pertaining to the inheritance of acquired characteristics. He later accepted that natural selection was one of the causes of biological evolution, coining the phrase "survival of the fittest."

However, he did not believe that natural selection or the inheritance of acquired characteristics was the key mechanism for evolution. He maintained that the ultimate mechanism of change was an "unknown and unknowable absolute force"[28] that acted continuously on the world, producing from "homogeneity" a "heterogeneity," or an increase of variety of forms (or lower complexity to higher complexity). He based his theory on his two laws of matter: the law of the "persistence of force"[29] and the law of the "multiplication of effects."[30] He believed that homogeneous matter did not remain the same when a continuous external force acted on it; instead, this force would affect some parts of the matter differently from others. This would result in an increase in the variety of matter, or a heterogeneity; hence a "multiplication of effects." This law of "multiplication of effects" he thought was the answer to the understanding of biological and cosmic development.

Spencer applied his evolutionary theory to the evolution of the universe, to human society, industry, language, art, and science, naming this application the "Synthetic Theory." In social evolution he claimed, for instance, that human societies evolve from undifferentiated groups into complex civilizations by means of an increased division of labor. He believed that this evolution was continuous and that there should be no interference, socially or economically, with the process. In other words, with human society there should be no interference from the state. Social evolution, he asserted, should lead to a free or capitalist state. Spencer in his *Man versus the State* (1884) main-

26 Spencer, *First Principles*, 358.

27 With Darwin's *OS* in 1859, Spencer "immediately saw reason to modify some of his own conclusions with reference to organic evolution, although it did not move him from his evolutionary position in regard to the development of the social organism, or in ethical and metaphysical matters." T. W. Hill, "Introduction," in Ibid., xi.

28 Ibid., 3–106.

29 Ibid., 165.

30 Ibid., 390.

tained that socialism allowed for the worst in society to survive and multiply. Although laissez-faire was seen to eliminate the unfit, it was also supposed to boost humanity to improve itself by its own volition; this approach clearly had Lamarckian undertones. The idea of exerting effort was popular and was also found in works such as Samuel Smiles' *Self-Help* (1859).

Spencer ultimately failed to "synthesize" the sciences under an umbrella of biological evolution; this evolution could only provide simplistic answers for the growing, complex, and divergent specialization of the sciences. Before the turn of the twentieth century his evolutionary philosophy was already dated. His law of "persistence of force" was based entirely on a metaphysical principle—that is, the "unknown and unknowable absolute force," which could not be validated by science. Spencer's evolutionary theory was widely accepted by many intellectuals who considered that he popularized Darwin's theory. Indeed, for a long time many intellectuals, such as those in the literary field, confused it with Darwin's theory of evolution. This is relevant to Greek writers such as Xenopoulos, where at times he appears to conflate Darwin's and Spencer's ideas. Thus in chapter 3 of this book I examine an interview in which Xenopoulos sees all societal structures in terms of this combination of theories.

Darwin's *DM* had shown that an individual's moral faculty was subject to evolution. So the ethical question considered was whether one could still maintain a free will and at the same time be directed by biological determinism. Intellectuals thus argued whether there was any point in fighting one's nature if one's fate was predestined.[31] The idea of biological determinism was widespread in the nineteenth and early twentieth centuries, and refers to the belief that the individual is shaped by biological factors, more than likely by heredity; in this sense, the term is interchangeable with hereditary or genetic determinism. Here environmental or social factors play no part in the individual's make-up. Taking this line of thought would mean that an individual's traits could not be improved by making changes to social conditions. Based on this, sociopolitical strategies could reduce or eliminate the numbers and the influence of those individuals who were considered by nature, or by heredity, as "unfit." The ideology of hereditary or genetic determinism would play a major role with the advent of eugenics.

31 Petros Vlastos takes this up with George Skliros, as will be discussed later in this chapter.

Francis Galton (1822–1911), influenced by his cousin Darwin's theory of natural selection, wrote *Hereditary Genius* (1869), which also impacted to a degree on Darwin's *DM*. Galton saw that this concept of selection could be applied in achieving the biological improvement of the human species, and in 1883 he called this approach "eugenics." Unlike Spencer's laissez-faire social Darwinism, which rejected state intervention, the eugenics movement declared that the state should control its members in society. Darwin had mentioned that in a civilized society natural selection did not eliminate the unfit anymore. So, based on the assumption that the poorest individuals are those with the highest proportion of inferior traits, Galton—prompted by the poverty he saw in English slums—argued that the state should intervene in the control of its extent.

Eugenics was an influencing factor in formulating concepts and ideas associated with nationalism, modernity, and race theories. Society was seen as being directed not only by social factors but also natural laws. Western countries saw themselves in light of the evolutionary process, with the more developed countries perceived as higher up the evolutionary scale and as an example to be copied by the underdeveloped countries.[32] Moreover, by the beginning of the twentieth century in Europe eugenics was to be "one of the most significant social applications of Darwinian and neo-Darwinian thought."[33] In terms of biological and cultural degeneration, many nations were experiencing "a crisis of the individual and a national 'degeneration' [and] wished to create 'a healthier' and more 'efficient society.'"[34] A similar trend was seen in the USA with emphasis on race.[35]

National regeneration was central to the eugenics movement, which aimed to improve the character of a race or nation predominantly by means of artificial selection. Galton's strategy was based on two mechanisms: positive eugenics, which encouraged the birth of the "fit" and was often in the form of a reward from the government if one complied; and negative eugenics—that is, selective breeding—where the birth or survival of the unfit to parents with inherited problems was discouraged.[36] In his book *Mod-*

32 Leeuwenburgh and Heide, "Darwin on Dutch Soil," 184–185.
33 Hermann and Šimůnek, "Between Science and Ideology," 206.
34 Ibid.
35 See further on eugenics in America Bowler, *Evolution* (2009), 311–312.
36 See the chronology associated with eugenics internationally, dating from 1859 to 1989, in Bashford and Levine, *Oxford Handbook of the History of Eugenics*, 559–567.

ernism and Eugenics Marius Turda clearly shows that in Europe during the period between 1870 and 1940 eugenics "became part of a larger biopolitical agenda," indicating that "eugenics was as diverse ideologically as it was spread geographically, both advocated and adhered to by professional and political elites across Europe irrespective of their political and intellectual affiliation."[37]

The concept of linear evolution contributed to the view of a hierarchy of the races of humanity. Racial theories based on this idea predated Darwin.[38] Unlike Darwinian evolution, which is branching and irregular, with all the human races arising simultaneously from a common origin, linear evolution showed the races arising at different points on the evolutionary line. Hence, "lower races" allegedly arose from earlier stages on this line compared to "higher races," which were perceived to be at later stages. Darwin's theory was still viewed in this way well into the twentieth century, and was used to validate these racist ideas.

The progress of a race or nation in terms of social and cultural evolution was often viewed as synonymous with biological evolution. This further resulted in the perception that the hierarchy of cultural evolution of a race or nation could parallel that of its mental evolution. Darwinism and Mendelism, both hereditarian, provided a scientific basis for eugenics in determining one's biological character. However, Darwin was cautious about eugenics. Though convinced of Galton's ideas on the inheritance of intelligence, he did not advocate any programs which had to do with eugenics or social Darwinism, both of which were at their peak after his death.

The prevalent trend of imperialism in the early twentieth century was associated with Darwinism—as in the struggle for existence between "races of humankind." The domination of a race or nation over others was thus considered a natural part of humanity's advancement. Still, the subtitle of Darwin's *OS, The Preservation of Favoured Races in the Struggle for Life,* has very often been misconstrued to refer to "human races," and thus controversially validate various ideologies such as imperialism and eugenics. Darwin's use of the word "races" was in reference to the word "species," from the naturalist's perspective, and will be dealt with later. Eugenic discourse will

37 Turda, *Modernism and Eugenics*, 2.

38 For further information on racial theories see Bowler, *Evolution* (2009), 292–297.

be examined in the work of Greek intellectuals in this chapter and in chapters 2, 4, and 6 of this book.

Darwinism in literature and criticism

This section examines the primarily thematic approach needed for the analysis of the writings of Greek intellectuals. The issue of what might be considered as Darwinian or post-Darwinian thought in literary work has been quite varied, resulting in contradictory analysis between commentators. What used to be regarded as pseudo-Darwinian was also referred to as "Darwinistic," a term which is not recognized now by scholars. In 1972, as noted by Michele L. Aldrich, Elton Miles categorized literary works that derived their ideas of evolution through Emile Zola (1840–1902), or through philosopher Hippolyte Taine (1828–1893) as pseudo-Darwinian.[39] Zola, an evolutionist himself, did draw on Darwin, but he was influenced mainly by Claude Bernard (1813–1878), whose major work *An Introduction to the Study of Experimental Medicine* (1865) dealt with the scientific method. William Irvine claimed that Taine was not the "chief exponent of Darwinism in literature," as many were to believe.[40] Irvine maintained that Taine's scientific methodology belonged to that of Charles Augustin Sainte-Beuve.[41]

William Leatherdale, agreeing with Morse Peckham, labelled ideas which were not from Darwin's theories as "Darwinistic," or as "surrogate Darwinism" when associated with some other figure that had influenced evolutionary thought. As an example, Leatherdale defined Spencer's ideas of evolution in literature as Darwinistic.[42] However, I am in agreement with the more recent view of George Levine, who maintains that "it is impossible to disentangle 'authentic' Darwinism from what is called Darwinisticism."[43] For instance, as noted by Oldroyd, "the movement of literary naturalism sprang from the same soil as nineteenth-century philosophical naturalism.

39 Aldrich, "United States," 222–223.

40 Irvine, "Influence of Darwin on Literature," 618.

41 Ibid., 616–617.

42 Peckham, "Darwinism and Darwinisticism," 19–40. See also Leatherdale, "Influences of Darwinism on English Literature and Literary Ideas," 2–3.

43 Levine, *Darwin and the Novelists*, 9.

And Darwin did much to fertilize this soil."[44] For this reason one would not be able to exclude Zola's work from having Darwinian elements.

To further determine what is Darwinian or is influenced by Darwinism in Greece, I refer to the multidisciplinary approach for my analysis of Xenopoulos' texts, and that of other modern Greek intellectuals' works. It is a broad approach used by a number of commentators who include Gillian Beer in her *Darwin's Plots* (1983), a benchmark study in the critical discourse of Darwinism in literature; she examines Darwin's *OS* and the literary works of nineteenth-century British writers such as George Eliot and Thomas Hardy.

This multidisciplinary approach derives to some degree from the "polyvalent"[45] nature of Darwinian ideas applied to various disciplines. Investigating the critical discourse on Darwinian and other evolutionary ideas in literature, I discovered a number of issues which need addressing. Darwin's theory, as stated by Beer, has "extraordinary hermeneutic potential."[46] His work has the potential to be interpreted and applied in varying and even contradictory ways. For instance, as I have mentioned earlier in this chapter, both socialism and capitalism have had intellectuals apply some form of Darwinian evolutionary theory to them, despite the fact that they are opposing ideas. For his theories, Darwin drew on numerous disciplines apart from just botany and zoology. For instance, for his *OS* he was influenced by animal breeding and geology. Particularly for his writing of the *DM* he drew from anthropology and craniology, and it is argued that his work was intertwined with Victorian culture and ideology. For his book *EE* he was influenced by the work of past physiognomists and anatomists.

When specifically investigating the influence of Darwinism in literary works, culture, and society, there are certain ideas which do not comply with the Darwinian tenets. Such ideas, as pointed out by Beer, include "stasis," "return," "absolute replication" (cloning is its contrary), "pure invariant cycle," or "constant equilibrium."[47] So that any ideas that implied immutability could not be Darwinian. In addition, what Beer refers to as no "return" and no "pure invariant cycle" could apply to Nietzsche's philosophical con-

44 Oldroyd, *Darwinian Impacts*, 328.

45 Beer refers to the "multivalency of evolutionary concepts." See Beer, *Darwin's Plots*, 12.

46 Ibid., 8.

47 Ibid.

cept of the "eternal return" (also known as the "eternal recurrence"), a physical concept involving the return of beings in the same form an infinite number of times. Unlike time in Darwin's evolution, which is linear, Nietzschean time is cyclical. Nietzsche's concept of eternal recurrence and Darwinian evolution can sometimes be confused. An example of this will be examined in chapter 2 in relation to Kostis Palamas' work. Indeed, Nikos Kazantzakis, a quintessential Darwinian, rejected "eternal recurrence" in his chapter on Nietzsche in *Report to Greco*.[48]

The trend in scholarship has been to avoid an overall categorization or formula for Darwinian themes. An attempt to create "a systematic representation of scientific ideas" in literary works would be misleading, argues Gillian Beer.[49] George Levine in *Darwin and the Novelists* attempts "a sort of gestalt of the Darwinian imagination, a gestalt detectable in novels as well as in science"; but he argues that "no simple list of 'Darwinian' ideas will quite suffice to evoke it."[50] Jina Politi also speaks of a "Darwinian text," which she says is drawn from political economy (Malthusian). She aligns "Darwinian text" to Spencer's social and psychological Darwinism; to literary Darwinism created in works of the Victorian and subsequent periods, such as Tennyson's "In Memoriam" (1849), which preceded the *OS*; to the post-Darwinian novels of Thomas Hardy and George Meredith; and with a discourse of anti-Darwinian argument from a religious point of view.[51] Peter Morton considers any attempt to systematize Darwinian influence in Victorian literature, as part of the cultural impact of Darwinian biology, an "opus magnum."[52] Regarding the *OS* he admits: "Critics and historians have been fundamentally divided in their interpretation of Darwinism *vis-à-vis* the literary imagination. ... As for Darwin's broader influence on the literary imagination, there has been no agreement at all—neither about the nature of that influence nor even (in the extreme view) whether it is to be found at all."[53]

Since 1995, Joseph Carroll and his subsequent advocates have proclaimed that all literature can be viewed from an evolutionary perspective,

48 Kazantzakis, *Report to Greco*, 317–339. See further on Kazantzakis and his book in chapter 2 of my study.
49 Beer, "Science in Literature," 796.
50 Levine, *Darwin and the Novelists*, 13. See further on his themes later in this chapter.
51 Politi, "Δαρβινικό κείμενο και *Η φόνισσα*,"155–156.
52 Morton, *Vital Science*, 3.
53 Ibid., 3–4.

naming this approach Literary Darwinism.[54] They believe that all aspects of human nature, including one's creative imagination, are directed by the tenets of evolution, their environmental and/or genetic roots, and can be analyzed accordingly.[55] There is a concern that this approach lends itself to an overapplication of evolutionary theory. Furthermore, one might question whether the science on human nature has developed sufficiently that we can confidently label everything as either reliant on nature or nurture.

Hence, there is no rigid formula or agreement by commentators on the categorization of such ideas and themes in literature that can be adapted to the modern Greek literary perspective. We can only be guided by the knowledge of scientific trends, theories, and concepts at the time, the background of the author, the literary trends of the period, and the social context. These are the factors which contribute to determining what is relevant in terms of Darwinian and post-Darwinian influences in literature.

The ubiquity and vast diversity of application of Darwinism and associated theories to literature necessitates, as I have mentioned, complex multidisciplinary and interdisciplinary exploration. The range of areas which will be investigated includes: the history of science, feminist literary scholarship, and social history, as well as the application of Darwinism, medicine, and science to literature, such as the degeneration theories.[56]

This leads to further discussion associated with the way scientific material such as Darwinism translates to the literary world. I take the approach that there cannot always be a neat correlation between scientific Darwinism and the way it is represented in literature. In her 1996 book *Open Fields,* in the chapter entitled "Translation or Transformation? The Relations of Literature and Science," Beer highlights "interchange rather than origins and transformations rather than translation."[57] She concludes that "the questioning of meaning in (and across) science and literature needs to be sustained without seeking always reconciliation."[58] Beer's advice to students in

54 I use the term "Literary Darwinism" in association only with the work of Carroll and his associates; whereas "literary Darwinism" has been used in general, to denote literature influenced by Darwinism, and is independent of Carroll's approach.

55 For further reading on Literary Darwinism see Carroll, *Evolution and Literary Theory*; Carroll, *Literary Darwinism*; Gottschall and Wilson, *Literary Animal.*

56 This is not an uncommon general approach to the analysis of literature that contains elements of a biological or evolutionary nature. See Greenslade, *Degeneration*; Chamberlain and Gilman, *Degeneration.*

57 Beer, *Open Fields*, 173.

58 Ibid., 195.

the field of "science and literature" is to avoid "one-to-one systemization."[59] It is worth pointing out that historically, scientific discourse such as that of Darwin's *OS* was easily accessible, particularly to the English and American people, and very readable, so that those without scientific training could communicate the information through creative literature.

In this book, investigations associated with the responses of writers to Darwinism and post-Darwinian theories and ideas are not limited to pro-Darwinian thought, but also include anti-Darwinian thought. So that, for instance, when I am saying that the Greek writer Alexandros Papadiamantis was influenced by Darwinian ideas in his novel *The murderess,* I am not stating that he is necessarily an advocate of Darwinism. What I am saying is that his work displays themes or ideas which can be argued as having some Darwinian roots or influence. Unless a writer is openly taking a stance for or against Darwinism or other theories, it is not always possible to extrapolate from one novel whether a writer himself had pro-Darwinian tendencies.

Also, as is often the case with ideologies, one's views may vary in intensity or change a number of times within a lifetime. A prime example is the writer Samuel Butler (1835–1902), who initially was a strong Darwin supporter at the time he wrote his utopian satire *Erewhon* (1872), but later converted to neo-Lamarckism when he wrote such novels as *The Way of All Flesh* (published posthumously in 1903), *Life and Habit* (1877), *Evolution Old and New* (1879), and *Luck or Cunning* (1879).[60] An exception perhaps was Xenopoulos' underlying support for Darwinism reflected throughout his "Athenian Letters," even at times of variable Darwinian sentiment in society. His views in the "Athenian Letters" are the topic of discussion in chapter 3.

Almost any issue that could be associated or fused with Darwinism or evolutionism became a theme or idea in some form, taken up in many countries by the writers of imaginative literature in the nineteenth and twentieth centuries. Darwin's *OS* and later the *DM* produced a wealth of ideas for creative writers and writers in general. Darwin's theories fuelled the long ongoing evolutionary debate in the scientific world, triggering other scientific theories and intellectual trends. It should be noted that evolutionary theory had been an issue in the scientific world and thence the literary world

59 Beer, "Science in Literature," 797.

60 Bowler, *Evolution* (2009), 238–239, 367. Note also that Butler started to write *The Way of All Flesh* about 1872 and worked on it intermittently until 1884.

even before Darwin explained his theory of evolution by means of his theory of natural selection.

With Darwin's *OS* came a new interpretation of progress. Evolution, as applied to society, was not synonymous with progress towards a desirable goal. The Darwinian theory of gradual, random evolution is a "neutral scientific conception," and so evolution can appear to be progressive or regressive.[61] It should be noted that biological evolution is only random in terms of genetic variance, not regarding the process of natural selection.[62] Along with the progressive/regressive perception and evolutionary theories like those of Herbert Spencer,[63] many in the scientific and literary worlds would continue to view progress as a natural consequence of evolution.

Intellectuals applied progressive and regressive evolution to social, biological, and psychological ideas which became the theme in novels. H. G. Wells' book *A Modern Utopia* (1905), although of a science fiction genre, deals with a republic where the two upper classes are categorized according to natural psychological differences: the "kinetic" or the "intellectual," and the "poietic" or "creative." This is interpreted as one which "involves the presentation not of a finally perfect state of *being*, but a hopefully ascending one of *becoming*."[64] This theme will be seen in Xenopoulos' *Rich and poor* in association with the idea of immobility of the two social classes of rich and poor, and also in relation to the concept of socialism.[65]

The regressive possibilities of Darwinian evolution only managed to fuel the theories of biological and social degeneration which were circulating even prior to the *OS* and up to the 1940s. I have already mentioned Lankester's theory of biological degeneration. Darwin's theories played a decisive role in the changing meaning of degeneration. In English literature, the concept of degeneration was an important component of the novel from the mid-nineteenth century to 1940.[66]

61 Bury, *Idea of Progress*, 335.

62 The selection process is not random but directed. Natural selection is the "nonrandom elimination" of "inferior individuals." Mayr, "Darwin's Influence on Modern Thought," 80.

63 His theories were based on the belief in progress.

64 Henkin, *Darwinism in the English Novel*, 244 (italics in original). See Henkin regarding the evolutionary ideas associated with these types of dichotomies.

65 Alden, *Social Mobility in the English Bildungsroman*. See her work on Gissing, Hardy, Bennett, and Lawrence which focuses on the social factors involved.

66 To place degeneration in the context of other ideas in English literature of the nineteenth century see Trotter, *English Novel in History*, 111–127.

At the turn of the twentieth century, degeneration of the individual and social class was extended to that of the race and nation. The pursuit of "national regeneration" and "racial improvement" was associated with eugenics in its various forms. The eugenics theme did not escape the literary world, which will be further discussed later in this chapter and in chapters 2, 5, and 6.[67]

There are a vast number of works which exemplify the ideas of degeneration in a post-Darwinian framework.[68] The fin de siècle brought with it a situation where the concept of "degeneration facilitated discourses ... between the normal and the abnormal, the healthy and morbid, the 'fit' and 'unfit,' the civilised and the primitive."[69] The devolutionary nature of women is exhibited in Arthur Conan Doyle's novel *The Parasite* (1894) with the "vocabulary of atavism," and where "images of reversion to lower states of animality signified the uncontrollable mystery of female desire."[70] Female sexuality was a force to be feared, and dramatist August Strindberg (1849–1912) makes a point in the preface of *Miss Julie* (1888) that the negative feelings he had for women had Darwinian roots.

The following are examples of various forms of degeneration from a post-Darwinian perspective: the reversion to a primitive state in Conrad's *Heart of Darkness* (1902);[71] the theory of recapitulation as found in Arthur Conan Doyle's short story "The Adventure of the Empty House";[72] the duality of man in terms of the repressed "other" represented as a subhuman beast, as in R. L. Stevenson's *The Strange Case of Dr Jekyll and Mr Hyde* (1885);[73] H. G. Wells' *The Island of Dr Moreau* (1896), where Moreau attempts to bridge the gap between humans and animals with a surgeon's knife;[74] the degenerate Clem Peckover in George Gissing's *The Nether World* (1889), who reveals

67 For literature influenced by eugenics see Childs, *Modernism and Eugenics*; Cuddy and Roche, *Evolution and Eugenics*.

68 Greenslade, *Degeneration*. Greenslade also looks at themes of the New Woman, race-regeneration and eugenics.

69 Ibid., 2.

70 Ibid., 18–19.

71 Conrad, *Heart of Darkness*.

72 As quoted in Greenslade, *Degeneration*, 68–69. Recapitulation refers to Ernst Haeckel's nineteenth-century biological theory, where the development of an embryo to adult was said to reflect the evolution of species from their primitive ancestors; this theory of "ontogeny recapitulates phylogeny" is now invalid.

73 Stevenson, *Strange Case of Dr Jekyll and Mr Hyde*; Greenslade, *Degeneration*, 85–86. See also chapter 5 of this book where the "other" theme is dealt with further.

74 Wells, *Island of Dr Moreau*.

atavistic traits of maleness and eroticism due to her "deviation from conventional womanhood";[75] Bram Stoker's *Dracula* (1897), which shows themes of atavism and parasitism, and picks up on the debate on the physiognomy of the born criminal.[76] To add to this, metamorphosis or transformation plays a role in some of the above and also in Gissing's *Demos* (1886)[77] and in Arthur Conan Doyle's *The Hound of the Baskervilles* (1902), where Lombrosian physiognomy is used to brand criminals as having "a defective evolutionary inheritance."[78] This type of degeneration will be demonstrated in chapter 5 with the devolution of Xenopoulos' character Tereza. The following passage serves to exemplify this with the atavistic and beastly character of Slimy in George Gissing's *The Unclassed* (1884): "Leaning on the counter, in one of the compartments, was something which a philanthropist might perhaps have had the courage to claim as a human being; a very tall creature, with bent shoulders, and head seeming to grow out of his chest; thick, grizzled hair hiding almost every feature, with the exception of one dreadful red eye, its fellow being dead and sightless. He had laid on the counter, with palms downward as if concealing something, two huge hairy paws."[79]

The negative representations of the New Woman in late nineteenth- and early twentieth-century novels were also associated with post-Darwinian degeneration, with the New Woman's alter ego, the "womanly" woman, seen as a theme of regeneration in novels. Commentator Sandra Siegel describes the degenerate form of the New Woman as follows: "The New Woman, like her mirror image, the new decadent, who was always male, confused what was essential to her nature. She not only moved in the public sphere, but behaved like a man, even as the new decadents, in their self-absorption and inaction, behaved like women, lost their masculine vigor. For the Victorians, any confusion of gender was bound to have implications for 'civilization.'"[80] Degenerate representations of the New Woman in

75 Gissing, *Nether World*; quoted in Greenslade, *Degeneration*, 75–76.

76 Stoker, *Dracula*. For a further discussion on the physiognomy in *Dracula* see Pick, *Faces of Degeneration*, 167–175.

77 Gissing, *Demos*; Greenslade, *Degeneration*, 79–83.

78 Doyle, *Hound of the Baskervilles*; Greenslade, *Degeneration*, 93–94. Also, Italian criminologist Cesare Lombroso (1836–1909) believed that physiognomically a criminal type was distinguishable from a normal person. Note that Lombroso's degenerate physical stigmata were also found in geniuses, artists, and prostitutes.

79 Gissing, *Unclassed*, 66; quoted in Greenslade, *Degeneration*, 52.

80 Siegel, "Literature and Degeneration," 209.

literature are seen in the works of August Strindberg, Henrik Ibsen (1828–1906), and Emile Zola. I will be investigating Xenopoulos' discussion of Strindberg's *Miss Julie* and his use of the play as a basis for his own novel, *The Three-Sided Woman,* in chapters 5 and 6 of this book. For further discussion on the New Woman novels see chapters 5 and 6, particularly on how the New Woman was perceived as a detriment to the building of the nation or race.

In 1959, on the centenary of the publication of the *OS,* William Irvine in his essay "The Influence of Darwin in Literature" highlighted the trend among late nineteenth-century commentators to use evolutionary ideas as an analogy for the development of literature.[81] He observed that French commentator Ferdinand Brunetière (1849–1906) in his first volume of *L'évolution des genres dans l'histoire de la littérature* (The evolution of genres in the history of literature) (1890) used Darwin's principles to explain, by biological analogy, the evolution of literary genres and the evolution of literary criticism.[82] Further to this in 1963, as an example, René Wellek indicated that Brunetière "parallels the history of genres with the history of human beings," uses "the analogy of the struggle for existence to describe the rivalry of genres, and argues that some genres are transmuted into other genres."[83] Both Irvine and Wellek maintained that Brunetière's applications did not always hold—for example, when he discussed the transformation of poetry to prose. Darwinian and other evolutionary ideas were applied in many areas such as the history of ideas, the history of poetry, and the evolution of style by William Bagehot.[84] Irvine believed that, above all else, this literary application of evolution was most important to "the concept of genetic development."[85] The Greek poet Palamas applied the concept of evolutionism to his essays on language, poetry, and the continuity of the Greek "soul," and this is dealt with in chapter 2.

81 Irvine, "The Influence of Darwin on Literature," 617–619.

82 Ibid., 617.

83 Wellek, *Concepts of Criticism,* 44.

84 Irvine, "The Influence of Darwin on Literature," 618–619. For further applications see Wellek, *Concepts of Criticism,* 42, on John Addington Symonds' application to the history of Elizabethan drama (1884); Richard Green Moulton applied it to Shakespeare (1885) and used it as late as 1915. Wellek also indicates that English and American studies of oral literature in these times were Darwinian based. Spencerian evolutionism also applied the law of development from the simple to the complex to the development of literature.

85 Ibid., 619.

Victorian and Edwardian novelists fused Darwinian theory with "allusions to much older texts" or with "evocations of ancient texts and wisdoms."[86] This is also exemplified in two novels by Xenopoulos which will be examined in the next chapters—specifically, by the Greek mythical characters Lamia in *Tereza Varma-Dacosta,* and Phoibos and Merope in *The night of degeneration.*

It is important to note that critical literary scholarship has also been based on interpretation of Darwin's *OS* as itself a literary work, and not just a scientific writing. One of the first studies to examine Darwin's monographs as literary works in their own right was Stanley Edgar Hyman's *The Tangled Bank* (1959).[87] Hyman saw the *OS* as a dramatic poem of tragedy and discussed Darwin's intended metaphoric nature of the terms "the struggle for existence," "natural selection," "great tree," "tangled bank"—metaphors for the evolution of life—as a theme running through the *OS*.[88]

Similarly, Gillian Beer examined the evolutionary narrative in Darwin's *OS* and has argued for the presence of Darwin's own literary imagination.[89] Also, she has argued that Darwin himself, when writing the *OS*, was greatly influenced not just by other scientific works, but by a vast range of disciplines, including creative literature. For instance, she demonstrates the presence in his book of numerous literary ideas, including various themes such as those of mythic metamorphosis and transformation.[90] The following is an example of the theme of transformation from the *OS*: "In North America the black bear was seen … swimming for hours with widely open mouth, thus catching like a whale, insects in the water. Even in so extreme a case as this, if the supply of insects were constant, and if better adapted competitors did not already exist in the country, I can see no difficulty in a race of bears being rendered, by natural selection, more and more aquatic in their structure and habits, with larger and larger mouths, till a creature was produced as monstrous as a whale."[91] Although this passage reflects creatively Darwin's theory of evolution, it does not reflect the extreme slowness of evolution, and as a result the situation appears absurd. Imaginative writ-

86 Hopkins, *Giants of the Past*, 11.

87 Hyman, *Tangled Bank*, 11–78. For further work on Darwin as a creative thinker see Culler, "Darwinian Revolution and Literary Form"; Manier, *The Young Darwin*; Beer, "Darwin's Reading," 543–588.

88 Hyman, *Tangled Bank*, 32–34. For Darwin's metaphorical interpretation of the terms see Darwin, *OS*, 116, 115, 171, 458.

89 Beer, *Darwin's Plots*, 8.

90 Ibid., 97–114.

91 Darwin, *OS*, 215.

ers were able to take up this type of description for their purposes. Beer cites Ellegård: "Anti-Darwinian writers seem to have deliberately played on the ambiguity of 'a bear' and 'a whale.' The word can be used to denote the individual, and to denote the class, or species."[92] In chapter 5 I will demonstrate how Xenopoulos uses this particular theme of transformation in illustrating changes in the main character Tereza in *Tereza Varma-Dacosta*. This creative quality in the *OS* allowed literary writers to take up features into their own work that were not purely scientific. My point here, then, is that the *OS* has elements of non-fiction science and also of creative literature, and that this shows an overlap of genres, not uncommon in cross-disciplinary fields such as those of science and literature.

The future of the race

At a time when scholarship on race and eugenics is growing, there is also a recent interest in modern Greece. Sevasti Trubeta, for instance, discusses eugenics in association with various Greek institutions and societies.[93] Other studies show that in the earlier part of the twentieth century in Greece eugenic thinking and reforms were triggered in response particularly to the social questions of population growth, high birth rates, and public health, and more specifically, that they were associated with the issue of improving the biological quality of the nation.[94] Trubeta adds further that in order to redistribute resources in society eugenicists "suggested reducing the percentage of the needy in the population by restricting their biological existence rather than by changing the conditions of inequality," so that "the emphasis on the 'quality' of the population and the omnipresent ghost of degeneration effectively shifted from social care to managing inequality through birth control and eugenic selection."[95] However, she points out that in this early period the Greek state was unable to create a scheme for eugenic birth control even though there was significant support from various fields, including medicine, academia, and politics. Trubeta further speculates that the reasons could be political or even due to the controversial nature

92 Ellegård, *Darwin and the General Reader*, 238–241; quoted in Beer, *Darwin's Plots*, 98.

93 Trubeta, *Physical Anthropology*.

94 Trubeta, "Eugenic Birth Control," 273; Theodorou and Karakatsani, "Eugenics and Puericulture," 299–303.

95 Trubeta, *Physical Anthropology*, 212.

of eugenics, but also due to the then nascent stage of Greece's infrastructure, which would probably have been unable to support such programs.[96] Various institutions and societies which had affiliations with eugenics, race, and population strategies would be established in the interwar period, such as the Greek Anthropological Society in 1924 and the Chair for Physical Anthropology at the University of Athens in 1925.[97]

Complementing this research, Vassiliki Theodorou and Despina Karakatsani have also indicated that in Greece in the 1930s eugenic approaches were mainly associated with educating people on positive eugenics in the pursuit of a healthy new generation.[98] Compared to other countries of Eastern Europe, eugenics in Greece became more developed after World War II.[99]

Internationally, nation-building has been an important matter for investigation from the nineteenth century to the present. Though much of the scholarship associated with modern Greece's nationhood and national identity is sociopolitical, there is another perspective which Greek intellectuals in the early period did voice. Towards the end of the nineteenth century and into the twentieth, many European scholars, including Greek, appropriated an evolutionary discourse to investigate not only the ongoing concern of the degeneration of the individual but also that of the nation.

After 1821, when gaining its independence from the Ottoman Empire, Greece pursued its 1844 irredentist concept of the "Great Idea," and also attempted to regain a pure Hellenic identity. A controversial counter-discourse to the idea of the continuous lineage of the modern to the ancient Greeks was that of the Austrian scholar Jakob Phillip Fallmerayer (1790–1861). In 1830, he declared that in the period between late antiquity and the early Middle Ages the ancient Hellenes had been completely wiped out and were replaced by Slavs; so that the modern Greeks, he claimed, were biologically not descended from the ancient race.[100]

---

96 Ibid., 296–297.

97 From the second decade on, the roles of certain Greek anthropological institutions would include an attempt to find scientific evidence for the biological continuity between the ancient and modern Greek races, and also to refute the theories which claimed Greek inferiority and racial impurity due to the presence of other ethnic groups.

98 Theodorou and Karakatsani, "Eugenics and Puericulture," 322–323.

99 This is according to the current work on the history of the Greek Eugenic Society carried out by Alexandra Barmpouti at Oxford Brookes University.

100 Other theories on the degeneration of the ancient Greeks were also in circulation—for instance, Charles Darwin commented in support of such a view in his *DM*, 1:177–178.

With its 1897 defeat by the Turks in Crete, Greece showed its inability to reclaim territory, thus foreshadowing the doom of its "Great Idea." In 1922 Greece's pursuit of this "Idea" to reestablish its Hellenism in the Near East became ill-fated with the defeat by the Turks, resulting in the Asia Minor catastrophe.

After the 1897 defeat to the Turks, with Greece at a low, Greece's intellectual elite would investigate how to steer the future direction of the nation. As occurred elsewhere in Europe, the evolutionary discourse in Greece would become a player in the development of the nationalist discourse. In the *DM* Darwin had propounded that not only the physical but also the mental attributes of humans were subject to the theory of evolution by natural selection. At a time when progress was depicted in an evolutionary manner, the physical and mental evolution of the individual and the nation were viewed as mechanisms in the survival of the fittest.

Any factor that might be perceived as hindering the progress of the nation or exhibiting no direct advantage to survival was potentially a "degenerate" element in a society. Such elements included the arts, philosophy, and aesthetics, when around the turn of the twentieth century those in the art world showed concerns that with the evolutionary Zeitgeist science would see the death of the arts and its associated fields. Racial eugenicist discourses among the elite were significant in medicine, science, literature, and politics. Other matters such as "the language question" of the Greek nation were expressed from a sociopolitical perspective and also biologically, where the demotic or spoken Greek language was considered by many purists as degenerate.

In Europe the question whether art and aesthetics were necessary tools for the development and survival of an individual also extended to that of a nation.[101] As will be discussed in the analysis surrounding Skliros' letter, Greek intellectuals certainly contributed to this controversy that was part

101 An example of this was the view taken by German ethnologist and philosopher Ernst Grosse (1862–1927) in his 1897 book *Beginnings of Art*, where he argued in support of art as a survival mechanism: "If art were indeed only idle play, then natural selection should have long ago rejected the peoples which wasted their force in so purposeless a way, in favor of other peoples with practical talents. ... Art ... is an indispensable social form, one of the most efficient weapons in the struggle for existence." Grosse, *Beginnings of Art*, 312–314. Contrary to this, Max Nordau (1849–1923) in his 1892 book *Degeneration* had expressed his view of the degeneracy of art as a detriment to society. Nordau's book was translated into Greek in 1911, subsequently influencing many intellectuals. For various other theories on the adaptive nature of the arts see Miller, "Arts of Seduction," 156–173.

of a bigger picture. Still this question is being discussed—that is, if art has an adaptive purpose in terms of promoting the survival of a nation or race.

George Skliros' letter in Grammata

In the first decade of the twentieth century, the eugenics movement continued its momentum with the formation of various committees and societies in Europe, America, and Australia. The first International Eugenics Congress was held in London in 1912. Intellectuals of various disciplines, politicians, and doctors would take part in an evolutionary discourse on nationhood and national regeneration.

The following analysis seeks to exemplify this trend in Greece also. In the period between 1913 and 1915, a discussion among intellectuals was published in the Greek periodical in Alexandria (Egypt), *Grammata* (*Γράμματα*), in response to a letter written on November 30, 1913, in Cairo, by George Skliros, entitled "Skepticism and our race."[102] The discussion appears to be representative of some of the first post-Darwinian and biological explorations of Greek nationhood. As one reads the letters it becomes clear what the title is alluding to—that is, the questioning of biological, social, and political factors potentially important for the progress of the Greek race.

The discussion hinges on these more experienced intellectuals attempting to advise the young contributors of the literary periodical as to the type of articles they should be writing. What these intellectuals have in mind is an ambitious view to reshape the mentality of future generations of Greek youth and, in general, to rectify the flaws of the Greeks; this all in the name of the social and biological evolution of the nation. Certain key aspects of the discussion are expressed within a post-Darwinian and, more generally, evolutionary framework. They cover various opinions on the role played by the arts, philosophy, aesthetics, science, and language as determining factors of a nation's progress. How some of these intellectuals wished to create a race fit for the future takes a hereditarian perspective, and at times a eugenic approach, in a Darwinian or Lamarckian form.

What transpires in these letters is a new discourse put forward by a number of Greek intellectuals who have adopted a wide range of post-Darwinian

102 Skliros, "Ο σκεπτικισμός και η φυλή μας," 289–291.

ideas, which they then situated within a national rhetoric on the progress of the nation. In his letter, Skliros commented on the mental and intellectual attributes of the young contributors of *Grammata,* and labeled them as the potential "regenerators of our race." He was very concerned that they too, like the Greek intellectuals to date, "have not been able to escape the relentless Fate of our Nation." This, Skliros believed, was due to their "negative, skeptical, and exclusively aesthetic tendency."[103] He warned that contributors of the periodical were not interested in what was happening in the real world, and asked of the young intellectuals to look to the example of Europe in terms of how to progress. The following is one of the key excerpts from Skliros' letter:

> Take a glance at Europe, the great guider of us all—because it has surpassed us by far in social evolution—and you will understand by yourselves why. What do we see there? Thousands of educated people who, even though they very well recognize skepticism and the other pessimistic systems, immerse themselves, without thinking too hard, in the volcano of everyday life, becoming enthusiastic, excited, affected by the various issues of the day—practical issues, economic, social, scientific, etc. Why is this? Because *enthusiasm is innate in their blood,* which needs expression and action and no philosophy and rationale can choke it and contain it. ... They love metaphysical philosophy also, but they leave it for their period of relaxation.[104]

Here the blood metaphor[105] implies an "inheritable" trait, one which, as Skliros will later explain, was unattainable by the Greeks: "Because—alas!—we are children of Anatolia drenched to the bone with the culture and mentality of Asia. And what is meant by Asia ... but a vast cemetery, with an endless deadness and silence in all its expressions of life, with a civilization whose only foundation is the material and only the material, the glorification, that is, of material success in life with all the legitimate and illegitimate means."[106] Such racial rhetoric would contribute to the argument for the purification of the nation of "unwanted" or "degenerate elements" which, as Skliros maintains here, was important for the future of the nation. Skliros then went on to

103 Skliros, "Ο σκεπτικισμός και η φυλή μας," 289.
104 Ibid., 290 (my italics).
105 Starr, *The Blood of the Nation,* 8.
106 Skliros, "Ο σκεπτικισμός και η φυλή μας," 290.

say that the right environment was necessary for society to produce children with superior mental attributes: "How can, I ask you, in such a traumatically dead and silent, traumatically material, and traumatically anti-human environment, for there to be born children with vital souls, with innate enthusiasm towards life around them and nature, and above all with innate concern and love for 'humanity' and its social destiny?"[107]

Is Skliros contradicting himself then with the following or was it a battle of nature versus nurture playing in his mind?: "This innate indifference towards humanity, no power, no rationale can change, unfortunately, because it is deeply rooted in our blood; it is a physiological necessity. We, therefore, will not do as the immediate positive workers of our race, because it is *physiologically impossible for us to change our blood* and only the grave will correct us—this great servant of evolution and progress."[108] This negative trait was to the detriment of the evolution and progress of the race. He saw extinction via death as the only way to rid the race of those carrying this trait: "Let us, however, have inside us one drop of sincerity, a little self-flagellating strength, and real love for our race; let us attempt at least to save our children, the new generation, which again and again is born and is therefore now forming its mental world; and with the nourishment we will give it, we, the *evolved* representatives of our race, will very much determine the form, the pace, and the direction which its mentality will take. Do not forget that *man is shaped in his childhood from which the rest of his life is dependent.*"[109]

Skliros' initial concern was that the young contributors of the periodical were wasting their time immersed in the arts and philosophies, particularly skepticism, which he believed would not help the Greek nation progress to the level of the rest of Europe. His by and large biological deterministic view does contain a Lamarckian element when he suggests that the new generation should be given "nourishment" to improve the race, and that "man is shaped in his childhood from which the rest of his life is dependent." In the early 1900s in Europe, neo-Lamarckism was one of the evolutionary theories determining the discourse on national character.[110] The view that

107 Ibid.

108 Ibid., 291 (my italics).

109 Ibid. (my italics).

110 Outside of Europe, the American educator and eugenicist Jordan David Starr (1851–1931) in 1901 would see a "double meaning" for progress: one was the advancement of civilization, dealing with promoting education and training, and the second meaning was race improvement. Starr, *The Blood of the Nation*, 14.

human traits acquired during one's lifetime could be inherited by offspring became the foundation for "national pedagogy."[111] The amelioration of the race could thus occur with the work of social scientists and psychologists—by changing a person's character and way of thinking.

Further to this, education of the nation's youth was of prime importance in the progress of the race. By taking on as a topic the development of the mental attributes of the Greek youth, the letters in *Grammata* also contributed to a broader discourse on "progress." In terms of education, the impetus came from progressive educationalists and demoticists who held that the proper education of the youth was a key factor in the development of the nation. Alexandros Delmouzos and others were instrumental from early in the twentieth century with the setting up of various educational programs and associations. With the difficult years in Greece (1898–1908), Delmouzos returned from his studies in Germany inspired to actively participate in the education movement. He saw it as "the ideal for the prosperity of the individual, society, and the national progress."[112] It became a priority for a number of new organizations such as the National Language Society (Η Εταιρεία η Εθνική Γλώσσα)[113] and the Education Association (Εκπαιδευτικός Όμιλος).[114] It was argued by the likes of Delmouzos that in order to improve the nation, the demotic language should be promoted in schools, resulting in the creation of progressive and free thinkers; such positive traits, it seemed, would culturally link the modern Greeks to the glorious ancient Greeks in an unbroken continuum.[115] One of his experiments was setting up in 1908 the progressive Higher Primary Girls' School (Ανώτερο Δημοτικό Παρθεναγωγείο) in Volos, which was subsequently closed down in 1911 because of its emphasis on the demotic language and the supposed teaching of Darwinism.[116]

Starr noted that education was not considered hereditarian, whereas "the change in the blood which is the essence of race progress ... finds it cause in selection only" (15-16). He argued that not only does "the blood of a nation determine its history," but that "the history of a nation determines its blood" (7).

111 Trencsényi, *The Politics of "National Character,"* 12.

112 Papanoutsos, *Α. Δελμούζος*, 13.

113 It was founded in 1904. Founding members included Ion Dragoumis, Kostis Palamas, Grigorios Xenopoulos, Elisaios Gianidis, and Andreas Karkavitsas.

114 It was founded in 1911. Founding members included Alexandros Delmouzos, Andreas Karkavitsas, and Nikos Kazantzakis.

115 Mackridge, *Language and National Identity in Modern Greece*, 284. Not all the intellectuals involved in the betterment of the nation would see this; instead, many felt that only change in the social structure of society could improve the nation.

116 See chapter 3 for a discussion by Xenopoulos in his "Athenian Letters" on this school.

Further to the discussion on racial discourse, Skliros envisaged that the nation's problems were caused by the mentality of the Eastern influence. He wanted the Greek people to be less insular, alluding again to the European model: "It would be a great shame and heartless for our race if we only looked at our emotional pleasure rolling hedonistically in our Anatolian nirvana, completely indifferent to external activity and life."[117]

Skliros' last words in the letter are reactionary: "It is about the regeneration of the mentality of our race. ... It remains for us to make the social and mental revolution, *purifying our race* from all the horrible Asian stigmata which till now have contaminated and withered it, condemning it to obscurity and barbarism."[118] The racial rhetoric is evident. His underlying aim was to harness a biological approach and rid the Greek race of degenerate elements by eugenic means, even though those means were not voiced. In accordance with the ideals of the "Great Idea," it is likely he was referring to removing the Turkish and any other "foreign" ethnic element occupying Greece, as a means to purify the race.

On the other hand, there are those eugenicists who were neo-Lamarckian, accepting that the environment could shape human nature and behavior, and also that acquired characteristics could be inherited. In spite of the prevailing synthesis of neo-Darwinism and Mendel's heredity, which suggested that the germ plasm was unaffected by the acquired traits of its host, neo-Lamarckian eugenics was still popular well into the 1920s and 1930s.[119] What follows here are the relevant aspects of subsequent letters by several intellectuals, published in the Greek periodical *Grammata* in response to Skliros' letter. In his response dated January 10, 1914, Nikolaos Giannios (1885–1958) largely agreed with Skliros; he believed the fault did not lie with the East, but rather that it was political—there being no adequate "driver" to direct the Greek nation. In light of hard hereditarianism he argued that if skepticism was a negative trait, heritable and irreparable—as Skliros believed—then it would not be possible to save the youth.[120]

The nationalist writer Ion Dragoumis (1879–1920) responded to Skliros in a letter from Vienna, dated January 31, 1914. He argued that aestheti-

117 Skliros, "Ο σκεπτικισμός και η φυλή μας," 291.
118 Ibid. (my italics).
119 See Turda and Gillette, *Latin Eugenics in Comparative Perspective.*
120 Giannios, "Ο σκεπτικισμός και η φυλή μας," 505–506.

cism and skepticism were transitional phases in a youth's life and could be an asset: "Skepticism is most valuable for the Greek youth of our time and shows sufficient maturity, as it bothers them to embrace and accept without objection and inquiry every old and new idea which the Greeks, Anatolians, or European civilizations present to them or want to impose on them. ... It is not that they are indifferent. ... For some of you ... thought will be fated as the only supplement of your life, because you are born this way."[121] He justified the status quo of the periodical by reasoning that its creation was a dynamic activity in itself: "Besides, I have great doubts if [the ability] to act can be sharply separated from the thought, with the two intertwined, so much so, that they cannot be untied. So that is why I say that the creation of *Grammata* is an act, that is, an action, since it required energy to result from a theory to an act, and since the ideas inspire energy in others."[122]

Showing further support for the youth as they are, his last comment in the following passage is conducive to Darwin's posit on the variability within species: "Mr. Skliros prefers that you are enthusiastic for practical, economic, social, and scientific questions. You however prefer to be enthusiastic about aesthetics, the arts, [and] philosophy. Good on you! ... Why should we be exclusive and all at the same level? How great is the world with its inequalities and varieties!"[123]

Unlike Skliros, Dragoumis did not believe that the Anatolian civilization was particularly materialistic or that it had "a deep disregard for every mental and moral expression."[124] In terms of bettering itself materialistically, he could not observe any difference between the Anatolian and European civilizations. Dragoumis asked the readers to refer to his 1913 essay "Neohellenic civilization" ("Νεοελληνικός πολιτισμός"), where he took an opposing view to Skliros' glorification of the European civilization and his condemnation of the East.[125] In order for Greece to move forward, Dragoumis implored the readers to nurture every aspect of Greek culture and tradition, and to turn their backs on trying to imitate other European cultures.

121 Dragoumis, "O σκεπτικισμός και η φυλή μας," 507. Dragoumis had several pseudonyms, including Anthropoid Ape (Ανθρωποϊδας πίθηκος), which is probably in the spirit of his Darwinian views. Anthropoid Ape might be a complex pun on his other pseudonym, Idas. He signs off as Idas in this letter.

122 Ibid., 508.

123 Ibid.

124 Ibid., 509.

125 Dragoumis, *Ελληνικός πολιτισμός*, 71–86.

Petros Vlastos (1879–1941) wrote in response to Skliros' letter on December 22, 1913, in Athens. Here they are in agreement that with nurture a lot could be achieved to correct the Greek race. Vlastos' concern was how the race would be shaped, revealing: "The essence of the race determines the life and the culture of every nation. As to how a race will improve and be cleansed, it is a complex matter and one which still requires a long study. A society can lay down some *negative barriers,* so that the blood of the race is not poisoned by those entirely rotten, but the essential job, in my opinion, only individuals can do it."[126]

His comments are eugenically based with reference to the implementation of negative eugenics, though a plan of reform does not appear forthcoming. If anything, his suggestion of how to remove undesirable elements from the race is ominous.[127] Vlastos' final point highlights the ongoing debate on the Greek "language question." In fact, what is seen in the following is his support for the pervasive perception among many Greek intellectuals that the state of the language is linked directly to the progress of the Greek race: "And please allow me to take to task Mr. Skliros that since he has so many ideas about the Greek race and its reformation, he must not write in the katharevousa but in the demotic. From where will the action start if not from us who see the truth? ..."[128]

Language purists, unlike the demoticist Vlastos, maintained that the modern Greek race had changed little from ancient times, and believed that the official modern Greek language should be the katharevousa—that is, a purist form with archaic expressions based on ancient Greek. They claimed that the demotic or contemporary spoken language of the people—a simpler form with its foreign borrowings—was a degenerate element in society, and that purifying the language would actually improve the Greek race.

Demoticists and liberal thinkers who were supporters of this common language of the people also assumed a connection between the ancient and modern Greek civilizations. Mathematician and demoticist Elisaios Gianidis (1865–1942) in his study *Language and life* (*Γλώσσα και ζωή*) (1908) argued that this newer Greek language "is not a degenerate end product, but

126 Vlastos, "Ο σκεπτικισμός και η φυλή μας," 509 (my italics).

127 Vlastos goes on to say there should be respect for atomic energy, restriction of "tyrannical socialism," and an understanding of the importance of not being ashamed of oneself. Ibid.

128 Ibid., 510.

a pure offspring of the ancient language, born according to all the laws of the science of language."[129]

Scholarship on the Greek "language question" has focused on the sociopolitical occurrences influencing the language issue.[130] In particular, Greek socialists such as Dragoumis believed that maintaining the demotic language of the people as the official language would direct the recognition of the rights of the people, and give them a voice politically. Katharevousa was used in all formal affairs and was the language of the educated that were in the minority.

For Vlastos, Skliros' interpretation of the Greek race, which had biological undertones, should have been written in the demotic. Vlastos' comment on the Greek language in relation to national regeneration is significant, as it marks the perceived view that the right outcome on the "language question" would contribute to Greece's progress. Skliros would later point out that the mixed form of the language—that is, a combination of the demotic and katharevousa—was not perfect, but would gradually evolve, resulting in this mixed form which he believed would suit the people.[131] He aligned the evolution of the language with the tenets of biological evolution: "It is necessary then for quite some time for some transitional mixed stages [to occur]. This after all is suggested by the simple rationale of *the gradual evolution of all things* and the experience of everyday life. ... If the structure of katharevousa is demolished, then from its ruins and from the material of the demotic, our language will *gradually* be created."[132]

In 1907, the converted and strong advocate for demoticism Kostis Palamas identified the language issue as an intellectual one, requiring an evolutionary approach. Believing it was not a political issue, he knew though that it would become a political tool.[133] At that time, the demotic or common language, particularly in its extreme form, was referred to as the "malliari" (μαλλιαρή), literally meaning hairy, but implying "a vulgar tongue."

129 Gianidis, *Γλώσσα και ζωή*, 612. Stamatios Stamatakis wrote under the pseudonym Elisaios Gianidis.

130 See Mackridge, *Language and National Identity in Greece.* Mackridge provides a thorough historical disquisition on the sociopolitical developments of the modern Greek language in relation to the nation.

131 Early on in the first decade of the twentieth century, Xenopoulos would also inform his readers, in his "Athenian Letters," that the best language to use was a mixed form, one that was "middle-of-the-road"—that is, neither the vulgar demotic nor the archaic katharevousa. Further on Xenopoulos' instructions to the youth on the topic of language see Malafantis, *"Αθηναϊκαί Επιστολαί,"* 136–147.

132 Skliros, *Τα σύγχρονα προβλήματα*, 14 (my italics).

133 See chapter 2 further on this.

The term was said to be derived from the word given to those demoticists in the late nineteenth century who wore their hair long. It is not inconceivable that this term could also have implied a perceived primitivism with respect to the appearance of the user of the language, and also with the language itself; in which case, one could consider that this was an example of how elitist intellectuals believed that the more developed or sophisticated a language, the same could be said about the nation or race.

In the *DM* Darwin suggested that languages, like organic beings, can follow the tenets of evolution.[134] He used the analogies of extinction, variability, and gradualism, and quoted Max Müller (1823–1900) as saying that: "A struggle for life is constantly going on amongst the words and grammatical forms in each language. ... Those words and grammatical forms that survive owe their success to their 'inherent virtue.'" To add to this, Darwin highlighted: "The survival or preservation of certain favoured words in the struggle for existence is natural selection."[135]

The German biologist Ernst Haeckel, a strong supporter of Darwin, played a significant role in the dissemination of Darwin's theories in Greece. Many Greek intellectuals, including George Skliros, having studied in Germany, followed Haeckel's lectures.[136] Contrary to Darwin, Haeckel maintained a theory of evolutionary polygenism, believing that human races were created separately. Further influenced by linguist August Schleicher (1821–1868), Haeckel believed that languages belonged to a hierarchy that paralleled the evolutionary hierarchy of the races. Later commentators would choose to assume that certain races, perceived as superior, would be associated with a superior language.

In the *DM* Darwin made an observation which he would use as an analogy to dispute ideas on the superiority of languages according to complexity and symmetry. He stated:

> With respect to perfection, the following illustration will best show how easily we may err: a Crinoid sometimes consists of no less than 150,000 pieces of shell, all arranged with perfect symmetry in radiating lines; but a natural-

134 Darwin, *DM*, 1:60. Darwin was influenced here by Charles Lyell, who in his 1863 *Geological Evidences of the Antiquity of Man* envisaged that the evolution of language resembled that of a species.

135 Ibid., 60–61.

136 Krimbas, *Δαρβινισμός και η ιστορία του*, 128.

> ist does not consider an animal of this kind as more perfect than a bilateral one with comparatively few parts, and with none of these alike, excepting on the opposite sides of the body. He justly considers the differentiation and specialisation of organs as the test of perfection. So with languages, the most symmetrical and complex ought not to be ranked above irregular, abbreviated, and bastardised languages, which have borrowed expressive words and useful forms of construction from various conquering, or conquered, or immigrant races.[137]

In the late nineteenth and early twentieth centuries, all manner of interpretation would have arisen from evolutionary discourse invoked to evaluate languages. Important to the Greek purists was to have a language devoid of the "degenerate" foreign influence, but which embraced the legacy of the supposed superior ancient Greek race. Contrary to this, the demoticists would have taken Darwin's view in the aforementioned citing. They would have valued the spoken language as a result of its modern diachronic development, with all its foreign borrowings appropriated from the Ottomans, Venetians, and other Europeans. Nonetheless, the Greek "language question," strongly associated with nationhood and Greek identity, would remain unresolved until well into the twentieth century.

Dimitris Zachariadis (1879–1922) was one of the founders and editors of the periodical *Grammata*; he was also anti-Marxist. In response to Skliros' letter, Zachariadis wrote: "Europe, that great guider, may know Marxism and other such stupid theories, but educated people there embrace daily life and indulge in art, philosophies, and scientific studies."[138]

In general, as noted by Raf de Bont, the application of evolutionism to society in the late nineteenth and early twentieth centuries was mainly by scientists with socialist or progressive liberal backgrounds, sociologists, pedagogues, and anthropologists.[139] So it is probable that Zachariadis did not accept such approaches when he mentioned that he viewed sociologists as pseudoscientists and rejected Marxist ideology.[140] Zachariadis maintained that in accordance with Marxist theory, Skliros saw artistic expression as lowly, and asserted: "Naïvely, the supporter of Marxism thinks that

137 Darwin, *DM*, 1:61–62.
138 Zachariadis, "Σκεπτικισμός, αισθητισμός και οι θεολόγοι," 511.
139 Bont, "'Foggy and Contradictory,'" 197.
140 Zachariadis, "Σκεπτικισμός, αισθητισμός και οι θεολόγοι," 514–515.

the poetic period of humanity has passed and that the scientific period has from now on superseded it. He did not notice, nor does he notice, that it is about two unbroken parallel and simultaneous ways of life; true, different in their expression, analogous however in the result. And all this misunderstanding because the scientist did not take care to determine that there are two types of truths in the world, the scientific and the poetic."[141]

Other Greek intellectuals would also contribute to the debate on science and the humanities. At that time, theorists speculated that Marx did not consider art as having contributed to the essential requisites for the historical progress towards a socialist future.[142] In 1857 Marx himself highlighted in his *Critique of Political Economy*: "As regards art, it is well known that some of its peaks by no means correspond to the general development of society; nor do they therefore to the material substructure, the skeleton as it were of its organisation. ... Certain important creations within the compass of art are only possible at an early stage in the development of art. If this is the case with regard to different branches of art within the sphere of art itself, it is not so remarkable that this should be also the case with regards to the entire sphere of art and its relation to the general development of society."[143]

Zachariadis noted that "Marxist sociology, the scientific one, ignores every poetic expression, since it [poetic expression] cannot be quantified, or it is considered to be primitive ravings."[144] He hailed the lofty nature of art and the artists' mission, which he felt did not coincide with a scientific mission. Further to this he mentioned: "Skliros thinks that he is allowed to intervene and to dissuade the artist from his mission in the name supposedly of humanity's interest. ... *Grammata* therefore justifies its program, and consciously follows its artistic aesthetic mission. Its expression is neither stained by Asian stigmata, nor characterized by retrogressiveness."[145]

Hence, he did not believe that the young contributors to the periodical upheld a degenerate perspective; and yet the "artistic aesthetic mission" would be seen by many as a deterrent to the progress of the nation. Regarding the view of the Greek civilization as opposed to the European civilization, Zachariadis endorsed arguments promoted by Dragoumis and Perikles Giannopoulos thus:

141 Ibid., 517.

142 Werckmeister, "Marx on Ideology and Art," 501–502.

143 Marx, *Contribution to the Critique of Political Economy.*

144 Zachariadis, "Σκεπτικισμός, αισθητισμός και οι θεολόγοι," 517.

145 Ibid., 519.

> The Greek civilization is subordinate to the European. But that occurs because the contemporary Greek civilization is simply an imitation of Europe. Every imitation signifies an inferior stance. Everyone who comes into contact with Europe tries to mimic it, making it more complete, more perfect, more alike to the prototype, till the final assimilation. ... Only when the moment of creation comes ... for the Greeks, when they depart from the point of imitation and come to the stage of self-existence, ... when our civilization becomes different to the European, and it departs from the common standard, the common criterion which ties it today with the European, only then will we be able to speak proudly about the Greek civilization, not as superior than the others, but different.[146]

Socialist Christos Christoulakis also wrote his response to Skliros' letter in August 1914. He asserted that passing the time thinking philosophically nourished the individual. Appropriating a biological discourse, he explained that Skliros' views were contradictory, and entered into the debate as follows: "If for us it is 'physiologically' impossible to change our blood, why is this possible for our children? ... Either we stay with the physiological theory of the blood, and then the same we are obliged to accept for our children; or, if we accept for them a possibility of a change in mentality, then the same possibility we are obliged to recognize in ourselves."[147] Christoulakis pointed out the common arguments for nature versus nurture. His philosophical question raised his awareness of freewill and race: "Can we modify our will, yes or no? ... Can my will, which I found to be, for example, Asian, ... be modified and made non-Asian?"[148] He indicated that much had been written on this, but rationalized that nothing had been satisfactorily answered.[149] His intention basically was "to restore the authority of philosophy which Mr. Skliros so mistreated in his letter, and to show that the most profound goal of philosophical observation and self-analysis is self-creation, the exact opposite of what Mr. Skliros supports—that is, 'for man it is physiologically impossible to change his blood.'"[150] Christoulakis maintained:

146 Ibid., 520.

147 Christoulakis, "Ο σκεπτικισμός και η φυλή μας," 81–82.

148 Ibid.

149 His reference to "determinisme," as he put it, shows that he received the concept from the French, and he would also invoke Bergson's philosophy.

150 Ibid., 92.

> There are individuals who internally influence their will, and create their lives to a degree. If such individuals are not found among us today, then this is due to chance, as it would be due to chance if such individuals existed. Neither Asianism nor Europeanism nor anything else plays any role. Of course, for a philosopher to be born, as we understand it, some complex conditions must assist with this; on no account do I believe [these complex conditions] to be social. More likely it seems that they are *hereditary*. ... Perhaps these [complex conditions] bring to reality an intense burning of vitality, which *naturally would require many centuries to bring about*. And this burning could be one of the primary prerequisites for vital gentleness, which allows the philosopher to contemplate and create his life.
>
> As for the Greek race, it seems to me there will be moments when, contrary to Mr. Skliros' opinion, it will naturally surpass all the others one day from this perspective. To make known, that is, ... exemplary self-creative individuals.[151]

Optimistic about the Greek race, Christoulakis felt that what he considered to be the inheritable trait of vitality, would for the Greeks require a long term process of evolving; and requiring a gradual and natural process of mental evolution analogous to biological evolution.

George Papandreou (1888–1968), a liberal who was to become the prime minister of Greece in 1944–1945 and 1963–1968, responded to Skliros on July 26, 1914, from Berlin. In agreement with Zachariadis, he viewed art and science as two independent disciplines, both with the right to exist and of equal value. He revealed his "hard" and "soft" hereditarian views: "There are certain temperaments—geniuses, talents—who are inevitably inclined towards a certain direction. They definitely will take their natural course, unrelated to the environment's influence—they will be the born Scientists, Artists, the Practical ones. ... In life, however, apart from them, there are also the people who are average—and they are many, the Called ones!—with a corresponding education and corresponding capabilities in all areas. Their road is dependent mainly on the trends of the environment."[152]

Noteworthy is how much emphasis these intellectuals placed on the value of the arts and philosophy when ascertaining whether or not they

151 Ibid., 92–93 (my italics).

152 Papandreou, "Ο σκεπτικισμός και η φυλή μας," 94. The original Greek text has "the Called ones" as *οι Κλητοί*. This would be in reference to the saying "Πολλοί μεν οι κλητοί, ολίγοι δε οι εκλεκτοί" (Many are called, but few are chosen).

were relevant to the future direction of the Greek race. This view is confirmed by Papandreou, who expressed his view on the scene in Greece at the time, indicating: "Because in Greece till now the movement (apart from the national mania, political mania, and trade) was mainly 'language' and then 'literary,' that is why the great majority of our youth occupied themselves with the language and the literary movement with unhealthy unilateralism and excess, whereas other forms of life in comparison did not take up their time."[153] He was referring here to the continuing "language question."

He also felt there was in Greece another group, of youth who were "half-educated and idle," a product of "the free State ... who tried to dress their idleness utilizing literary means, so as to justify their idleness to the world and to themselves."[154] This situation, he claimed, was the "sick" state of Greece, and he did not believe it had anything to do with the supposed "immutable character of the race, which preserves Asian stigmata," but on the contrary was due to the "mutable historical conditions."[155] He explained that the reason for this was that Greece was still in its relatively infant stages as an independent state. Apart from its established parliamentary system, he envisaged that Greece "did not have the prerequisite political tides of economics ... (with economic evolution not having as yet created sufficient opposition among the classes) and, in general, 'social' trends. That is why the 'intellectual' battles dominated, and for reasons again historical, these were first and foremost the language and literary interests."[156] Papandreou thought that as the economic situation and the general evolution of Greece progressed, the aspirations of the Greek youth would diversify to other areas.[157]

Skliros responded primarily to Zachariadis' commentary, in a letter dated June 1, 1915. At this point, Skliros defensively admitted that initially he had actually been invited by the editors at *Grammata* to write something on sociology, freely expressing his opinion. He had not expected Zachariadis' negative and heated response, which emphasized that ultimately the periodical would remain purely literary.[158] Skliros highlighted that "sociology with all its imperfections would spread everywhere because it represent[ed]

153 Ibid.
154 Ibid.
155 Ibid.
156 Ibid.
157 Ibid., 94–95.
158 Skliros, "Ο σκεπτικισμός και η φυλή μας," 95.

the real essence of life."[159] He wrote that he had made no mention of Marxism, and that Zachariadis was avoiding the real argument—most likely, that of addressing the flaws in the Greek race. In addition, Skliros claimed that Zachariadis did not understand that life should be viewed from a biological perspective, as well as social: "Man as a biological [and] sociological unit, rammed in the cauldron of life; whether he wants it or not, he is obliged to spend his life biologically, in accordance not only with the imperfect, related human prism of his perceptions in general, but also, in particular, with the material of his epoch, rolling in the ideas, in the beliefs, in science, and in the arts of his epoch. ... And fortunately that it happens like this, because humanity owes its progress to this exactly."[160]

Skliros' final words—as he put it, on how to correct the faulty state of the Greek society— were:

> That only with two approaches could there be rectification of this flawed social situation: with the development of economic, industrial, and social condition of life on the one hand, and with the intellectual activity of the *chosen* element of our race on the other; activity which would be broad and would embrace all the disciplines of life—and not only poetry and literature—filling the minds of the children, who are now growing with rich and complex depictions for all the expressions of life, and not only one-sidedly with art, as occurs in the feudal societies, and as would occur out of necessity till now with us also, in accordance with our poor feudal environment.[161]

Undoubtedly, Skliros saw the arts as being of less value to society compared to the sciences, though he did concede that both are necessary. He responded to Zachariadis in relation to Marxism and the arts, defending Marxists. He believed they did not wish to abolish the arts, but simply to popularize them so that people of all classes could enjoy them.[162]

The closing comments came from the editors of the periodical *Grammata,* though it is not clear if Zachariadis, as part of management, was involved in creating these comments. The editors felt that for the social evolution of a nation to continue, the arts and aesthetics were necessary:

159 Ibid.
160 Ibid., 96–97.
161 Ibid., 99–100 (my italics).
162 Ibid., 102–103.

"Our sociologists come and, by naively simplifying the philosophical and aesthetic problems, they want to subjugate every intellectual expression in view of the populace. They say: 'We want to popularize the arts.' With one rushed oligarchy they believe in the objective knowledge of the sociological laws of evolution, baptizing it [evolution] as the 'warm embrace of life,' when dealing with the concerns associated with social and political problems; and sociologists name those who pursue the emotion of the beautiful as senselessly drenched 'in icy skepticism,' when it comes to negativity in the environment."[163] Again this confirms the opposition received by the arts world on the application of evolutionary ideas in the social sciences. Despite all that was said and done, the final consensus among those involved with the periodical was that it would remain at a purely literary level.

Worth noting here is Dragoumis' closing paragraph of his 1913 essay "Neohellenic civilization," which reveals his eugenic approach to ameliorating the Greek race: "A civilization is not only a community of thought, senses, and wants, but a basis for it to elevate the *superior species of man,* the exceptional blooms, the great men of every kind of human capability. *Civilization has the ability to create the exceptional individuals,* who include and represent all the attributes and capabilities of the civilization and nation which gave birth to them."[164] Interestingly, it is not mentioned how the "superior species of man" is to be created. Perhaps this was too controversial at the time to be expressed publicly.

Similar to Dragoumis, the staunch nationalist Perikles Giannopoulos (1869–1910) in his 1902 essay "Contemporary picture" emphasized Greece's loss of identity and its effect on the nation, as follows: "Europeanism, as an imitation, is not progress—instead, it is regressive."[165] In his quest for true Hellenism, most likely referring to the "Great Idea," he remarked: "Let the few take courage, the many will follow them fatefully, when we all help to *purify the horizons of ideas and physical things,* which blocked from us the good road towards the Greek Rebirth."[166] His disdain for any foreign influence is found further in several of his essays in the same book.[167]

---

163 Unnamed, "Ο σκεπτικισμός και η φυλή μας," 109.
164 Dragoumis, *Ελληνικός πολιτισμός,* 86 (my italics).
165 Giannopoulos, "Η σύγχρονος ζωγραφική," 60.
166 Ibid., 65 (my italics).
167 See "Η ξενομανία" (Foreign mania) (1903) and "Ελληνική γραμμή/Ελληνικό χρώμα" (Greek line/Greek color) (1904).

It is feasible to say that this type of discourse, with its evolutionary and eugenic undertones among highly influential Greeks of that period, would have played a role in the sociopolitical events which were to follow in Greece. This analysis is indicative of a discourse which was not exclusive to the intellectual elite of Greece, but pervasive in Europe and America. Greek scholars have tended to omit or minimize commentary on eugenic and other post-Darwinian discourse that surrounded matters such as the "Great Idea," the "language question," Greek identity, and nationalism. This would imply potentially incomplete evaluations of these issues.

To some Greek authors, Darwin's theory of evolution became associated with a fear of the loss of the arts. This was a concern strongly expressed by Zachariadis, but which has also been observed by others, such as Petros Apostolodis. In 1895, Apostolidis elaborated with reference to the negative first-wave reception of Darwin's theory by many literary writers: "The delicate world of the daydreamers and the hypersensitive constituted the crusade's lead—those who saw the poetry of life extinguished by the breath of the scientific geniuses. Indeed, a little later, Mr. Achilleus Paraschos, poet representing the hypersensitive of that period, did not omit demonstrating the culminating wrath in fiery verses."[168] Apostolidis attempted to explain the reasons for this, also revealing that this was an issue not just specific to the Greeks: "This symptom in general is nature. The fear of science and truth on behalf of all those [afraid]—that is, those confused who have an idea about the great and complex problems of life with a misguided perspective of the laws, that is, the phenomena of the governing of nature and life—did not occur only in our embryonic society. Everywhere and always, poets and artists, the believers in the mystery and transcendental character of the arts, intimidated by scientific discovery and who did not want to understand that science designed the mysticism by the theory of good, proved however that the arts have their roots in those recesses of human life that constitutes the noblest bloom."[169]

It is important to note that Skliros also wrote about his perception of the mentality of the Greek race in his book *Our contemporary problems* (1919):

168 Apostolidis, "Επιστήμη και Τέχνη," 25.
169 Ibid., 25–26.

> This mentality still has not grasped well the idea of evolution and has not become accustomed to the dialectical method. ... Only these can today keep an individual and a society in step with the vibrant environment of this century of ours. This mentality which with all of today's progress in science and philosophical thinking means to stay stuck in the pedantic, metaphysical way in which it thinks; which supposes things are *static, immutable,* and *separate* one from the other; which supposes the ancient Greek civilization as something disconnected, without previous sources and influences, after all, like some miracle; which was outside common human laws; and which supposedly said the last word in all things: the arts, science, philosophy, civilization, etc. And so, it is claimed, following only from the lights of that civilization, people can even today perfect themselves in all areas and be able to resolve all their questions and all the problems which concern them. Needless to say that this theory is literally freakish and not only can one most easily quash it historically with innumerable arguments, but even theoretically it constitutes serious hubris against the *theory of evolution* and the *gradual progress* of all humanity, from which today's scientific mind cannot and must not be distanced.[170]

In this passage, Skliros is scathing in his view maintaining that the cultural progress of modern Greece at that time was hindered by its ancient Greek essentialist notion of constancy and discontinuity. Rejecting metaphysics, he appropriated a pro-Darwinian approach in terms of how society should evolve in all aspects of society. Skliros argued against a "static" and "immutable" mentality, concepts which are counter to the Darwinian tenet of non-constancy. The view that things are "separate one from the other" and that the ancient Greek civilization is "disconnected without previous sources and influences" also goes against the Darwinian view of a common ancestral heritage for all life. In fact, Skliros praised evolutionary theory along with its tenet "gradual progress" of all humanity, a tenet which exemplifies "no breaks or discontinuities."

It should be said that this form of Platonian essentialism was a view held by many European philosophers and other intellectuals right up to the twentieth century. The essentialist view was that variation was an imperfect occurrence of the *eide,* or form, or essence. Darwin's theory of evolution

170 Skliros, *Τα σύγχρονα προβλήματα,* 46 (my italics). This passage has also been discussed in Zarimis, "Darwinian Reception and Selected Responses," 653–654.

which required variation of species for evolution to occur would go against the centuries-old essentialist view of the static, immutable, and separate.

Skliros' comments here contribute to the discourse on continuity which had its roots in nineteenth-century Europe and which became biological in the post-Darwinian period. From the Greek perspective, the theme of continuity—whether biological, and if not, cultural—was a key concept in debates about nationhood, race, Greek identity, and the Greek language; a theme which, as we have seen here, greatly influenced intellectuals in the early twentieth century.

CHAPTER II

DARWIN AND MODERN GREEK WRITERS

I often bless all novelists.
CHARLES DARWIN[1]

Jina Politi has once remarked that in 1877 "Darwinism constituted a field of intense questioning in Greece," concluding that "literary writers should not have remained unaffected."[2] Literary commentary on Darwinian ideas in the work of modern Greek writers has only been recent; it is disjointed, sporadic, and views on a particular piece of work can often be quite variable and contradictory. In my introductory chapter I discussed the possible factors contributing to the gap in literary scholarship, which were historical and also due to the way Darwinism was received.

Complementing Politi's views are those of Eftychia Amilitou, who argues for the absence of Darwin's scientific method in Greek literature, particularly in naturalism. According to Amilitou, the Greek expression of Zola's naturalism was restricted to "certain images of life of the common class and of their animal world, with the erotic instincts, the presentation of the body, [and] the realistic version of the characters' life."[3] She indicates that in Greece all these images were received from the French literary trend in an unclear and selective manner: "The *experimental novel* and the experimental method were never implemented in the Greek domain as a structural system of theories. The transference of the scientific method of Darwin, Claude

1 Darwin, *Autobiography of Charles Darwin and Selected Letters*, 54. Darwin wrote this in the context of what he thoroughly enjoyed reading. In his later years he much preferred novels to poetry or Shakespeare. Prior to writing the *OS*, he had been very interested in poetry and Shakespeare. Indeed he was highly influenced by them when writing the *OS*.

2 Politi, *Συνομιλώντας με τα κείμενα*, 160.

3 Amilitou, *Νικόλαος Σιγαλός*, 50.

Bernard, and Dr. Lucas to literature; the interaction of character traits in a family; heredity as natural law and the genealogical tree as a hermeneutic form; the aberrant behavior, and pathology determined by the combination of the nerves and blood in the body; the observation and the medical viewpoint are issues which never engaged Greek writers with the claims of a scientific approach that Zola propagandized for naturalism."[4] Although Amilitou initially makes a broad statement covering all Greek literature, it appears that she is focusing only on the naturalist period. However, she does not consider Darwinian thought in the later period of the realists, which is the period that I examine in this book.

My primary aim in this chapter is to explore a selection of the modern Greek writers' literary responses to Darwinism and to review any critical work already associated with these responses. I will be examining two poems regarding the impact of Darwinism in Greece in the 1880s. Furthermore, I have also chosen to examine a range of writers who are known to have read Darwin's work within the first wave of Darwinism.

The Greek writers I will be examining are the novelist, essayist, and short story writer Emmanuel Roidis, the poet Kostis Palamas, and the novelist, dramatist, poet, and essayist Nikos Kazantzakis. My analysis of each writer will be essentially by means of a selection of his own critical or creative work. I have also included a commentary on the much discussed novel *The murderess* (*Η φόνισσα*) (1903) by Alexandros Papadiamantis, as a primary example of how commentary associated with Darwinian ideas can vary so much. The themes arising from the works of these writers will provide a context for Xenopoulos' work, which will be dealt with in chapters 3, 4, 5, and 6. Other Greek writers will be examined along the way, such as Kalliroi Parren, feminist leader of the Women's Movement in Greece around the turn of the twentieth century, in chapters 5 and 6, and also Andreas Karkavitsas.

As noted in the introduction, the interest in the sciences in Greece was not as intense as in other Western countries, nor was the application of the sciences, in particular Darwinism, to literature a common trend among Greek writers. However, throughout this study I have noted that those writers who had some affiliation with the sciences or who had exposure to French, German, or English versions of Darwin's work were responsive to

4 Ibid., 50–51 (italics in original).

Darwinian ideas in their literary work. Investigations will show that Darwinian and post-Darwinian thought did influence Greek writers both in the period when naturalism dominated, at the end of the nineteenth century, and also in the period up to at least the 1930s.

"Dock" and "voltaire"

A significant poem but unknown among commentators is a thirteen-stanza poem entitled "Darwin" ("Δαρβίνος"), written in the demotic language—or the spoken Greek. The poet signs off with the pseudonym actually written in English as "Dock"; one could assume that it means Doc., short for doctor. It is a pseudonym that is still anonymous to the literary world.[5] The poem was published in May 1882 in the Greek satirical literary periodical *Don't Get Lost* (*Μη Χάνεσαι*) (see the appendix for the poem and my English translation).[6] Charles Darwin died on April 19, 1882, so the poem appears to be a memorial to him. Due to the pro-Darwinian views reflected in the poem, Dock could have been involved in the controversy between the University of Athens Medical School and the theologians in the 1880s.[7] In other countries too, some of the anonymous poems soon after the publication of the *OS* were by individuals who were affiliated with the sciences.[8]

The poem, which responds (though a little delayed compared to the British responses) to the aftermath of the publication of the *OS*, is very likely to be one of the earliest pieces of Greek literature to cover aspects of Darwinism extensively. It presents certain ideas in a satirical manner, hence evoking controversy and questioning in relation to Darwinism. The poet, through a first person narrator, is in awe of Darwin and presumably his theory of evolution. However, as noted in the first stanza, he finds Darwin and what he represents quite confronting, to the point that he wishes he could return to pre-Darwinian times. In stanzas 2, 3, and 4, overwhelmed by what Darwin may represent, he tries to fathom how Darwin could possibly have managed

5 My research on this poem is also in Zarimis, "Dock and the Darwin Debate," 369–384. The paper only refers to Pavlos Nirvanas as the possible Dock. In the present study the research is revised and extended to include several further possibilities for the identity of the poet.

6 Dock [pseud.], "Δαρβίνος," 4–5.

7 See the section on the reception of Darwinism in Greece in the introductory chapter of this book, showing that the periodical *Μη Χάνεσαι* housed a number of articles on Darwinism.

8 Beer, "Satire, Voice, and Nineteenth-Century Science," 273.

to work out the creation of the human species from a primordial organic and, possibly, inorganic world. He refers to it as a conjuring act which, in a way popular in those times, demotes Darwin's theory to that of the obsolete theory of spontaneous generation—the theory refuted by Pasteur in 1859, the year Darwin first published his book on the origin of species.

In stanza 2 the poet speaks to Darwin, yet it appears that he identifies him with his mechanism of evolution, a mechanism which involves the whole world. It may be that he is implying a pantheistic concept of the world, embracing the god as nature—an ideology taken up by those who had their religious faiths shaken post-*OS*. In the face of his religious childhood teachings, the poet compares Darwinian creation with traditional theistic creation. When he writes that god creates from "nothingness" and Darwin from "life," he brings attention to the fact that this evolution does not explain the cosmic process; it is rather a concept that describes the consequences of that process. Yet, on the other hand, the poet speaks to Darwin in the last two lines of stanza 6, declaring that the two versions of creation are quite similar in one particular, important aspect—that is, as he believes, they both include a creator, a god: "Ah philosopher, and if within, you don't hide a god, / all the same you truly do reveal him!!" It is possible that the poet here is referring to Darwin's initial exclusion in his *OS* of a creator, which reads: "There is grandeur in this view of life, with its several powers, having been originally breathed into a few forms or into one; ... from so simple a beginning endless forms most beautiful and most wonderful have been, and are being evolved."[9] The seeming exclusion of a creator in this passage caused a backlash from the Church, resulting in Darwin changing the sentence in the aforementioned passage to "originally breathed by the Creator" in his subsequent editions. Further to this, the poet speculates that Darwin's theory implies the presence of a creator who would have fashioned the early "forms" before the process of evolution actually took place.

The evolution to "perfection," as seen in stanza 7, was a misguided view held by philosophers in the eighteenth century and by some in the nineteenth century, such as Herbert Spencer. It would be a concept accepted as part of the evolutionary process and a theme appropriated by many post-Darwinians.

9 Darwin, *OS*, 459–460.

As displayed in stanzas 7 and 10, a key aspect of the poem is its ability to satirize aspects of Darwinian theory, such as the process of evolution of living things. Satire in the Darwinian works of literary writers became a tool for questioning and probing the ideas. It also served to voice the initial response to ideas which appeared absurd. While this was a very common technique used in both prose and poetry immediately post-*OS*, it was also popular in the visual arts, as can be seen in figure 2.1, "Man Is but a Worm" from the British *Punch Almanack* (1882). Considered to be satirizing Darwin's *Formation of Vegetable Mould through the Action of Worms, with Observations on Their Habits* (1881), it shows a rather refined gentleman who has evolved from the worm. The man appears to be in reverence of Darwin, who is somewhat reminiscent of God in the Creation of Michelangelo's Sistine Chapel.

That human beings arose from one or a few cells linking them to the remainder of the animal kingdom was a difficult premise to accept. Even though Darwin's evolution was branching, it was not unusual for writers, or as in this case poets like Dock, to link humanity unilinearly to other animals in an absurd manner, as is observed in the following excerpt from the poem:

> And now how can you marvel, if from the worm
> Was born the goby fish, if from it tortoises,
> From tortoises the eagles, from eagles the wolves,
> From wolves the reptiles, then rockmelons,
> In our skies the birds, the blossoms in the earth's soil,
> And from blossoms and birds even man was born?
> (stanza 10)

In stanzas 8 and 9 the poet attempts to appease the anti-Darwinian sentiment and show that Darwin's theory may have appeared absurd at that time, but in years to come society would come to accept it; in the same way, accepted technological advances of the day might have been considered unfathomable fifty years prior.

Stanza 11 continues with the theme of change, introducing the Darwinian concepts of the struggle for existence, extinction, variation of species, with the implication of natural selection, the latter of which is described as "the stronger ones remain, the finest ones remain." The poet also notes: "Nature resembles an endless factory / where you are created so you can

Figure 2.1. "Man Is but a Worm," *Punch's Almanack for 1882*, London, December 6, 1881, no page numbers. Courtesy of the State Library of NSW.

vanish, you die so you can live..." These last two lines thematically touch on the popular ideas of continuity and immortality, used as metaphors for the continuity of the human race and society. Here the poet highlights that only through the part of "you" which is the hereditary material does continuity occur, and the body dies. These ideas would foreshadow Weismann's theory of the continuity of germ cells.

The poet in stanza 12 reveals a feature of himself—he is in a position to view life under a microscope, so it is likely that he has a scientific or medical background. To further demonstrate the validity of evolution, he directs the reader to examine through a microscope the living world found in a drop of water or a crumb of cheese. Revealed is the Darwinian microcosm of organisms such as protozoa, exhibiting a struggle for existence, variation, and extinction, identical to that of the human world.

The theme of progress which was discussed in chapter 1 is emphasized in stanza 13. Following from stanza 12 the poet implies that progress is synonymous with evolution and is always advancing to a higher goal. He also extrapolates that as a universal facet of life, progress can be applied to everything, including technology (stanzas 8 and 9). This theme of progress was very popular with the literary world.

The poem shows that Dock agreed with most Darwinian concepts. However, it appears that Dock could not entirely exclude the idea of a god who created the world from nothing (stanza 6). Dock believed that once the world was created by a divine power then life was able to evolve according to Darwinian theory—that is, by way of natural selection. Like many intellectuals post-*OS* who could not come to terms with the absence of a creator, he embraced a pantheistic approach. The poem has all the hallmarks associated with the initial responses to Darwinism, exemplifying the type of poem which was seen in other Western countries up to the end of the twentieth century.[10]

Investigations into the identity of the poet have led to several possibilities. According to commentator Giorgos Valetas, the poet, dramatist, novelist, and doctor Pavlos Nirvanas was writing in the radical and pro-Darwinian periodical *Don't Get Lost* in 1884 under the pseudonym of "Haholos" (Χάχολος).[11] This could not be found. His other pseudonyms include "Doctor," "Mr. Doctor," "Dr. A," and "Dr. Ap" (Ιατρός, Herr Doctor, Δρ Α, and Δρ Απ).[12] Investigations show that there are no entries in this periodical under these pseudonyms; nor is there any work by Nirvanas or Apostolidis. It is

10 See Stevenson, *Darwin among the Poets*, 43–345; Roppen, *Evolution and Poetic Belief*, 175–464.

11 Valetas, *Η γενιά του '80*, 8. Valetas has probably made a mistake with his date of 1884 because *Μη Χάνεσαι* was published from January 16, 1880, until October 24, 1883. Note that the word "Haholos" means Russian peasant and also can mean one who is awkward in the way he stands and walks; the word could be translated into English as a "gawk" or "bumpkin." It is significant also that Nirvanas was born in Russia.

12 Delopoulos, *Νεοελληνικά φιλολογικά ψευδώνυμα*, 157. The pseudonym Dock is not in Delopoulos' book. Note that the use of pseudonyms, often more than one, was common practice by Greek writers. The name Pavlos Nirvanas, by which this writer is generally known, is itself a pseudonym; his real name is Petros Apostolidis. Note also that in Delopoulos' extensive list of pseudonyms Apostolidis is the only writer with pseudonyms that allude to a medical background.

possible that Nirvanas may have written this poem at the age of sixteen or seventeen, when he was about to become a medical student at the University of Athens. Despite his age, he was well entrenched in the poetry circles of the time. As noted by Valetas, "Nirvanas ... took part ... in the great poetic and intellectual revolution of '80, even though he was somewhat younger than the others."[13] Nirvanas' medical training would have given him the distinct appreciation of the microscopic world referred to in stanza 12.

This poem is representative of the poetry of the New Athenian School (Η Αθηναϊκή Σχολή), also known as the "literary generation of 1880" ("η λογοτεχνική γενιά του 1880"). The poets of the generation wrote in such periodicals as *Don't Get Lost* about social, political, and scientific themes in the demotic language; till then Greek poetry was written predominantly in the katharevousa or puristic Greek.

Interestingly, Dock only appears this once in the whole of the periodical *Don't Get Lost*. He could have also written under other pseudonyms, of course. George Souris (1852–1919), satirist poet and later humorist, also belonged to the New Athenian School. He did not have a scientific background and it is not known if he had read Darwin. Unlike Nirvanas, he wrote prolifically in the periodical from 1880 to 1882 under one of his known pseudonyms, where he actually signs off in English as "Souris." In his writings he generally used the demotic and at times the puristic, or even a mixture of both. He satirized the political and social matters of the day, and so many of his poems were quite ephemeral. Apart from politics, his works show that he was very well-read in all areas, to include feminism, philosophy, and foreign literature, such as Shakespeare.

It is also possible that another poet of the New Athenian School who had read Darwin and was interested in science could have written this poem. One such poet that stands out is Kostis Palamas, who also wrote prolifically under various known pseudonyms in *Don't Get Lost* at that time. It cannot go unnoticed that in that period several celebrated scientists in Greece who were Darwinians also wrote poetry. The Hellenized botanist Theodor von Heldreich (1822–1902), who corresponded with Charles Darwin writing to him in French, on one occasion sent Darwin an essay and poem that he had written. This we know from an 1880 letter sent by Dar-

13 Valetas, *Η γενιά του '80*, 6.

win to Heldreich thanking him for his essay and poem.[14] Botanist Theodoros Orfanidis (1817–1886) wrote satirical poetry for which he was well-known.

An even earlier Greek poem on the theme of Darwinism, though much shorter, is the three stanza poem entitled "Does it change place?" ("Τόπο αλλάζει;"), published in 1880 in the periodical *Rabagas* (*Ραμπαγάς*) (see the appendix for this poem and my English translation).[15] It is by Aristotelis Kourtidis, known for his children's stories.[16] He wrote this poem under the pseudonym "Voltaire," using an extreme colloquial form of the Greek demotic language. It is a highly satirical response by the poet's first person narrator, a rather crudely spoken student, who attends a lecture on Darwin's theory on descent. The poem indicates that the lecture was given by the Darwinian Ioannis Zochios (1840–1912), who was professor of physiology and anatomy in the School of Medicine at the University of Athens. As mentioned in the introduction, due to pressure by the Church, Zochios was actually reprimanded by the University for his confronting lectures on Darwin's theories. Following from this controversial matter, I would say that this poem was written in the spirit of that period, where the general community would have been horrified to hear that they, as humans, could have been descendent from the apes. And it is in this mocking and scoffing style that the poem has been written.

It is interesting that Kourtidis should choose in stanza 1 and 2 to align the transformation of the monkey to a human with that of an acorn to an oak tree. At the time, Darwinians such as Thomas Huxley utilized the transformation of an acorn to an oak tree to highlight their theory on development. Huxley argued, for instance: "It is a truth of very wide, if not universal, application, that every living creature commences its existence under a form different from, and simpler than that which it eventually attains. The oak is a more complex thing than the little rudimentary plant contained in the acorn; the caterpillar is more complex than the egg; the butterfly than

14 Krimbas, "Ο δαρβινισμός στην Ελλάδα," 91.

15 Voltaire [Aristotelis Kourtidis], "Τόπο αλλάζει;" 2. Also my investigations show that Kleon Ragkavis (1842–1917) in the "Notes" at the end of his 1877 epic drama *Ιουλιάνος ο Παραβάτης* (Julian the apostate) questions the Creation, invoking Darwin. See Ragkavis, *Ιουλιάνος ο Παραβάτης*, 497.

16 He was editor-in-chief of the magazine *Children's Guidance* from its beginnings in 1879 until 1895, after which Xenopoulos took over the position until 1947. Kourtidis' works include, as well as poems, articles in newspapers and periodicals (including in the *Children's Guidance*), short stories, novels, and translations of foreign literature.

the caterpillar; and each of these beings, in passing from its rudimentary to its perfect condition, runs through a series of changes, the sum of which is called its Development."[17] Huxley goes on to refer to the higher animals in a similar manner of metamorphosis and transformation. These transformations continue to fascinate us even today, just as they did in Darwin's time; so imagine the incredulous sentiments at that time when one was told that, in a similar manner, humans were said to be descendants of the apes.

In the third stanza of the poem the poet seems to almost come to terms with, as he sees it, the transformation of monkeys to humans. However, he cannot come to terms with the monkey's tail, the glaring point of difference between the two species. He satirizes as to where it might have been repositioned in humans, or perhaps that it might have withered away at some point. We will see next that Roidis would also find the issue about the tail a problem in view of the comparisons between humans and apes.

Emmanuel Roidis

Extensive satire on Darwinian theory can be found in the works of Emmanuel Roidis (1836–1904),[18] which he wrote in katharevousa. His formal background was literature and philosophy, which he studied in Berlin. In 1871, in an essay Roidis mentioned that he was reading Darwin when he received a collection of theatrical comedies from Angelos Vlachos (1838–1920).[19] In 1863 he declared:

> I happened to have become immersed, I don't know how, in the theories of the Darwinian school. For two months already my inseparable companions had been Vogt, Darwin, Büchner, Lamarck, Moleschott, and others who aspire to prove Solomon was right in asserting that *"man has in no way surpassed the*

17 Huxley, *Man's Place in Nature*, 81–82.

18 A significant amount of the content on Roidis can be found in Zarimis, "Influence of Darwinian Greek Literary Writers," 177–190. Since this article was published, the work has been revised and extended here in this study.

19 Note that in 1877 Angelos Vlachos and Roidis will battle regarding two issues: the first was whether "a poet is born or made"; the second was the status of Greek poetry, whether it had a future or not. The first issue touches on the work of scientists such as the Darwinian Francis Galton. This controversy will be discussed further in chapter 4 with reference to Xenopoulos' novel *Rich and poor.* The issue of the status of Greek poetry was part of a much broader discussion, to which many Greek writers contributed. Palamas, as noted in the next section, also talks about Greek poetry using Darwinian and other evolutionary ideas.

> *animals," and that there is no substantial difference between us and our ancestors the apes apart from the tail.* The arguments of these gentlemen convinced me at times, but always grieved me; and my anxiety came to a head whenever I leafed through a dictionary of natural history and found man defined there as "a vertebrate animal, belonging to the genus of mammals and to the species of the bimanous, broad-nailed, omnivorous, etc." Faced with this insulting classification or rather condemnation, it was impossible for me to bow my head in silence, and night and day I devoted myself in pursuit of some distinguishing characteristic, separating man from the remaining mammals. I searched first of all in the works of writers who claimed to have disproved the Darwinian and Büchnerian theories, namely Quatrefages, Janet, and the Cardinal Bonnechose, but in them I could find nothing beyond impassioned rhetoric in support of faith and fatherland, *the immutability of species,* the responsibility of man, and the temporal power of the pope.[20]

Roidis would have read a copy of the *DM,* where volume one dealt with "The Evidence of the Descent of Man from Some Lower Form," the "Comparison of the Mental Powers of Man and the Lower Animals," and "The Manner of Development of Man from Some Lower Form."[21] In the passage Roidis claims to have felt an exaggerated anguish when he read Darwin and the other advocates of evolution, which if interpreted as satire is typically his form of writing. The view that man was no different from the ape, minus the tail, caused Roidis endless searching for characteristics which would differentiate man from other mammals. The searching only leads him to the arguments of anti-Darwinians who saw their beliefs threatened.[22] The article continues with Roidis searching for answers to the differences between man and beast. This issue was widely debated and researched in this period

20 Roidis, "Αγγέλου Βλάχου κωμωδίαι," 28 (my italics). The natural scientists Carl Vogt (1817–1895), Ludwig Büchner (1824–1899), and Jacob Moleschott (1822–1893) were known as the scientific German materialists. Vogt, a Darwinian, is cited by Darwin in *DM* on a number of occasions, in his introduction. Jean Louis Armand de Quatrefages de Bréau was a French natural scientist who corresponded with Darwin, who refers to some of his studies in his *DM*. Quatrefages, though, opposed Darwin's evolutionary theory. Henri-Marie-Gaston Boisnormand de Bonnechose (1800–1883) was a French cardinal who between 1869 and 1870 participated in the First Vatican Council.

21 Darwin, *DM,* 1:9–157. As Roidis completed his schooling at a Greek-American high school and then studied philosophy and literature in Berlin, thereby knowing English and German, one would argue that he read both the *OS* and the *DM* in one of these languages. In German, the *OS* was first published in 1860 and the *DM* in 1871.

22 Roidis' distaste for papal secular power is reflected in his novel *Pope Joan,* which is a wicked satire about a woman who disguises herself as the Pope.

by scientists and other intellectuals, as seen through their literary works.[23] I will be discussing later in this section works in Greek literature which addressed this issue, including a short story by Roidis.

At various stages in his writing, as Georganta and Bezas have noted, Roidis pondered on this issue.[24] The comparison of animal and man was a universal theme of satire, which in Roidis' case was bolstered by the Darwinian ideas of the time. Further to this, Georganta has indicated that Darwinian theory "occupies a special place in the work of Roidis ... and gives one more link for its connection with naturalism."[25] In her article "From Aristotle to Darwin," Georganta discusses a study on laughter that Roidis incorporated into his critique of the comedies by Angelos Vlachos. So it is highly relevant that Roidis was reading Darwin when he received Vlachos' work. It appears that Roidis' study was prompted by his own Darwinian readings, leading to his investigation of those distinguishing characteristics in man which make him human. Roidis synthesized the theories of various ancient philosophers with trends in science. An example showing this included Aristotle, who believed that man is the only animal capable of laughter. Roidis concluded that: "Man 'laughs' and animals don't 'laugh.' In laughter, then, and there only, lies the difference between man and beast."[26] Roidis maintained: "In all sobriety and seriousness I will confess that for me what makes man superior to the animals is his capacity to laugh at himself and at everything which he aspires to and worships in moments of moral inebriation."[27] He also reproduced the classical tropes of European racism: "Some peoples of the Orient, the Ethiopians for instance, and the Fellahin, although in many ways better than us, lack this gift of mockery; but also their brain, according to the latest anthropological research, scarcely differs from that of the apes."[28]

The passage reveals Roidis' knowledge of the anthropological studies of his time, although the source of his information, or rather his misinformation, regarding the Ethiopians and the Fellahin[29] is unknown. Also it is not clear how seriously one should take him. With its origins in Lamarck and

23 For literary works dealing with this theme see Henkin, *Darwinism in the English Novel*, 83–111.

24 See Georganta, *Εμμανουήλ Ροΐδης*, 274; Georganta, "Από τον Αριστοτέλη στον Δαρβίνο," 43; Bezas, "Εμμανουήλ Ροΐδης," 26–27.

25 Georganta, *Εμμανουήλ Ροΐδης*, 274.

26 Roidis, "Αγγέλου Βλάχου κωμωδίαι," 30.

27 Ibid.

28 Ibid.

29 The Fellahin are peasants or agricultural workers in Egypt.

Darwin, the racial biology of the human race was a study which provided information on "racial variation in physical and psychological traits in the human species."[30] Internationally, numerous social and scientific theories emerged in the 1800s to highlight the inequality of races within the human race, such as that of Louis Agassiz (1807–1873). Scientists believed that due to similarities between man and animal, human races like animal species tended to localize in certain areas in the world. "Signs of inferiority" within a race were taken to indicate that the race was degenerate.[31] In the passage, Roidis could again be satirizing the idea that the scope of one's intelligence was determined by race and by certain features of one's cranium (as perceived by the popular craniological studies of the time).

These ideas were also seen in the *DM*. Here Darwin dedicates two chapters to the "Comparison of the Mental Powers of Man and the Lower Animals," where he writes on issues such as the emotions, imitation, and language. It should come as no surprise then that having read Darwin and his supporter Carl Vogt, whose craniological studies Darwin refers to in the *DM*, Roidis was able to make these comments, though they are somewhat different from those of Darwin.[32]

Roidis would also have known that in the last paragraph of Darwin's introduction to the *DM*, Darwin mentions that he had wanted to add his study on the "expressions of the various emotions by man and lower animals."[33] Darwin was drawn to the work of the prominent anatomist Sir Charles Bell. In Darwin's words, Bell "maintain[ed] that man is endowed with certain muscles solely for the sake of expressing his emotions."[34] This view was "obviously opposed to the belief that man is descended from some other and lower form"; so Darwin needed time to consider it. He also wanted to "ascertain how far the emotions are expressed in the same manner by different races of man."[35] For these two reasons, he excluded the study from the *DM*, and the following year, in 1872, he published the *EE*. Anticipation of this next book of Darwin's would have sent the scholarly world buzzing. I would imagine that Roidis' disquisition on laughter was his contribution to such discourse.

30 Stepan, "Biological Degeneration," 97.
31 Ibid., 98–99.
32 Darwin refers to Vogt's study "Memoire sur les Microcéphales" of 1867. Darwin, *DM*, 1:44, 56–57, 121.
33 Ibid., 5.
34 Ibid.
35 Ibid.

Figure 2.2. "The Lion of the Sea," *Punch*, May 25, 1861, 213.
Courtesy of the State Library of NSW.

Following from this would appear Roidis' undated short story "Story of an ape" ("Ιστορία ενός πιθήκου").[36] In 1918, Simos Menardos viewed it as a synopsis of Darwinism.[37] More recently, Georganta and Bezas acknowledge Menardos' comment, but neither elaborates on the short story.[38] Neverthe-

36 Roidis, "Ιστορία ενός πιθήκου," 347–353. His *Collected works* show it to have been first published between 1894 and 1904. Note that the short story *Η δίκη των πιθήκων* (The trial of the apes) published in 1881 by Zisimos Typaldos (1839–1907) is a satire along similar lines.

37 Menardos, *Εμμανουήλ Ροΐδης*, 20. Menardos comments that in the story Roidis "is attempting to remove all the differences man has by portraying Thomas the ape as a servant and as a comical librarian of a high-spirited nobleman."

38 Georganta, "Από τον Αριστοτέλη στον Δαρβίνο," 43; Bezas, "Εμμανουήλ Ροΐδης," 27.

less, it is a satire based on Roidis' presumed readings of the *DM*, dealing with the comparisons and differences between man and ape. It is likely that Roidis drew from the *DM* to satirize Darwin's comments on "the principle of *Imitation*" as being strong in man, and so monkeys which are close to man in kinship are known to be "ridiculous mockers."[39] The story touches also on the concept of mutability of species as described in the *OS*.

Iakovos Pitsipios' satire *The ape Xouth* (*Ο πίθηκος Ξουθ*) (1848) was possibly one stimulus in the realization of Roidis' short story in question.[40] Yet Roidis' story is reminiscent of other satires, such as Thomas Love Peacock's novel *Melincourt* (1817), where an ape named Sir Oran Haut-ton is introduced to polite society.[41] The relationship between man and ape was also satirized in the visual arts. *Punch* magazine shows a picture of a rather alarmed Victorian aristocrat alongside a gorilla dressed in a tuxedo (see fig. 2.2). Darwin's *OS* implied humanity's close connection with the ape, rather than being created in God's image. The scholarship on evolution prior to Darwin's work was confirmed with the *OS*.

As one of the few key Greek prose satirists of the nineteenth century, the following analysis will attempt to place his "Story of an ape" within the context of European works of the nineteenth century.[42] As suggested by Gillian Beer, at the end of the nineteenth century "language ... is taken as the distinguishing feature between the human and the animal."[43] This is one of the main features that are satirized in Roidis' short story. She also notes the "usual" theme of mimicry of man by ape in satire, indicating that "the

39 Darwin, *DM*, 1:44 (italics in original). Darwin theorized that the monkey's strong tendency to imitate facilitated the origin of articulate language: "It does not appear altogether incredible, that some unusually wise ape-like animal should have thought of imitating the growl of a beast of prey, so as to indicate to his fellow monkeys the nature of the expected danger. And this would have been a first step in the formation of a language." Ibid., 1:57.

40 Bezas, "Εμμανουήλ Ροΐδης," 27. See also O'Neill, "The Unfinished Ape," 67–110.

41 Peacock, "Melincourt," in *The Works of Thomas Love Peacock*. *Melincourt* was translated into French soon after its publication. William Hauff wrote a short story, "The Young Englishman", with a similar idea.

42 Beaton, *Introduction to Modern Greek Literature*, 52. On Roidis' one novel, his satirical masterpiece *Pope Joan*, there has been substantial commentary.

43 Beer, "Satire, Voice, and Nineteenth-Century Science," 271. In this paper, it is worth noting that Beer examines a poem entitled "Monkeyana" and an associated picture of a gorilla wearing a placard which says "Am I a man and a brother?" The poem was published in the English magazine *Punch* in 1861, and although the poet was initially anonymous (he signed the poem Gorilla), it was later discovered that it was written and illustrated by geologist and taxonomist Philip Egerton (1806–1881). It is a satire of an ape who, naturally, is unable to speak, but is able to write. In the poem the ape is questioning his status on the evolutionary ladder, particularly in relation to man. Ibid., 273–276.

human family tree" was commonly satirized, revealing "a queasy uneasy anxiety about humanity's unique status" with regard to any resemblance of the ape to man.[44]

This is reflected in Roidis' story when admirers liken Thomas the ape's portrait to not just any person, but to members of the aristocracy. In "Story of an ape," Roidis through a third person narrator, as omniscient observer, tells the clever and wickedly funny tale of Thomas, an ape of the species "Chimpagni" (a spoof on biological names), who is highly revered and lives with his owner, the banker and archaeologist Baron Dimitrios Kuste. The story highlights the fact that the ape cannot talk but that he can imitate humans in his behavior. The baron has Thomas painted by the famous artist Izola (presumably a play on the word Zola), who like the French naturalist writer aimed in his work to "paint" life in microscopic detail, no matter how ugly it was. Thomas, who the narrator admits "completely resembled all apes,"[45] has admirers of his portrait believing he resembled individual members of the aristocracy, such as a countess and a female poet.

When sick with a cold, he has the best Italian doctor, "so as to hydrate his monkey with chamomile, theriac, and warm wine."[46] The baron invites those who do not believe that Thomas is worthy of such treatment to his mansion for a lavish celebration of the ape's name-day, St. Thomas' day (the patron saint of doubters). At the reception Thomas' physical, moral, and spiritual attributes are wryly observed. The baron attempts to make Thomas appear human. Thomas mimics his owner in the way he dresses, and with his aristocratic airs: "In the evening, when we arrived at Kuste's mansion, the master of the house hurried up to welcome us, and with him the ape Thomas, who extended his hand to each of us in imitation of his master. ... He had donned the uniform of his office, a frock coat of blue velvet, red breeches, a fine necktie, and a camellia in his buttonhole. ... He was as handsome as an ape could possibly be. If I was not afraid of being considered contradictory, I would say that he was handsome with an ugly snout. And these were merely his external features; Thomas' moral and intellectual qualities were far superior to these."[47]

---

44 Ibid., 278–279.

45 Roidis, "Ιστορία ενός πιθήκου," 348.

46 Ibid.

47 Ibid., 349–350.

The baron, determined to prove to his guests Thomas' deserved "status," eagerly asks Thomas to relay to them how his previous owner abandoned him on the street. What follows is a comic "dialogue" between the baron and Thomas. Thomas proceeds to carry out a complicated mime of the relevant events leading to his abandonment. The baron, who does not appear fazed by the fact that the ape cannot speak, improvises by speaking for the ape.

Roidis satirizes the ape's intellectual proximity to man by presenting Thomas as the baron's "fine" librarian. Thomas' library duties are described, which in reality are only capabilities within the scope of a trained ape. His owner has trained him to imitate the stance and gait of those visitors known to them. So that when a known visitor arrives to use the library, Thomas is able to mimic that person in front of the baron, who then decides whether or not this person is allowed to enter. Using sound association and a long cane pointer, the baron is able to direct the ape to the book that he wants retrieved. The very high bookshelves are not an access problem for Thomas who, according to the narrator, was "probably educated in his younger years in the art of climbing."[48] The ape's duties are mocked further by the narrator with this final comment on the issue: "Indeed the alacrity with which he fetches the book and his shaking off the dust and the clatter of his feet are the first and essential qualities worthy of study, and I challenge the director of the Lavrion Library to show me a single librarian comparable with mine."[49]

The chief librarian in the municipality would have held a rather important and dignified position. Calling on the chief to find a finer librarian than Thomas is mocking the chief himself. Apparently, Roidis was able to avenge opponents at the national library with this satire.[50] Undoubtedly, he does achieve this in the story by undermining the intelligence of the library staff, by proclaiming Thomas' superiority as a librarian even though he really can only perform rudimentary tasks. Roidis confirms in a footnote in his story that despite Thomas' intelligence, "this type of association of a particular sound with a particular object does not exceed the intelligence of such domestic animals according to the observations of Darwin, etc."[51]

48 Ibid., 352.
49 Ibid.
50 Menardos, *Εμμανουήλ Ροΐδης*, 20.
51 Roidis, *Αφηγηματικά κείμενα*, 647.

The satire's culminating scene occurs when the baron overindulges in food and wine, and suffers severe indigestion for three days. The doctor arrives with a vial of medicine that, when taken orally, will cause the patient to shake violently, and so facilitate digestion, thereby avoiding death. According to the doctor, there is a danger that the patient may not be able to survive the shaking. What then occurs is seen in the following passage:

> The ape grabbed the medicine from the table; climbed onto a cabinet and started to examine it carefully; he would smell it and then avert his face in disgust; looking at the patient and shaking his head as if to advise him not to take the foul-smelling medicine. Finally he started to shake the bottle up and down, as if he had read the stock formula "Shake the bottle before taking" on its label. The pantomime was so expressive and so extraordinarily comic that two of the patient's friends standing nearby were overcome by uncontrollable, convulsive laughter. This laughter exploded like a bomb, spread like fire, took hold of the servants who had come up, and finally also the dying baron himself; it was so intense and uncontrollable that it was sufficient to bring on the shock which was the purpose of the potion, and hence save the patient's life.[52]

The baron and other observers of the ape's buffoonery finally see the ape's ability, as an animal, to mock man by imitating him. However, the response of laughter only serves to highlight, at that time, the perceived absurdity of man's close kinship with the ape. Roidis juxtaposes cleverly what he sees as a unique attribute of man—that is, his ability to laugh—with the ape's ability to imitate man and its inability to speak. The concept of mimicry is directly linked to and elaborated in Darwin's *DM*. Roidis argues in his study that only man is capable of laughter, and that the more the laughter, the more human one becomes.[53] As indicated earlier, the concept of expressions is mentioned in anticipation of the *EE*, which will include the expression of laughter in man and the ape.

The short story does also relate to another aspect of Darwin's concepts, which arises in the *OS*—namely, the mutability of the species. The narrator

52 Roidis, "Ιστορία ενός πιθήκου," 353.
53 Roidis, "Άγγελου Βλάχου κωμωδίαι," 30.

makes a point of describing the baron's zoological museum, which did not house stuffed animals, but "rather a real Noah's ark, because the animals were alive."[54] He goes on to describe the animals:

> He had twenty-seven dogs of every breed and size, from the Apennine sheepdog to the Maltese terrier and the spaniel of the court of King Charles of England, as well as innumerable cats from Ankara, from Tarma, Spanish tabbies, and golden-haired Persians. Kuste could be considered the forerunner of today's flourishing lovers of cats. But most remarkable of all was his unique collection of all kinds of exotic birds in an annex to the glass greenhouse, from two imposing ostriches to the dazzling, microscopic humming birds, which resemble flying emeralds and sapphires. This entire feathered, downy, and hairy domain was fed and serviced through a special account, amounting to several thousand pounds, presented each month to His Excellency Baron Kuste by a special employee, bearing the title of "Curator of animals."[55]

Likening it to Noah's ark alludes to the biblical creation of species. He proceeds to describe the twenty-seven canines of every species and size and the countless varieties of cats contained in it. The literal biblical version of creation (creationism) adheres to the premise that all the species of the animal kingdom were created through separate acts and that they are immutable; hence, according to the Bible Noah's ark housed all the species of the animal kingdom (and presumably the plant kingdom). By comparing the baron's museum to Noah's ark, Roidis mocks the biblical creation. By emphasizing the numbers and naming some of the species of animals in the baron's "ark" and also highlighting the huge expense of feeding these animals, Roidis attempts to show the absurdity of keeping so many animal species in one space. On the whole, Roidis attacks the biblical story of the ark—a story which discounts Darwin's theory of the origin of life arising from one or a few cells, from which all species evolved gradually. In other words, according to creationism, all the different species known today have always existed, exactly as they are now, since the time of creation. As noted by Henkin:

54 Roidis, "Ιστορία ενός πιθήκου," 349.
55 Ibid.

> Trouble arose ... from the mere number and size of the species that came to be known with the progress of zoology. The numbers now known amount to millions, and if two of each of these kinds, "every creeping thing that creepeth upon the earth, and every fowl," were in the ark, it was felt that the ark must have been rather overcrowded. Theology's views on animal life were uncompromising. In those days the doctrine of the Special Creation of species was an idiom that a child acquired before he was ten years old; even a naturalist would no more suppose that one species or sub-kingdom gave rise to another than that Italy gave rise to Spain. Men believed that every existing animal, however numerous in kind, belonged to species represented in Noah's ark; that all those extinct, discoverable in fossil form, had been with few exceptions drowned in the Noachian Deluge.[56]

Roidis' Noah's ark has only one ape, Thomas. In the narration, Roidis places Thomas at the end, after the list of other animals, hence highlighting the ape's place in the evolutionary continuum of life. In addition, the ape is referred to in a paragraph on its own, isolating it from the other animals and implying a closer proximity to man. The Curator of the animals of the "ark" seems to be an omnipotent being whose identity is not discussed any further. However, the owner of the animals and the "ark" is Kuste who, as the human, represents man's self-proclaimed superiority over all animals. Roidis in an essay said it aptly: "It is true that man received or rather gave to himself the name of king of the animals."[57]

From his *Collected works* (*Άπαντα*) around 1900 it can be seen that Roidis did not consider it offensive to be called an animal; he maintained that whatever differences exist between man and animals do not prove man's superiority: "From what I have been fortunate or unfortunate enough to discover, I am, I believe, the only person who, if they named me animal, would not consider it an insult. The more I mix with animals, the more I am rather convinced, [not that] there is no difference between them and humans, as some lovers of paradoxes have wished to assert, but only that the things in which we differ from animals do not all establish human superiority. The primary difference between them and us is that they received from humans everything good humans possess, and they avoided imitating all that is useless, harmful, and ridiculous."[58]

56 Henkin, *Darwinism in the English Novel*, 19–20.

57 Roidis, "Σκαλάθυρμα," 190.

58 Roidis, "Ιστορία ορνιθώνος," 218.

Roidis goes on to give instances of humanity's negative habits, which he claims are not found in animals. Perhaps this could be compared to Darwin's final comments in the *DM*: "For my part I would as soon be descended from that brave little monkey, who braved his dreaded enemy in order to save the life of his keeper; or from that old baboon, who, descending from the mountains, carried away in triumph his young comrade from a crowd of astonished dogs—as from a savage who delights to torture his enemies, offers up bloody sacrifices, practises infanticide without remorse, treats his wives like slaves, knows no decency, and is haunted by the grossest superstitions."[59]

This exploration has shown Roidis' strong literary response to Darwin's *OS* and *DM*. He appears to have been quite disillusioned initially by the idea that man and beast have an evolutionary kinship. Despite this, he was quite accepting of the mutability of species and presumably of Darwinism. Initially, Roidis' disquisition on laughter, a reaction to Darwin's observations in the *DM*, led him to believe that man is unique due to his ability to laugh. However, this is prior to the 1872 publication of the *EE*, where Darwin showed evidence of smiling and laughter in apes.[60] The "Story of an ape" highlights Roidis' ideas on man's ability to laugh at himself, by simultaneously observing the ape's ability to *imitate* man. Eventually, in his later responses Roidis declared that man was not superior to other animals and that there were no essential differences between them. This view, of course, would have agitated the Church, a response which he had also received with his novel *Pope Joan*.[61]

Kostis Palamas

While the eminent Greek poet Kostis Palamas (1859–1943) did not have any formal training in the medical or biological sciences, his personal library contains a comprehensive range of scientific books from his era, reflecting the scientific interest he had throughout his life. Alongside copies of Darwin's works he also possessed works by Darwinian and post-Darwin-

59 Darwin, *DM*, 2:404–405.

60 Darwin, *EE*, 130–135.

61 See further Roidis' essay on issues regarding man's descent and the definition of man: Roidis, "Η εορτή του Όνου," 9–10. Also, his preoccupation with animals is seen in his other work: "Ιστορία μιας γάτας" (1893), "Ιστορία ενός σκύλου" (1893), "Ιστορία ενός αλόγου" (1894), and "Ιστορία ορνιθώνος" (1897).

ian authors such as Haeckel, Nordau, and Lombroso. His scientific interests can be seen in the constant references he made in his literary criticisms and in his poetry. In his sixteen-volume *Collected works* he made numerous references to Darwin and Darwinism.[62] He held Darwin in high esteem, referring to him, along with Engels and Rousseau, as one of the "heads of intellect [who] released great and fertile ideas to those aloof in the vulgar worlds."[63] Palamas maintained the nineteenth century was "the century of Science."[64] Similarly, in an essay which commences with the words "Science gives birth to Poetry," he named Darwin and Spencer (among others) as major influences in poetry: "Philosophers and naturalists, Darwin and Spencer, the Democrituses and the Spinozas, astronomical neighbors of the mansions of the Imagination, first designers and founders!"[65]

In 1917, he referred to the Darwinian theory of the development and the evolution of living organisms as the "great idea which marks modern science and thought in general, which has spread to all areas of knowledge and which interprets ... the philosophy of history."[66] Then again, he also believed that evolution was an idea taken by Darwin from an ancient philosopher (not named).[67]

An investigation of Palamas' personal library has shown that it contains two different copies of the *OS* in French. One copy has no publication date and no added markings or notes, whereas in the other, published in 1866, there are pencil markings and added words throughout the preface, presumably made by Palamas. This preface was written by the authorized French translator of the *OS*, natural scientist and feminist Clémence Royer (1830–1902), and it is her personal interpretation of Darwin's theory of evolution. In it she attacks Christianity and suggests that in order to progress the weak and sick should be eliminated. Darwin was immediately enraged by this preface, which had become highly disputed, and he maintained that

62 Palamas, *Άπαντα*, 16 vols. For the relevant references see the index volume Savvidis and Kehagioglou, eds., "Τα ευρετήρια," 503, 571. Note that the date of some of his work is not known.

63 Palamas, "Ένας πατριώτης," 160.

64 Palamas, "Η εθνική γλώσσα," 245.

65 Palamas, "Μετά την παράστασιν του Άμλετ," 146.

66 Palamas, "Ένας μεγάλος φίλος των Ελλήνων," 441–442.

67 Thales (ca. 600 BC), Empedocles (ca. 490–460 BC), and Plato (ca. 428–348) are often referred to as having had evolutionary concepts in their philosophies. Although Aristotle founded natural history, he believed that all species possessed an immutable essence; so his ideas were in conflict with those of the modern concept of evolution that is based on variation of species.

Royer had overstepped her role as translator.[68] Palamas would have been quite aware of the notoriety of this preface, and so would have been keen to read it. In an essay published in 1911 he actually does refer to Royer's translations and in particular to her preface, which he refers to as "learned and exciting prolegomena."[69] Also found in Palamas' library is a French edition of the *DM* circa 1873, which does not contain any added markings.

In the late 1800s to early 1900s, it appears that Palamas was a strong advocate of evolutionary ideas and their application to other fields, particularly in literature and the language issue in Greece. He made constant references to the "Darwinian theory of unravelling of all" (δαρβινική θεωρία του ξετυλιμού των όλων), "the idea of unravelling" (η ιδέα του ξετυλιμού), "the theory of unravelling" (η θεωρία του ξετυλιμού), and other forms such as "it evolves" (ξετυλίγεται) and "while evolving" (ξετυλίγοντας). Note also that Xenopoulos refers to Darwin's theory in this manner in one of his "Athenian Letters," as will be further discussed in chapter 3. Henceforth I will be referring to "ξετυλιμός" and "ξετυλιγμός" as "evolution."[70]

For many intellectuals of the nineteenth century who dealt with evolutionism, such as Goethe, the development theory of the world was "a process analogous to the gradual unfolding or unravelling of a seed."[71] Further to this, Darwin's fanatical advocate Thomas Huxley in his famous Romanes Lecture on "Evolution and Ethics" in 1893 made reference to the children's story of "Jack and the Bean-Stalk." He used the development of the unravelling bean as an analogy for the concept of evolution.[72]

In accordance with Venetia Apostolidou's observations, the evolution of literature, as an analogy of Darwinian evolution, is a concept that Palamas believed in and repeatedly mentioned in his works.[73] In a 1907 note to Pavlos Nirvanas, who did not accept this view on the "evolution of poetry," Palamas wrote:

68 Darwin was also infuriated by Royer's addition of her own extensive footnotes throughout the *OS* French translation. See Clémence Royer, "Préface de la première édition," in Darwin, *De l'origine des espèces par sélection naturelle*, xv–xxxviii. There are more pages to this preface but in Palamas' copy they are damaged and not readable.

69 Palamas, "Το νέο βιβλίο της Κυρίας Δέλτα," 87.

70 The Greek word for "evolution" used now is "εξέλιξη" which, as in English, can mean "development."

71 Oldroyd, *Darwinian Impacts*, 311.

72 Huxley and Huxley, *Evolution and Ethics*, 60–62.

73 Apostolidou, *Ο Κωστής Παλαμάς*, 94.

> I would have very much desired a broader explanation for my belief in "the *evolution of poetry.*"… The only thing which I can note here is that we must give this *idea of evolution* some sense analogous only to the scientific meaning of the word; as almost always occurs with terms borrowed from the natural sciences by the historical disciplines, which always claim, quite rightly, that they too—that is, the historical sciences—obey and are subjected to the same laws. … I don't remember very well where and how I supported the *idea of evolution*; of course speaking about the evolution of poetry I must have meant some kind of more *complex, gradual build-up,* a certain regular *progression and ongoing change.* Even though I don't have any difficulty believing and proclaiming evolution everywhere, from the more natural of creation to the most subtle and barely definable objects of the moral world. I would have been encouraged by the example of a great philosopher, Spencer, and from a great critic, Brunetière.[74]

Palamas acknowledged the interdisciplinary nature of the ideas that Brunetière used to describe various fields, such as language and literature, in terms of biological evolution. He also noted the evolution of poetry to be a *gradual* process to a *more complex* form, which results in *progress.*[75] Palamas appears to be informed on developments associated with scientific ideas. In fact, this passage shows that he was attuned to the general scientific and also to the evolutionary discourse of his time: "It is true that lately some anti-Darwinian cases have started to surface in the broad outer scientific circles. They tell us that a naturalist has proven, shall we call it, the sudden bursting forth of new species of animals. So alongside the one-act player Lyell with his theory that the great changes happen *bit by bit* and unnoticeably, Cuvier's ghost, with his cataclysms, is reestablished."[76] He is most likely referring to the Dutch botanist and geneticist Hugo de Vries (1848–1935), who in his "Mutation Theory" of 1901–1903 believed that the evening primrose (Oeno-

74 Palamas, "Σημείωμα για τον Παύλο Νιρβάνα," 536–537 (my italics). Ferdinand Brunetière likened literary ideas to biological ideas and the history of ideas to the history of humanity. For a further discussion on the discourse between Palamas and Nirvanas see Apostolidou, *Ο Κωστής Παλαμάς*, 94–96.

75 The theme of gradualness will be discussed further in chapter 3 in relation to Xenopoulos' work primarily, and also to that of Karkavitsas.

76 Palamas, "Σημείωμα για τον Παύλο Νιρβάνα," 537 (my italics). Palamas refers to the British geologist Charles Lyell (1797–1875), who published the *Principles of Geology* (1830–1833) and who propounded the theory of gradualism in the earth's formation, and to the French naturalist Georges Cuvier (1769–1832), who believed that the extinction of species was caused by some local catastrophe. A cursory read of Darwin's correspondence shows that there were many claims to spontaneous generation by biologists; all have been refuted by science.

thera lamarckiana) could suddenly produce mutated forms of the species.[77] This spontaneous production, known as saltationism, where new species were considered to originate suddenly from existing ones without transition or any visible preparation, goes against Darwin's gradualism. In his *OS* Darwin mentions the adage known in natural history "Natura non facit saltum," that is "Nature takes no leap," used to explain the slow and gradual evolution of bodily organs and instinct.[78] By 1910, evolutionists realized that de Vries' "mutations" did not breed true, and they were shown to be only hybrids of the original plant.

In 1908 Palamas wrote about "the war of ideas," which is the introductory phrase of an essay in his *Collected works*, reflecting the essence of the essay. He touched on what he considered, at that time, to be the "myriad ideas and powers [which] are fighting in the Europes and the Americas for victory and domination."[79] He drew attention to the different perceptions of various ideas: "Positivists here, and elsewhere idealists, positivists of idealism, idealists of positivism, in ethics, in the arts, in philosophy, the same names used in this or that intellectual circle take on different meanings. Pragmatism in literature is different from pragmatism in metaphysics."[80] The passage shows that his knowledge covered a wide area of disciplines. Clearly, positivism was an area he was interested in. The advantageous effect that positivism had on science and its negative effect on metaphysics were acknowledged by Palamas in the next passage, which follows from the above. He heralds:

> Darwinists and Lamarckists, together with the followers of the law that shows things which remain stable by the law against evolution and transformation. Just when we are to believe the great hypotheses to be true, we hear some commanding voice calling out to us. Don't believe them! They have problems also! Just when, through the influence of the new scientific ideology hostile to mysticism, we've got used to regarding man as noth-

77 Mayr, *Growth of Biological Thought*, 546. In addition, de Vries was made a foreign member of the Academy of Athens (the only foreign evolutionist since the Academy began). Xenopoulos was a member from 1931 until his death.

78 Darwin, *OS*, 223–224, 237, 263. Referring to organs he argues: "Why should not Nature have taken a leap from structure to structure? On the theory of natural selection, we can clearly understand why she should not: for natural selection can act only by taking advantage of slight successive variations; she can never take a leap, but must advance by the shortest and slowest steps." Ibid., 223–224.

79 Palamas, "Σημειώματα στο περιθώριο," 90.

80 Ibid.

> ing more than a being subject to natural law, a single drop in the centerless and boundless whole, somebody suddenly turns up with a quiet whisper to set trembling once more all those things we had thought were firmly established within us. Is man a supplement to nature, bringing nature to perfection, accomplishment, or is it the beginning of a new world, or a new step on the endless ladder of things? Science is a giant on Olympus and metaphysics is Antaeus.[81]

Antaeus, a giant in Greek mythology, gained his strength when touching the ground. If he was lifted high off the ground for a period of time, he would lose this strength and could die. Presumably, Palamas meant that science, which had attained its prestige in the nineteenth and early twentieth centuries, was superior to metaphysics, which had been rejected by the materialists, Comte, and Darwinian theory. This was part of the ongoing discourse where those opposed to the sciences, particularly evolutionary science, feared that any truth that such scientific discoveries as Darwin's theory would reveal, would put an end to the perceptions of life's mysticism.

In 1909, while pondering over the trends of French literature at the time, Palamas commented on his belief that literary symbolism was not dead, a view he said was not held by others. Instead this literary trend, he argued, evolved into a different form: "It merely *evolves* and is *transformed*. What is true for Science, we can see also in the Arts: each idea which has sown its seed and had completed its life cycle withdraws, stands aside, makes itself small, and gives its place to other, newly arrived, blossoming ideas. It appears to die. *And yet it is not lost. Its markings remain*. With these, new ideas are re-created, and occasionally the idea itself pops up again and is reworked, though under a different name, so as not to appear as the ghost of the earlier one, which would terrify people."[82] He argued here for the appreciation of both science and the arts, as did Petros Apostolidis.[83] In the passage there appears to be a neo-Darwinian concept emerging from his writing and, as the next passage will demonstrate, Palamas was well informed on the tenets of neo-Darwinism. Not only did he refer again to an idea evolving and metamorphosing, but he also argued that what occurs

81 Ibid., 90-91.

82 Palamas, "Ο ποιητής René Arcos," 91 (my italics).

83 See the introductory chapter and chapter 1 in relation to Apostolidis.

in science is also true for the arts. Consequently, he proceeded to elaborate that an idea is like a plant which gives the seed. The idea which at the end of its life cycle shrivels up makes way for the seed which will generate new ideas. Palamas made it clear that the idea is not lost and that traces of the idea remain. This dealt with the theory of the continuity of the germ plasm developed in 1885 by Weismann, the pioneer of neo-Darwinism, which was applied not only as a scientific theory, but also metaphorically to humans and society.[84] The implication of this is that it excludes the inheritance of acquired characteristics, or any inheritance influenced by external factors to the germ plasm, hence a hard inheritance. In other words, Palamas has applied this theory to the arts, where the germ plasms, as the "markings," are the ideas in the arts and which, he maintained, in time essentially do not change with the environment.

At this point Palamas also wrote about the French poet René Arcos (1881–1959), who he had met. In an essay he mentions Arcos' Darwinian approach in the writing of "scientific poetry" (επιστημονική ποίηση): "In a turbulent language, overloaded with neologisms borrowed mainly from the terminology of the sciences and philosophy, an advocate of *evolutionary monism,* he is intoxicated by *dreams of Darwinian cosmogonies* and filled with scientific ideology. 'Put some heart into your brains,' shout some conservative critics to these neo-Darwinists. 'We are metaphysicists,' they answer, without concealing a certain contempt towards these elegiac egotists, as they call them (fanaticism and exclusiveness is characteristic of the young), 'but our metaphysics is warmed by poetical emotion. Our heart is in our thought.'"[85] Palamas reflects here on a period when scientific writing, embracing hard inheritance, was perceived to have taken away the metaphysical aspects of life and also the warmth of such writing. He foreshadowed the type of discourse in Greece surrounding the science-arts rhetoric exemplified by the letters in the periodical *Grammata* from 1913–1915 and discussed in the previous chapter. The world of science after Darwin's theory of evolution would create

84 As mentioned in chapter 1, biologist August Weismann in 1885 proposed a theory of the continuity of the germ plasm, whereby the "germ track" (or genotype) is inherited independently of any influence from the body (or phenotype). His theory asserted that the "germ plasm enshrines the notion of 'hard' heredity, the belief that heredity cannot incorporate responses that the body makes to its environment" and that evolution is totally via natural selection. The theory was the initiator of neo-Darwinism.

85 Palamas, "Ο ποιητής René Arcos," 91 (my italics).

a reality counter to the world of the imagination, metaphysics, and the unknown. Palamas went on to state that scientific poetry first appeared in the works of the Greek "poet philosophers." Presumably, he was thinking of Presocratics such as Xenophanes and Empedocles. At the time he wrote this essay, though, he mentioned: "I feel how much, here too, fertile ideas are trampled on through ignorance, or barely recognized through narrow-mindedness."[86] He was open to this writing and went on to say that although he was unsure as to which direction Arcos' type of poetry was heading, he stood with "some special attention and was moved by verses and opinions like those of the young poet René Arcos."[87]

Palamas referred to one of his own famous poems, "The Twelve Words of the Gypsy" ("Ο Δωδεκάλογος του Γύφτου") (written 1899, published 1907), as an example of what he classed as "scientific poetry" that glorifies science.[88] This highly symbolic poem about the philosophical journey of a revolutionary gypsy examines the gypsy's views on work, love, his country, religion, science, and other matters in the historical context of the "Great Idea."[89] In the prologue (written 1905), regarding his inspiration for the poem he wrote in Darwinian terms: "Who knows which Gypsy spoke to my heart—so many years ago!—seductive words, and so sprouted from my heart were the first four lines. This was *the protoplasm from which my creation would gradually have evolved later.*"[90] Palamas explained the evolution or formulation of his poem using the concept of biological evolution. This is reinforced here by the other language that he uses, such as "protoplasm," "gradually," and "creation." There are some evolutionary undertones, such as those of artificial selection in this poem:

86 Ibid., 92.

87 Ibid.

88 Palamas, "Ο ποιητής René Arcos," 92. For the poem see Palamas, "Ο Δωδεκάλογος του Γύφτου," 285–479.

89 Greece was in the midst of debate on the "Great Idea" (Η Μεγάλη Ιδέα), which was about the formation of a greater and more glorious Greece, associated with a rejuvenation and geographical expansion of the new Greek state.

90 For the original Greek text see Palamas, "Ο Δωδεκάλογος του Γύφτου," 289. I have used my translation in the main text. The following published translation does not appreciate the full meaning of the word "ξετυλιγόνταν" to mean evolve: "Who knows what Gypsy woman may have spoken to my heart so many years ago. Thrilling words! And thus the first four verses grew from my heart; the nucleus around which my poem gradually developed." Palamas, *Twelve Words of the Gypsy*, xxiii. The original Greek uses "αγάλια αγάλια," translated as gradually.

And from us shall start the lineage
Of a new unblemished race;
Children who shall bring forth others
Like themselves; while every sore,
Canker, ugliness and evil,
Will have ebbed for evermore.[91]

Here is a utopian eugenic subtext, in the context of literature, associated with the idea of the selection of "a new unblemished" race, which will be ridden of the weak and imperfect. It mirrors an approach also held by other Greek literary writers of that period, such as Nikos Kazantzakis. In 1908, with Greece, nationhood, and the "Great Idea" in mind, Kazantzakis wrote the following under the pseudonym Akritas: "Individuals and nations must unite among themselves, gather their forces, become hard and egotistical, trample the corpses of their opponents. ... And the voice of our ancestors joins with the voice of science which is also shouting from the depths of time: to be beautiful and strong, to hate our enemies, to strengthen the nerves of our bodies and souls and be released from the Christian ideals of the crippled and diseased, to hold high the Selfhood of the race."[92] In accordance with Peter Bien, the two voices mentioned in the passage could well be those of the Spartans and of Darwinism respectively.[93] Alternatively, Kazantzakis could be referring to the voice of our primordial ancestors or even our ancestors in general, as he does on various occasions in his writings.[94] Whichever ancestors he is referring to, one could extrapolate further and state that both Kazantzakis' voices expose eugenic implications. These were part of a wider eugenic intellectual discourse that emerged in Greece about this time, a perspective which is discussed in more detail in chapter 1. This theme will be explored later in the work of Papadiamantis and Xenopoulos.

Palamas' views of evolution in the arts as well as language, as Apostolidou observes, are interwoven with the meaning of "rhythm"[95]: "One rhythm governs everywhere. When phenomena become ripe they decline. Following a

91 Palamas, *Twelve Words of the Gypsy*, 57. For the original Greek text see Palamas, "Ο Δωδεκάλογος του Γύφτου," 331.
92 Kazantzakis, "Επιστολαί από το Παρίσι," 1285.
93 Bien, *Kazantzakis: Politics of the Spirit*, 1:13.
94 For references to the primordial ancestors see Kazantzakis, *Last Temptation*, 7; for references to the "innumerable ancestors" and the "race of men" see Kazantzakis, *Spiritual Exercises*, 69, 70.
95 Apostolidou, *Ο Κωστής Παλαμάς*, 96–97.

rhythm they fruit again and then subside, dominate and then withdraw, and are forgotten and disappear, only to return and be remembered again, however changed they may appear in many ways. *Deep down they are the same.*"[96] Apostolidou indicates that Palamas' views of the evolution of poetry were Darwinian up to 1907, and that in the 1920s his ideas on this became Nietzschean, specifically with the Nietzschean idea of the "eternal return," as in the previous passage.[97] One could also add that these later ideas were possibly also influenced by neo-Darwinian ideas of a Weismannian nature—that is, a sense of the immortal. Note that Nietzsche was strongly influenced by Darwinism, so it is possible the neo-Darwinian and Nietzchean ideas here have been conflated in Palamas' work. He does continue to use an evolutionary lexicon, such as "unravelled" (ξετυλιγμένο) when he is talking about biological evolution in 1907 and 1909. In 1917, Palamas commented on the poem entitled "The death of a daughter" ("Ο θάνατος της κόρης"), written by an anonymous Athenian poet in 1892: "Within the boundless world of the imagination ... a certain positivist philosophy which glorifies the development of species, the progress of beings from the simpler to the more complex, is matched with a certain theosophical pantheism which wants everything to be alive, divine, ascending, becoming perfect, in a vision of the transmigration of the souls, a poetical interpretation of Darwinism."[98] In this interpretation of the poem Palamas aligned positivism with Darwinism and pantheism. Pantheism was not uncommon in post-Darwinism. Indeed, Xenopoulos too talked of taking a pantheistic approach to life after reading the German materialists. On Xenopoulos this will be discussed further in chapter 3.

Palamas believed that all great ideas "are not born suddenly and at once, but are made, evolve, return, never the same, but always varied; but in their essence, the same. And one such example is Darwinism."[99] The emphasis here is on the last part of the quote. Again he aligned the development of an idea with the biological evolution of Darwin, a concept which, as I have mentioned, he drew from Brunetière.

Palamas acknowledged the impact of scientific ideas, particularly, in his words, the "Darwinian truth," which as he indicated, thirty years prior had

96 Palamas, "Αντιρρωμαντικό στάσιμο," 503–504 (my italics). See also Palamas, "Ο λυρισμός του εμείς," 501.
97 Apostolidou, *Ο Κωστής Παλαμάς*, 96–99.
98 Palamas, "Πώς τραγουδούμε," 389.
99 Palamas, "Δυο παλιοτράγουδα," 450.

commenced to permeate even poetry. He identified the poet Tennyson as one who was influenced by such ideas.[100] In 1907, Palamas made further references to the evolutionary nature of literary concepts and ideas: "Literally genres do not die everlasting deaths. ... [The] resurrection of these genres always occurs more often in their essence rather than in form, and ... is not at all contrary to the laws *which through evolution transform things.*"[101]

The following passage from an essay dated 1907 reflects Palamas' views on the continuity of the Greek race and on the Greek language:

> At a time *when the theory of immortality, under the Darwinian evolution of beings,* departs from the theological concept and takes on the *light of science*—Greek Myth, at least in the hands of and under the gaze of [Nikolaos] Politis, shows itself like the ancient Proteus, with thousands of changes over the centuries, yet unchanged. Out of our various folktales and legends the *primeval soul of the Greeks* emerges alive and triumphant, alluring, with its *core characteristics.* And so the immortality of the Greek soul is revealed most clearly through Politis' interpretations of myth ... just as the immortality of the Greek language is revealed in a more lively way ... in the living, uncultivated language of the people. [102]

Palamas refers here to the Greek pioneer folklorist Nikolaos Politis (1852–1921), who in his 1871 study *Modern Greek mythology* (*Νεοελληνική μυθολογία*)[103] held the view that the continuity of the Greek culture from ancient to modern Greece was preserved through the traditional ways of the Greeks.[104] In his study Politis fervently would seek the kinship between modern-day Greeks and the ancient Hellenes in relation to their manners and customs.

In the aforementioned passage Palamas argues how the continuity of the Greek civilization (or as he says, "the immortality of the Greek soul") is expressed firstly in modern folkloric or popular culture with its origins in ancient times, as studied by Politis, and secondly it is apparent in the spoken

100 Palamas, "Η ποίηση στον περασμένο τον αιώνα," 195.

101 Palamas, "Από την αφορμή ενός λόγου," 510 (my italics). In the Greek original, the italicized phrase reads "που ξετυλίγοντας μεταμορφώνουν τα πράγματα."

102 Palamas, "Από την αφορμή ενός λόγου," 494 (my italics).

103 Politis, *Νεοελληνική μυθολογία.*

104 On folklore, its relation to Greek identity and nationhood, and also in terms of the European context see Herzfeld, *Ours Once More.*

language, the demotic (or as he says, through "the immortality of the Greek language"). In order to validate the continuity of the Greek race and also its language, Palamas appropriated metaphorically August Weismann's biological theory of the continuity of the germ plasm (or gene as we know it today) of living things, to highlight what he is referring to as the post-Darwinian theory of "immortality." Since Weismann rejected Lamarck's concept of acquired inheritance—that is, the inheritance of environmental factors—the germ plasm would be considered to direct the development of the whole organism and could not be modified by environmental influences when passed on.

So, the "primeval soul of the Greeks," a metaphor for Weismann's germ plasm, had not changed from one generation to the next and remained unaffected by the "environment"; environment here would suggest the prior Ottoman occupation of Greece and other foreign invasions. Invoking the ever-changing mythological Proteus, Palamas referred to the thousands of changes, presumably historical, which affected Greece; and as Palamas would have it, allusions to Weismann's theory would dispel any possibility of these changes being passed from one generation to the next—they were not part of the "germ plasm," that is the inheritable "core characteristics," as he said. At a time when social and cultural evolution was aligned with biological evolution, one could infer Palamas was also implying that the Greek race was biologically continuous from ancient times to modern. Note that while he mentioned human immortality according to traditional religion, presumably that of the afterlife, Palamas also asserted an ideological shift where he viewed "immortality," or a perpetual continuity, through the genetic component transmitted from one generation to the next.

In the previous quoted passage Palamas also linked the continuity of the Greek language to the continuity of the Greek race. He would view the demotic as the "living, uncultivated language of the people," utilizing evolutionary rhetoric in other passages. As one of the strongest advocates in his time for the demotic language, Palamas would also contribute to the dialogue on the "language question," using the same words he used to describe the evolution of the arts and poetry. His writings are in accordance with those of George Skliros when describing how the Greek language should develop. Palamas argued that the language issue was an ideological one and that it was very important for the future progress of Greece. He foreshadowed the fact that the language issue would become a political issue, much to his concern, and showed his sarcasm about the future: "The language issue is first of all an intellectual issue; from the outset it is not political nor is it purely social. It comes to politics and to society *gradually* as it *evolves* and spreads and grows and matures. ... It is not the concern of a few pen-pushers and a pretext for a certain circle of literates to fight each other. It is to show concern and to shake the Nation forward. Because even countries are not just political phenomena; they are phenomena, more so intellectual. ... The order of things will make it enter one day in Parliament and become a political party ... it will become a catchcry for the great social revolution which we long for."[105]

In one form or another, the debate on the "language question" continued well into the twentieth century, until the demotic was finally made the official language of Greece in 1976. Palamas appropriated the Darwinian discourse of gradualism, adaptation, transformation, evolutionary continuity, and immortality to discuss his ideas on the development or evolution of poetry, the continuity of the Greek psyche and race, and the development of the Greek language. This was part of an intellectual discourse associated with the nation-building of the Greek state, the issue of race continuity from the ancient Greeks to the modern Greeks, and the Greek language issue associated with the demotic and the katharevousa.[106] Indeed this study reveals the makings of a eugenic discussion in the literary field in the early 1900s.

There is no doubt that Palamas was indeed greatly influenced by Darwin through his readings. This influence is reflected in his essays, which include his literary criticism, and in his poetic works. It is worth noting that even as late as 1930 Dimitris Glinos (1882–1943), writer, politician, and demoticist, declared that Darwinism would still play a major part in the "intellectual" sciences, highlighting its role in the Greek language and literature: "The Darwinian theory of evolution, indeed with the all-embracing quality which was given by Spencer, will be the connecting link between nature and intellect. This tendency is apparent also in the language and in literature, where on the

105 Palamas, "Βιβλία για τα παιδιά," 545–546 (my italics). Note also that in the early 1900s application of the theory of evolution was also made to the development of the Arabic language. See Bezirgan, "The Islamic World," 386.

106 The different opinions on the "continuity of the Greek race" involve a vast discourse ranging from the nineteenth century to the present in an attempt to answer the underlying question of the individual's and nation's identity. The discourse has been multidisciplinary; historically, it was mainly sociopolitical, but also emerging is the recognition that Greek intellectuals did aspire to exploring a biological aspect.

one hand, the language was considered like a phenomenon, and its laws, natural laws; and on the other hand, each and every intellectual creation independent of every other critique has its historical value, its place in the unbroken chain of evolution, in the continuity of life and its transformation."[107]

As I will show in chapters 3 and 4, Xenopoulos used gradualism in his works for quite different issues, such as socialism and the evolution of individuals and groups.

Nikos Kazantzakis

It is well documented that literary figure Nikos Kazantzakis (1883–1957)[108] was influenced in his writing by Nietzsche, Marx, Bergson, Buddha, and Darwin.[109] However, the aim of this section is firstly to draw attention to the effect that Darwin's 1859 theory had on Kazantzakis; and secondly, to examine post-Darwinian and also Darwinian aspects of his famous 1927 essay *Askitiki: Salvatores Dei.*

In 1907, at the age of twenty-four, he wrote in a letter to his friend, Charilaos Stefanidis, that he had just read Darwin: "These days, I've read on the one hand Darwin's system, also Büchner, on the other hand something about spiritualism. Quite literally, I am all mixed up. I don't know what to believe."[110] In the period from 1907 to 1909 he studied in Paris, focusing on Nietzsche's philosophy and attending Bergson's lectures. It is after his publication of the first translation of the *OS* into Greek, in 1915, that he will acquire an intimate knowledge of Darwin's theory of evolution by natural selection.[111] Even much later, in his novel *Kapetan Mihalis* (*Ο Καπετάν Μιχάλης*) (1955), he would mention the *OS* as "the new English book which has turned the world upside-down and which demonstrates that man originated from the apes."[112]

107 Glinos, "Γιάννης Ψυχάρης: βιογραφικό και κριτικό σημείωμα," 19–20. I have taken the liberty of translating the word "πνεύμα" (or "πνέμα" as he writes it) as "intellect."

108 Aspects of this study on Kazantzakis are also found in Zarimis, "Darwinian Reception and Selected Responses," 656–660.

109 Key works which study Kazantzakis' writing are Bien, *Kazantzakis: Politics of the Spirit*; Bien, *Selected Letters of Nikos Kazantzakis*.

110 Bien, *Selected Letters of Nikos Kazantzakis*, 34. For the original Greek see Aposkitou-Alexiou, "Τριάντα τέσσερα άγνωστα γράμματα," 135.

111 Kazantzakis also translated Büchner's *Kraft und Stoff* (1855) into Greek, as *Δύναμις και ύλη* (1915).

112 Kazantzakis, *Ο Καπετάν Μιχάλης*, 184.

If we are to accept the account in his novelistic autobiography *Report to Greco* (*Αναφορά στον Γκρέκο*) (1961), his first exposure to Darwinism greatly affected him, and this led to the resolute rejection of his faith of Orthodox Christianity.[113] This is observed in the following passage from the book, where he learns from his physics teacher of the Copernican heliocentric system (which states that the earth is not the center of the universe). Most importantly, the teacher introduces him to Darwinism, clearly having a crushing effect on him. The time would be around 1900 in Crete:

> In short, what was this fairy tale our teachers had shamelessly prated about until now—that God supposedly created the sun and moon as ornaments for the earth, and hung the starry heavens above us as a chandelier to give light! This was the first wound. The second was that man is not God's darling, his privileged creature. The Lord God did not breathe into his nostrils the breath of life, did not give him an immortal soul. Like all other creatures, he is a rung in the infinite chain of animals, a grandson or great-grandson of the ape. If you scratch our hide a little, if you scratch our soul a little, beneath it you will find our grandmother the monkey!
> My bitterness and indignation were insupportable.[114]

Coincidentally, the statement that one's grandmother could be a monkey alludes to an actual event almost a century before, on June 30, 1860, the year after Darwin's *OS* publication. At a lecture in the Oxford University Museum of Natural History, Bishop Samuel Wilberforce asked the Darwinian Thomas Huxley if it was from his grandfather's or grandmother's side that he was descended from an ape. The response by Huxley and others is said to have "waxed hot."[115] The thought then that humans could conceivably have descended from apes was considered outrageous. Later, Kazantzakis' struggle to also accept this information leads him to anguish and finally

113 Kazantzakis, *Αναφορά στον Γκρέκο*. He states in his introduction that it is not an autobiography of his personal life but an account of his spiritual or philosophical ascent. He wrote it in 1955, just before he died. It was published posthumously in 1961 in Greek, and in 1965 and 1966 in English. Clear references to the loss of his faith due to Darwinism are found in Middleton and Bien, *God's Struggler*, 3, 8, 9, 114. As noted by Middleton and Bien, his early exposure to Darwin at school does not mean that he had actually read about him then (114).

114 Kazantzakis, *Report to Greco*, 115. (Kazantzakis, *Αναφορά στον Γκρέκο*, 139.)

115 There are discrepancies as to what followed after this question was posed. See the following for an account by botanist and confidant of Darwin, Joseph Dalton Hooker (1817–1911), who claimed that Huxley's retort was inconsequential, as it could not be heard clearly. Desmond and Moore, *Darwin*, 495–497.

to rebel over the Church's orthodox views. It appears that he never actually recovered from what he perceived was a lie created by the Church: "My disillusion and indignation endured for months. Who knows, perhaps they endure even now. On one side of the abyss stood the ape, on the other the archimandrite. A string was stretched between them over chaos, and I was balancing on this string with terror."[116]

At this time in his last year at school, his attempts and those of his friends to "enlighten" fellow Cretans about this evolutionary theory, had a negative reception: "We endured much in our effort to enlighten mankind on the succeeding days as we roamed through Megalo Kastro or toured the villages. We were called atheists, Freemasons, hirelings. Little by little we began to be hooted and barraged with lemon peels wherever we went, but we held ourselves proudly erect and pressed on through the insults and peels, content in the knowledge that we were witnessing and enduring martyrdom for the sake of the Truth."[117]

The naïve archimandrite (the school's religious instructor) is unable to answer Kazantzakis' question for proof of God's omnipotence. He becomes irate with Kazantzakis' persistence with the question and so expels him from school. In one village at the festivities after a child's baptism, one of Kazantzakis' friends, also an advocate of Darwinism, seizes the opportunity to "enlighten" the villagers. He preaches to them about "the origin of man," declaring that man's progenitor was the monkey. In addition, he states that man must not be so conceited as to believe in his supposed status as a privileged being created by God. The village priest watches this spectacle and speaks to the orator: "Excuse me, my boy, for staring at you all the time you were speaking. It's possible, as you say, that all men are descended from the monkey. As for yourself, however, forgive me for saying so, but you are a lineal descendent of the ass."[118]

Despite Kazantzakis' initial struggle with his Orthodox Christian faith, which led to its definitive loss, he continued the quest throughout his life for a belief and a purpose. He studied the teachings of philosophy and spiritualism, which were very important to his long-term development. In accordance with Nikos Pouliopoulos, Kazantzakis had Darwinism as a basic starting

116 Kazantzakis, *Report to Greco*, 119. (Kazantzakis, *Αναφορά στον Γκρέκο*, 144.)
117 Ibid., 123. (Kazantzakis, *Αναφορά στον Γκρέκο*, 149.)
118 Ibid.

point in his cosmological beliefs.[119] In fact, as indicated by Bien, Darwinism contributed to Kazantzakis' "doctrine of the transitional age," which influenced his writings.[120] Bien describes this early period in Kazantzakis' life: "The trouble with our times, Kazantzakis argued, is that we are caught up in the middle. On the one hand we have lost our spontaneous appreciation of this world's beauty; on the other we have lost our faith in the heavens above. We cannot be pagans because Christianity has poisoned our attitude toward material things: we cannot be Christians because Darwinism has destroyed the perfect spiritual world, which is the necessary basis of Christian behavior. We are thus the melancholy victims of a transitional age."[121]

The transitional period of literature between the Victorian and modern ages can be aligned with the "Age of evolutionism."[122] Tom Gibbons cites Bertrand Russell, who in 1914 saw evolutionary thought as a dominant force in literature and highlighted that certain post-Darwinian ideas cannot be viewed outside the realms of evolutionary thought: "Evolutionism, in one form or another, is the prevailing creed of our time. It dominates our politics, our literature, and not least our philosophy. Nietzsche, pragmatism, Bergson, are phases in its philosophic development, and their popularity far beyond the circles of professional philosophers shows its consonance with the spirit of the age."[123] Kazantzakis' exploration of Bergson's theory (based on evolution), and of Nietzschean theory, was to later expand this doctrine of the transitional age to a complete cosmology.[124] In 1908 in Paris, Kazantzakis wrote to a friend indicating that he wanted to elaborate on a personal philosophy and cosmology, which he could apply to his writing.[125] This would later come in the form of his essay *Askitiki*.

My focus now turns to Kazantzakis' essay *Askitiki: Salvatores Dei* (*Ασκητική: Salvatores Dei*), completed in 1923 and published in 1927;[126] it was translated into English as *The Saviors of God: Spiritual Exercises* in 1960.

119 Pouliopoulos, *O Νίκος Καζαντζάκης*, 306.

120 Bien, *Nikos Kazantzakis*, 8–9. Examples of work that express his idea of a transitional age include the novella *Serpent and Lily* (1909) and the novel *Day Is Breaking* (1907).

121 Bien, *Nikos Kazantzakis*, 8.

122 Gibbons, *Rooms in the Darwin Hotel*, 1–2.

123 Russell, *Our Knowledge of the External World*, cited in Gibbons, *Rooms in the Darwin Hotel*, 1–2.

124 Bien, *Nikos Kazantzakis*, 11–12.

125 Middleton and Bien, *God's Struggler*, 8–9.

126 Kazantzakis reversed the original title and subtitle in his 1945 second revised edition henceforth. The edition used here is his 1971 fifth edition. The translator Kimon Friar used the original form in his title as is seen here.

Much has been written about this essay, but my intention here is only to discuss its evolutionary phase. It is well established that *Askitiki* is central to Kazantzakis' writing. When first published, Kazantzakis referred to this work as the product of his intellectual and life experiences.[127] He was alarmed that it was interpreted as a work of art rather than as a "cry for searching and terror."[128] Yet his essay is so heavily influenced by the transitional period that it has been treated as a philosophical, religious, and even political work.

My intention is to explore the biological theme of continuity, prevalent in literature at that time, and which was mainly attributed to Darwin's theory of evolution and also Weismann's theory of continuity. Kazantzakis' Christian perspective of the afterlife is shattered post-Darwin, consequently resulting in his fear of the potential futility of life. *Askitiki* becomes his new way of dealing with the hope of some sort of immortality which, as will be seen here, is treated in parallel to the biological theme of continuity. The essay, excluding a short prologue, is divided into five phases: "The Preparation," "The March," "The Vision," "The Action," and "The Silence."[129] It is a philosophical quest which deals with birth, life, and death. At the beginning Kazantzakis declares that: "We come from a dark abyss, we end in a dark abyss, and we call the luminous interval life."[130] By means of his spiritual instructions he endeavors to enlighten the readers about this life and to understand this immortality.

We know from Kazantzakis' unpublished notes that he was continually reading philosophical literature in which Darwin and other scientists were discussed. In 1923, for example, Kazantzakis pondered the various world-views on the creation of the world and the philosophy of biology.[131] This work is an insight to the level of science that Kazantzakis was grappling with. Attuned to the scientific theories of his time, Kazantzakis may have given consideration to the Bergsonian perspective, but it is also likely he knew

127 Bien, *Kazantzakis: Politics of the Spirit,* 1:67.

128 Ibid., 78. I quote Bien.

129 "The Silence" is only made a separate section in his revised editions.

130 Kazantzakis, *Saviors of God*, 43. (Kazantzakis, *Ασκητική*, 9.)

131 My gratitude goes to Peter Bien for the copy of a page from Kazantzakis' notebook. The page is labelled "Διάφορες W.a. (Tat 1923 σ. 500)" [Various world views. (date 1923 page 500)], and is from a section labelled "Φιλοσοφία" (Philosophy). Although this comes from a section of Kazantzakis' handbook, one can only speculate as to whether these are his own words or whether he transcribed or summarized them from another text.

well that the mystical *élan vital* had commonalities with the force associated with the mechanistic "continuity of the germ plasm"; and later, by the 1930s and 1940s, he would have appreciated the modern evolutionary synthesis.

While the focus of this analysis is on the phase "The March," the questions on life that he is unable to answer ("Where do we come from? Where are we going? What is the meaning of life?") are explicitly stated in the last phase, "The Silence."[132] It is the section "The March" on "race" and "mankind" which we are interested in here. On race, the following passage is not so different from Palamas' Darwinian perspective on the theory of immortality of the Greek soul and race, discussed earlier in this chapter. Here the "meat" is the "germ plasm," which is passed on from one generation to the next and which does not change:

> The cry is not yours. It is not talking, but innumerable ancestors talking with your mouth. It is not you who desire, but innumerable generations of descendants longing with your heart. ...
> You are not one; you are a body of troops. One of your faces lights up for a moment under the sun. Then suddenly it vanishes, and another, a younger one, lights up behind you.
> The race of men from which you come is a huge body of the past, the present, and the future. It is the face itself; you are a passing expression. *You are the shadow; it is the meat.*[133]

Again, the following reflects the hereditarian continuity of evolution which transcends the generations of humanity: "You are not free. Myriad invisible hands hold your hands and direct them. When you rise in anger, a great-grandfather froths at your mouth; when you make love, an ancestral caveman growls with lust; when you sleep, tombs open in your memory till your skull brims with ghosts."[134] He highlights that the continuity of humanity relies on humanity itself:

> "Do not die that we may not die," the dead cry within you. ...
> "Finish our work! Finish our work! All day and night we come and go through your body, and we cry out. No, we have not gone, we have not detached our-

132 Kazantzakis, *Saviors of God*, 128. (Kazantzakis, *Ασκητική*, 91.)
133 Ibid., 69–70. (Kazantzakis, *Ασκητική*, 33–40.)
134 Ibid., 70. (Kazantzakis, *Ασκητική*, 34.)

> selves from you, we have not descended onto the earth. Deep in your entrails we continue the struggle. Deliver us!"
> It is not enough to hear the tumult of ancestors within you. It is not enough to feel them battling at the threshold of your mind. All rush to clutch your warm brain and to climb once more into the light of day.[135]

In the following, for the sake of humanity, he appears to be instructing the readers to select genetically favorable partners, with a strong eugenic message to discard what is not so favorable:

> But you must choose with care whom to hurl down again into the chasms of your blood, and whom you shall permit to mount once more into the light and the earth. ...
> Do not pity them. Keep vigil over the bottomless gulf of your heart, and choose. You shall say: "This shade is humble, dark like a beast: send him away! This one is silent and flaming, more living than I: let him drink all my blood!"[136]

Kazantzakis seeks solace in the continuity of life through the primordial and more recent ancestors, along with the potential of the future descendants:

> When you shake with fear, your terror branches out into innumerable generations, and you degrade innumerable souls before and behind you. When you rise to a valorous deed, all of your race rises with you and turns valorous. "I am not alone! I am not alone!" Let this vision inflame you at every moment. You are not a miserable and momentary body; behind your fleeting mask of clay, a thousand-year-old face lies in ambush. Your passions and your thoughts are older than your heart or brain.[137]

In the following passage, the "invisible body" is the "germ plasm" or genetic component of life, and the visible body is no doubt, in Weismannian terms, the somatic component which, viewed here, is the current and transient life. Added to this is the hint of further eugenic rhetoric on the improvement of mind and body:

135 Ibid., 71. (Kazantzakis, *Ασκητική*, 34-35.)
136 Ibid. (Kazantzakis, *Ασκητική*, 35.)
137 Ibid., 72–73. (Kazantzakis, *Ασκητική*, 36–37.)

> Your invisible body is your dread ancestors and your unborn descendants. Your visible body is the living men, women, and children of your race. ...
> Fight that all of your bodies may become strong, lean, prepared, that their minds may become enlightened, that their flaming, manly, and restless hearts may throb. [138]

The continuity of humanity is further highlighted along with the Darwinian premise of the fittest that are able to perpetuate the human race by passing on their favorable human traits to their offspring: "Your first duty, in completing your service to your race, is to feel within you all your ancestors. Your second duty is to throw light on their onrush and to continue their work. Your third duty is to pass on to your son the great mandate to surpass you."[139] Kazantzakis applies the theme of continuity on humanity and he focuses not on the human as such, rather he highlights the germ plasm's ability to remain immortal:

> It is not you talking. Nor is it your race only which shouts within you, for all the innumerable races of mankind shout and rush within you: white, yellow, black. ...
> We carry a torch and run. Our faces light up for a moment, but hurriedly we surrender the torch to our son, and then suddenly vanish and descend into Hades. [140]

Again the "invisible" is a metaphor for the genetic component which powerfully and perpetually transcends the generations of families:

> The mother looks ahead, toward her daughter; the daughter in turn looks ahead, beyond the body of her husband, toward her son—this is how the Invisible proceeds on earth.
> We all look directly before us, ruthlessly, driven by dark, enormous, infallible powers behind us.[141]

With the transitional period's multitude of ideas and ideologies which were interwoven at that time, while we have the established scholarship

138 Ibid., 73. (Kazantzakis, *Ασκητική*, 37.)
139 Ibid., 74. (Kazantzakis, *Ασκητική*, 38.)
140 Ibid., 75–76. (Kazantzakis, *Ασκητική*, 40–41.)
141 Ibid., 76–77. (Kazantzakis, *Ασκητική*, 41.)

that acknowledges in *Askitiki* Bergson's creative "primordial energy" in the human race, it is also Darwin's theory of evolution through natural selection and Weismann's continuity of the germ plasm which hold for the aforementioned aspects of Kazantzakis' essay.

The *élan vital* was Bergson's metaphysical interpretation of Darwin's mechanism of evolution. Utilizing this "vital impetus" offered a softer blow to the post-Darwinian mechanistic view of life, and was popularly taken up by literary writers, such as Kazantzakis, in the early decades of the twentieth century. Darwinism and Weismann's neo-Darwinian theory would later, in the form of the modern evolutionary synthesis, take over Bergson's *élan vital* by means of the force of genetics. It is apparent though that Kazantzakis' skillful use of words has allowed his essay to be viewed in various ways, and also in terms of the theme of continuity in evolutionary terms, which I maintain he knew of in 1922, when he started to write *Askitiki*.

Kazantzakis also speaks of the evolution of the "earth" and, as noted also by Pouliopoulos, the following passage in this phase of "The March" exemplifies Darwinian nature, encompassing a description of the creation of the world and a Darwinian evolution of man:[142]

> I recall an endless desert of infinite and flaming matter. I am burning! I pass through immeasurable, unorganized time, completely alone, despairing, crying in the wilderness.
> And slowly the flame subsides, the womb of matter grows cool, the stone comes alive, breaks open, and a small green leaf uncurls into the air, trembling. It clutches the soil, steadies itself, raises its head and hands, grasps the air, the water, the light, and sucks at the universe.
> It sucks at the universe and wants to pass it through its body—thin as a thread—to turn it into flower, fruit, seed. *To make it deathless.*
> The sea shudders and is torn in two: out of its muddy depths a voracious, restless, and eyeless worm ascends.
> The weight of matter is conquered, the slab of death heaves high, and armies of trees and beasts emerge filled with lust and hunger.
> I gaze upon Earth with her muddy brain, and I shudder as I relive the peril. I might have sunk and vanished amid these roots that suck at the mud blissfully; I might have smothered in this tough and many-wrinkled hide; or I

142 See Pouliopoulos, *O Νίκος Καζαντζάκης*, 306.

> might have twitched eternally within the bloody, dark skull of the primordial ancestor.
> But I was saved. I passed beyond the thick-leaved plants, I passed beyond the fishes, the birds the beasts, the apes. I created man.[143]

Again, "To make it deathless"—that is, immortal—reflects Kazantzakis' knowledge of Weismann's theory of the continuity of the germ plasm.[144] Further to this, the narrative is written in the first person, which appears to be a perpetual genetic agent in the evolutionary transformation. The passage demonstrates Kazantzakis' use of the concept of the struggle for existence, most likely a reflection of his own spiritual struggle in life. It also highlights natural selection and the concept of the transmutation of lower species, resulting in the human species. There is however after all this some sense of purpose to this evolution:

> Only now, as we feel the onslaught behind us, do we begin dimly to apprehend why the animals fought, begot, and died; and behind them the plants; and behind these the huge reserve of inorganic forces. ...
> We also toil with the same delight, agony, and exaltation for the sake of Someone Else who with every courageous deed of ours proceeds one step farther.[145]

Although Kazantzakis turned to Darwinism early in his life, as mentioned at the start of this section, he experimented with other intellectual and spiritual ideas as well. However, it should be noted that it was Kazantzakis' exposure to Darwinism which caused him to lose his Christian faith. Knowledge of Darwinism was to contribute to a central experience in modernity—that is, the loss of religious belief, which would be felt by many intellectuals.

143 Kazantzakis, *Saviors of God*, 82–83 (my italics). (Kazantzakis, *Ασκητική*, 47–48.)

144 By 1923 Weismann's concept of the continuity of the germ plasm or germ cell was a basic concept, well-known among those intellectuals who had followed the investigations on biological evolution. Kazantzakis' unpublished page which I mentioned earlier reflects such knowledge. It displays several references to the germ plasm and its continuity, and this is in relation to a discussion of Hans Dreisch's embryology experiments which caused anti-Darwinian sentiments among embryologists at the time. See further on the science Mayr, *Growth of Biological Thought*, 117–118.

145 Kazantzakis, *Saviors of God*, 83–84. (Kazantzakis, *Ασκητική*, 49–50.)

Alexandros Papadiamantis

After leaving the monastic life, Alexandros Papadiamantis (1851–1911) lived in Athens. Self-educated in English and French, he was presumably well-read on international issues. It is likely that he read Darwin, but it is not known for certain. Still, his knowledge of English and French, his journalistic work with foreign media, and his translating background would point towards an exposure to intellectual trends, including Darwinian and post-Darwinian ideas.

Papadiamantis' *Murderess* (*Η φόνισσα*) (1903), considered an important novel in Greek literature, is about the old woman Frankojannou (also known by other names, including Hadoula). She lives on the Greek island of Skiathos and she takes it upon herself to kill young females, from infants to little girls, who are either poor or physically feeble. She believes that these females are disadvantaged socially, that they never marry, and so remain a financial burden to their families. The strong tradition, then, that a girl's family must provide her with a substantial dowry to secure the marriage, plays a crucial role in Frankojannou's motives; for this reason poor families would not be able to provide their daughters with dowries. Sickly females would also not be suitable for marriage and basically, as Frankojannou thought, such a child "existed to be tormented and to torment us."[146]

In one situation Frankojannou scrutinizes the appearance and growth of two young sisters. She sees the perpetual growth of young girls as an uncontrollable problem: "'How they grow, my God!' thought Frankojannou. 'What garden, what meadow, what springtime produces this plant! And how it sprouts and flowers and puts out leaves and grows thick! And all these shoots, all the budding plants, will one day become greens, thickets, gardens? And this is how it will continue forever?'"[147] The point to be made here is of a similar Darwinian style in Karkavitsas' novel *The fair maid* (*Η λυγερή*) (1891), where he describes the young pregnant woman Anthi metaphorically transforming into a thriving flowering plant. See chapter 3 for this description.

This novel has generated substantial scholarship, particularly in the last decades. Commentators Jina Politi[148] and Dimitri Tziovas[149] appear to agree

146 Papadiamantis, *Η φόνισσα*, 76.
147 Ibid., 86.
148 Politi, "Δαρβινικό κείμενο και *Η φόνισσα*," 166.
149 Tziovas, *Other Self*, 100.

essentially that Frankojannou acts as a force of natural selection to reduce the female population. Politi maintains that the novel takes on a Darwinian worldview,[150] while Tziovas argues that the novel is antisocial and anti-evolutionary.[151] Tziovas maintains that the novel is "part of the intellectual climate of the end of the nineteenth century" with regard to such issues as progress, reform, and degeneration in evolutionary terms.[152] Further to this, he sees that "through natural selection Frankojannou believes that social inequality and human weakness can be confronted and obliterated."[153] He maintains that Papadiamantis' deep religious faith is inconsistent with a belief in Darwin's theories, "unless the emphasis on natural law is interpreted as a return to nature."[154] Katerina Kitsi-Mitakou views Frankojannou as the personification of Darwinian law and Christian law.[155] In addition, she addresses the "surplus female population" on Skiathos and claims that Papadiamantis was influenced by T. R. Malthus' *Essay on the Principle of Population* (1797) which, she indicates, later contributed to Darwin's work.[156]

My intention here is to situate the novel socially in Greece and to place it in the context of the European intellectual climate of that time. It is well documented that at the time Papadiamantis wrote the novel, the post-Darwinian ideology of eugenics was a world issue, as discussed in chapter 1. Eugenics had been popularized by Darwin's first cousin Francis Galton in his *Inquiries into Human Faculty and Its Development* (1883), coining the word eugenics to mean "good in stock."[157] Galton stated: "Man has already furthered evolution very considerably, half unconsciously, and for his own personal advantages, but he has not yet risen to the conviction that it is his religious duty to do so deliberately and systematically."[158] So it became a religious quest for Galton to see the improvement of humanity. Prior to Galton's work, in 1871 Darwin in his *DM* had discussed the issue of infanticide,

150 Politi, "Δαρβινικό κείμενο και *Η φόνισσα*," 164.

151 Tziovas, *Other Self*, 100.

152 Ibid., 99.

153 Ibid., 100.

154 Ibid.

155 Kitsi-Mitakou, "Aquatic Spaces and Women's Places," 192–193.

156 Ibid., 190–193. Malthus observed in primitive tribes that the geometric progression of a population would exceed the arithmetic rate of food resources, causing a "struggle for existence." In chapter 1 I argue that although Malthus' struggle for existence did inspire Darwin, it was limited in meaning, and Darwin's natural selection was a development way beyond what Malthus saw in primitive tribes.

157 Galton, *Inquiries into Human Faculty*, 17.

158 Ibid., 198.

in particular female infanticide in primitive cultures; this was a form of artificial selection for economic reasons:

> Infanticide—This practice is now very common throughout the world, and there is reason to believe that it prevailed much more extensively during former times. Barbarians find it difficult to support themselves and their children, and it is a simple plan to kill their infants. In the Polynesian Islands women have been known to kill from four to five to even ten of their children. ... Wherever infanticide prevails the struggle for existence will be in so far less severe, and all the members of the tribe will have an almost equally good chance of rearing their few surviving children. In most cases a larger number of female than of male infants are destroyed, for it is obvious that the latter are of most value to the tribe, as they will when grown up aid in defending it, and can support themselves.[159]

In fact, various sectors of society in general were preoccupied with eugenics as a consequence of the fear of the fin-de-siècle degeneration of society. Papadiamantis referred to his work as a social novel, which would direct us to the social context in Greece at that time. As I discussed in chapter 1, recent research has shown that in the early part of the twentieth century the problems of "overpopulation" and "social disease" in Greece were part of the "social question"; and aside from the sociopolitical strategies, this "social question" became a biological one, enmeshed in the minds and reforms of Greek eugenicists.[160] In Greece, poverty—seen as a symptom of degeneration—was widespread, and this was reflected in the high rates of abandonment of babies at orphanages (βρεφοκομεία) in Athens, in particular of female babies; babies were given up mainly due to the inability of needy parents to keep them.[161] The higher rate of abandonment of female babies as compared to male was due to "the disadvantaged position of women, and primarily of the poor, in Greece in the nineteenth century."[162] Greek islands such as Skiathos would not have had the infrastructure to accommodate such babies.

Episodes of female child infanticide were known to be prevalent at the end of the nineteenth century in Greece. A letter on female infanticide attrib-

159 Darwin, *DM*, 2:364.
160 See the work of Trubeta, Theodorou and Karakatsani.
161 Korasidou, *Οι άθλιοι των Αθηνών*, 101.
162 Ibid., 114.

uted to mothers was published in 1887 in *The Ladies' Newspaper* (*Εφημερίς των Κυριών*), whose editor-in-chief was the famous Greek feminist Kalliroi Parren. The anonymous writer of the letter indicated that from birth women in Greece were considered inferior and a burden to the family. The writer disclosed that for this reason "often and with not the slightest pang of conscience, the first or second day after the birth many of them are sent to their Deaths by their Christian mothers, so sparing them of their heavy burden."[163] Indicating that the numbers of infanticide were increasing, the writer pleaded for government intervention and also for the Church to assist by making these women understand that such actions were considered a sin against God. The letter was in response to an article in an earlier edition of the newspaper.[164] Also anonymous, this writer implored the reigning Queen of Greece Olga Konstantinovna, known for her philanthropy, to investigate incognito the conditions of the poor rural woman during her organized tour of rural Greece. The article reveals the horrors of women beaten by their husbands for what the men considered the "sacrilege"[165]—giving birth to female babies. The graphic accounts of the desperate measures used to kill these babies are confronting: hanging the infant by a rope on a tree, bleeding its head, and choking.[166] What is more, if the females were spared they were not sent to school—that is, as the article reveals, if schools existed in their area. This was the plight of the poverty-stricken rural female in Greece at that time.

In a 1917 lecture in Greece on eugenics, doctor Michalis Kairis described the first eugenicists as the ancient Greeks:

> The first example of an attempt at eugenics that we find in history is furnished by our ancestors the ancient Greeks, who invented and perfected everything. With the aesthetic sense and the concern for the state which distinguished them, they could not tolerate the birth of ugly, sickly, and imperfectly formed children. For this we see in the history of the ancient Spartans how these imperfect and ill-formed creatures were put to death by being thrown into the so-called Apothetai—a ravine adjoining Taygetus—or into

163 Parren, ed., "Μητέρες παιδοκτόνοι," 4.
164 Parren, ed., "Η Βασίλισσα περιοδεύει," 3-4.
165 Ibid., 3.
166 Ibid.

> Kaiadas, another ravine near the rock of present-day Mistra, or according to other authorities, into the river Eurotas, as soon as they were born, for they were regarded as useless burdens on the earth![167]

While he marvels at the ancient Greeks and their achievements in perfection, he is careful in his lecture not to condone the existing negative eugenics in various countries in modern times.[168] A significant part of his lecture is on infanticide[169] and on the high intake of children in orphanages. He writes that the one and only orphanage in the municipality of Athens in 1915 was recorded as having 1,916 abandoned babies and children, with his statistics showing that in those years the annual mortality rates of infants under the age of one in that orphanage were as high as 73%.[170]

In the early 1900s the eugenicist ideas were seen by some as a panacea in the reduction of poverty and also in the improvement of the human race. Eugenic feminists felt it their responsibility to educate on the reduction of the births among the poor, who were perceived to be overpopulating the nation and sending it into decline.

Creative literature reflected society's preoccupation with positive and negative eugenics and their application. In anticipation of Galton's eugenic ideas, probably influenced by Darwin's ideas of artificial selection, there is Robert Ellis Dudgeon's novel *Colymbia* (1873), where the practice of infanticide is used to eliminate the unfit babies.[171] Views adapting Galton's eugenics, degeneration, and the counter-discourse of regeneration may very well have been a major inspiration in Papadiamantis' novel at a time when others in Greece and the rest of Europe were discussing these ideas in relation to improving the race or nation.

---

167 Kairis, *Η ευγονία*, 4–5. Kairis uses also the word "η καλλιγονία." In both cases he is referring to "eugenics," as he cites Galton as the creator of the term "ευγονία." The word "η ευγονική" is the more common word for "eugenics."

168 Positive eugenics promotes the birth of the "fit," whereas negative eugenics aims at preventing the birth or survival of the "unfit."

169 He calls infanticide not only the killing of a child after birth but also that of the unborn fetus of pregnant women.

170 Ibid., 27–28. He suggests positive reforms be put in place to include for instance some sort of financial payment from the state to support the mother, so that she does not have to work while the child is very young, a system he indicted existed in America.

171 For an analysis on the novel see Henkin, *Darwinism in the English Novel*, 183–186. Dudgeon was Samuel Butler's medical adviser and close friend. The novel is little-known but appears to be one of the earliest English novels dealing with infanticide.

It is worth exploring the argument that Papadiamantis also portrays Frankojannou as a masculinized form of the New Woman (a literary negative representation of the members of the Women's Movement), who eugenically controlled the female population via artificial selection. She is described in the story as having a strong character and build, with a "masculine air" and "a thin moustache above her lip."[172] The New Woman theme is examined in chapters 5 and 6, and the eugenics theme in chapter 6.

Papadiamantis does allude to the ancient Greek style of eugenics when Frankojannou reveals what she believes should be the fate of a sick little girl: "If Death does not want to take it, then throw it over the precipice."[173] Frankojannou finally meets her own death. As she is chased by the authorities for her crimes, she drowns, a death some would call quite befitting considering that this was how victims of hers perished. The last passage of the novel leaves us with the moral dilemma of whether Frankojannou's drowning was justice served: "The old woman Hadoula found death at the crossing of Saint Sosti, at the neck which connected the outcrop of the hermitage with the land, halfway on the road, between divine and human justice."[174] Interestingly, just before she drowns she sees at a distance her dowry of land; it is the absence of a dowry that makes the young girls contenders for the killing. One could also add that the sentiment perpetuated by Galton in his time—that is, that it is the human species' "religious duty" to "deliberately and systematically" improve humanity—is an ideology that Papadiamantis juxtaposed on Frankojannou; an ideology that Papadiamantis speculated as teetering precariously between societal and divine laws, that is the ideology formed by the raw Darwinian natural laws, a view he wanted his readers to ponder.

172 Papadiamantis, *Η φόνισσα*, 60.
173 Ibid., 77. What she declares in Greek is the saying "Δεν είναι χάρος, δεν είναι βράχος."
174 Ibid., 227.

CHAPTER III

DARWINIAN REFLECTIONS: *CHILDREN'S GUIDANCE*

It has always seemed to me that for an Omnipotent & omniscient Creator to foresee is the same as to preordain; but then when I come to think over this I get into an uncomfortable puzzle something analogous with "necessity & Free-will" or the "Origin of evil," or other such subject quite beyond the scope of human intellect.
CHARLES DARWIN[1]

The popular Greek weekly magazine *Children's Guidance* (*Η Διάπλασις των Παίδων*), published 1879–1947, albeit a children's magazine, was widely read.[2] It epitomizes the advent of nineteenth-century periodical literature, which flourished as an important medium for communicating ideas to the wider educated public. Serialized fiction, scientific, and nonscientific issues were to be found side by side; so that among other ideas, scientific matters were discussed in popular weekly periodicals and intellectual quarterlies. Particularly with the British press, this type of periodical has been documented as having played a major role in the public debate which followed after the publication of the *OS*.[3]

1 From a letter to Asa Gray, dated February 24, 1860. See Burkhardt et al., *Correspondence of Charles Darwin, 1860*, 8:106 (punctuation and italics in original).

2 The entire collection was digitized in 2001 by the Hellenic Literary and Historical Archive (E.L.I.A.) in Athens, for the fifty-year commemoration of Xenopoulos' death. Part of the content in this chapter is based on the conference proceedings Zarimis, "Darwinian Thought in Grigorios Xenopoulos," 245–260. However, it should be noted that the research has been revised and expanded since then, and this is embodied in this chapter.

3 See Ellegård, *Darwin and the General Reader*, 18–61; Oldroyd, *Darwinian Impacts*, 193–203; Cantor and Shuttleworth, *Science Serialized*.

The aim of this chapter is to show that the *Children's Guidance* also played a significant role in the dissemination in Greece of Darwin's theories and other evolutionary ideas, reflecting the impact they had in many fields, including religion. The investigation will be viewed through the writings of Grigorios Xenopoulos who, as editor-in-chief from 1896 to 1947, wrote about numerous topics, including Darwin and associated matters, under the pseudonym "Faidon" in a section of the *Children's Guidance* entitled "Athenian Letters" ("Αθηναϊκαί Επιστολαί").[4]

From 1896, Xenopoulos would publish the magazine in a simplified katharevousa, making it easier for the youth to understand than its more difficult form. As the years progressed, he would utilize an even simpler form, thus slowly incorporating the demotic—which by 1917 had also been introduced into the primary schools. Along with other intellectuals, Xenopoulos had advocated the importance of the demotic language in educating the Greek youth, who were considered the future of the Greek nation. The liberal demoticists and progressive educationalists argued that only in this spoken language could the youth think freely and learn properly, without the pressures of having to learn in a synthetic language, the katharevousa.

Xenopoulos has tended not to be taken very seriously by literary historians and many commentators, as is often the case with very prolific writers. With reference to the relevant "Letters," the autobiography in his *Collected works*,[5] and other sources, an insight to his intellectual background will be revealed, specifically establishing his strong interest in Darwinism and other evolutionary ideas. This chapter will provide an appreciation of several of the post-Darwinian themes and motifs that Xenopoulos expressed in his writings.

When researching Xenopoulos' writing, one is limited to published works. The night of December 16, 1944, along with his family he abandoned his house in Athens, soon after it had been bombed by unknown perpetra-

4 In relation to the "Athenian Letters" see Malafantis, *Οι "Αθηναϊκαί Επιστολαί."* His study of the 2,000 "Athenian Letters" divides the letters into twenty-three categories, according to the main subject of each letter. The main categories are literature and children's education, but also listed are science and religion, which discuss a number of intellectual trends. A number of the letters I have researched are categorized in Malafantis' book. Some he has only mentioned by title. In some cases he has cited sections of the letters and for others he has also provided specific commentary. Further on the "Athenian Letters" see Malafantis, "Ο Γρηγόριος Ξενόπουλος," 494–501. Malafantis' research in this later article on *Children's Guidance* shows that it was not just read by Greek youth, but it was generally read by young and old.

5 His *Collected works* is actually only a selection of his works.

tors.[5] Xenopoulos' office and many of his unpublished works were destroyed. Therefore it would be impossible to quantitate to what extent these works were influenced by Darwinian ideas. This unfortunately has hindered a better picture of Xenopoulos. It is possible that some of the works that were destroyed were unpublished because of the evolutionary content, which he might have perceived as controversial, but this is only speculation. He does, however, refer to at least one paper on evolutionary theory that was in the fire. This will be further discussed later in the chapter.

Science and Positivism

From the autobiographical novel *My life as a novel* (*Η ζωή μου σαν μυθιστόρημα*), which Xenopoulos first published in serial form in 1939, we can obtain information about his intellectual sources. Even though Xenopoulos admitted in the novel that he could be more open about his views on various issues, it appears that he was still very careful regarding certain issues, such as those of religion.[7]

Xenopoulos' university education and his insatiable appetite for reading in Greek, French, German, and apparently some English exposed him to the intellectual trends of post-Darwinism. It is worth noting that Xenopoulos was one of only a few Greek creative writers of the early twentieth century to have had some formal university training in the sciences. Others include Pavlos Nirvanas, Andreas Karkavitsas (1866–1922), and Konstantinos Christomanos (1867–1911).[8] In 1883, Xenopoulos commenced his degree enrolling in the natural sciences at the University of Athens. Here his studies included botany, physics, mathematics, astronomy, and mineralogy. Outside these courses, of his own volition he also followed classes in physiology, history, literature, and philosophy.

It is important to note that around 1880 pro-Darwinian lecturers from the Schools of Philosophy and Medicine had been embroiled in disputes with anti-Darwinian theologians from the Theological School. One of Xenopoulos' favorite lecturers was the Darwinian Ioannis Zochios, who

6 Malafantis, *Οι "Αθηναϊκαί Επιστολαί,"* 327.

7 Xenopoulos, *Η ζωή μου*, 136.

8 There were also trained scientists, such as Theodor von Heldreich and Theodoros Orfanidis, who wrote poetry on the side.

had been a central figure in the controversies over Darwinism. Zochios taught him physiology, and Xenopoulos was so drawn to his lectures that he attended them for two years. On Zochios, Xenopoulos revealed: "He had ... the knowledge of his time; he passionately followed the sciences; what he said he knew well, and he said it in a charming manner."[9] It is very likely that Xenopoulos received much of his initial inspiration on Darwinism from Zochios in those two years. After five years at university, he did not complete his tertiary studies. He had said that he would have preferred a more general university education, excluding physics and mathematics and including linguistics, archaeology, psychology, and education. In 1895, he edited the prestigious periodical *Estia*, published in the demotic. Though not a specialist journal, its main goal was to popularize science, and it included articles on Charles Darwin.[10]

In his autobiography he indicates that as a schoolboy he had read Daudet, Flaubert, Zola, Maupassant, Dickens, Turgenev, Dostoevsky, and that he continued to read non-Greek literary masterpieces through to adulthood.[11] Interestingly, in his childhood he had read very few Greek books. As a teenager, he read the materialist Ludwig Büchner's *Kraft und Stoff* (*Force and Matter*) (1855), which adversely affected his faith.[12] This is reflected in his novel *Kosmakis 1: The first awakening* (*Ο Κοσμάκης Α΄: Το πρωτοξύπνημα*) (1923), where Xenopoulos aligned his own philosophical experiences in his early adolescence to the novel's main character Dionysis Alibrandi.[13]

Reading the work of astronomer, philosopher, poet, and popularizer of science Camille Flammarion (1842–1925) led him to become a "pantheist." Xenopoulos claims it took many years of reading numerous scientific and philosophical books to return to his original faith.[14] In 1925 Xenopoulos wrote an "Athenian Letter" as a tribute to Camille Flammarion, extolling

9 Xenopoulos, *Η ζωή μου*, 161.

10 See further the introduction of this book on *Estia*. In 1927, Xenopoulos founded and was editor-in-chief of the literary periodical *Nea Estia* (*Νέα Εστία*), which is still published. It has been mainly written in the demotic.

11 Xenopoulos, *Η ζωή μου*, 128.

12 Ibid., 132–133.

13 The four-volume novel *Kosmakis* was published in serial form in the Athenian newspaper *Ethnos* from March 26, 1923, to November 8, 1923. The four volumes are: *Α΄: το πρωτοξύπνημα* (1: The first awakening); *Β΄–το κέντρον* (2: The center); *Γ΄: τελευταία όνειρα* (3: Last dreams); and *Δ΄: ο γυρισμός* (4: The return). In 1930, it was also published in book form by Kollaros in Athens, with the subtitle *Ιστορία ενός φυσιολογικού αρρώστου* (Story of a physiologically sick person). It was published by the Vlassis Brothers in 1984.

14 Xenopoulos, *Η ζωή μου*, 132–136.

all aspects of his work.[15] The "Letter" reflects the admiration that Xenopoulos had for Flammarion, and it appears that his approach to his work and his philosophy had a strong influence on Xenopoulos. Further to this, it is also likely that Xenopoulos may have tried to mimic Flammarion's approach to popularizing science through his own endeavors to educate the readers of the "Athenian Letters" on scientific topics such as Darwinism.

During his university life, apart from physics and mathematics he read works by Aristotle, Bacon, Descartes, Kant, Schopenhauer, and almost all of Darwin's works.[16] His "Athenian Letters" reflect the high esteem in which he held science. In the following "Letter," he believed that those entering the sciences should not be mercenary:[17] "The reason for which one studies a science is generally the desire for knowledge, that deep and sleepless desire which makes us go on learning forever... . Unfortunately, here in Greece, we don't yet realize this great truth; we have not yet comprehended the lofty purpose of Science. ... We think that it is a simple bread-winning ticket. The true scientists, the ones dedicated to Science as their purpose in life, and to work only for it—can be counted on one hand."[18]

He considered the Greek mathematician and philosopher Giorgakis Hairetis, with whom he had discussions on socialism and positivism, as having been a major influence in his life. He attributed *Rich and poor* to the socialist teachings of Hairetis.[19] During his university years, he had read Auguste Comte's philosophy of positivism, which relies on science and scientific method as the only sources of knowledge, and rejects metaphysics and theism; so the world was to be viewed through science via observable phenomena. Xenopoulos believed that positivism, unlike materialism, allowed him freedom in his ideological thinking. He believed positivism did not sanction metaphysics, but also it was not able to confirm certain things.[20] Although he lost interest in socialism early in his life, he confessed in his

15 Xenopoulos, "Κάμιλλος Φλαμμαριόν," 236.

16 Xenopoulos, *Η ζωή μου*, 162.

17 The word "επιστήμη" (science) can have a much wider meaning in Greek to denote, for example, general academia. However, in this particular letter where there is some ambiguity, I see it in the same manner that Malafantis has in his study—that is, with the capital first letter as "Επιστήμη" to mean science. See Malafantis, *Οι "Αθηναϊκαί Επιστολαί,"* 216–217.

18 Xenopoulos, "Η επιστήμη," 291.

19 The influences of socialism and positivism in *Rich and poor*, in relation to Darwinian thought, will be examined in chapter 4.

20 Xenopoulos, *Η ζωή μου*, 261.

autobiography that: "A positivist I remained deep down forever. Comte's philosophy influenced my work also; from then on it became more realistic, more positive."[21] It is likely that Comte's positivist philosophy was an influencing factor when Darwin was formulating his ideas on evolution, and that it contributed to his oscillation between atheism and agnosticism. With Xenopoulos' strong support for Comte, one would imagine that his faith would have been affected. However, Xenopoulos wrote at a time in Greece when the establishment was solidly committed to the Greek Orthodox faith. His departure from that would have been perceived negatively, jeopardizing his professional writing career. He showed no obvious skepticism in his faith. On the contrary, in his "Athenian Letters" to the young readers of *Children's Guidance* he constantly instructed them that their faith was of utmost importance.

His university background in the sciences, though incomplete, appears to have translated to his literary work. While at university, his fellow students would tease him, referring to him as the "mathematician/physical scientist, philologist, and a bit ... the doctor."[22] This is reflected in Xenopoulos' work in this chapter and the forthcoming chapters of this book. Commentator Giannis Kouhtsoglou, who interviewed Xenopoulos, described him as follows: "Xenopoulos studied mathematics due to an innate tendency to classification and logical clarity. Essentially, his work is regulated (in its internal inclination, in its form, and in its architectural development) by an imperceptible law of positive method. The same goes for his critical functions where his penetrating—and sometimes revealing—sharp-sightedness and sensitivity have a constancy likened to the combinations and correlations of 'mathematical operations.' It is not an exaggeration if we say that Xenopoulos had a scientific mind."[23]

Xenopoulos "mathematically" applied certain evolutionary ideas, such as gradualism, to some of his "Athenian Letters" and novels. This will be examined in this chapter and in chapters 4 and 5. He was a strong believer in the naturalist movement, believing that a writer has the right to present whatever happens in life, whatever it may be.[24] He felt that prior to read-

21 Ibid.

22 Xenopoulos, *Η ζωή μου*, 162 (suspension points in original).

23 Kouhtsoglou, "Ο 'σπενσεριστής' Ξενόπουλος," 1015.

24 Xenopoulos, *Η ζωή μου*, 344.

ing Freud he had formulated his own beliefs on sexuality.[25] The wakening of the sexual instinct was one of his favorite themes.[26] The association with Darwinism, in particular the *DM*, can be likened to some of Xenopoulos' work, as will be seen in this study. In the prologue of his autobiography he mentioned it was only at the age of seventy that he was able to be more open about certain issues.[27] Even at this age, sexuality was a difficult subject for him to discuss, though he felt it was important in understanding humanity.

Xenopoulos demonstrated strong views on heredity and the environment in his writings of the twentieth century that show elements of post-Darwinian scientific knowledge.[28] Around 1925, he wrote in a letter to the Alexandrian literary critic Timos Malanos about the biological reasons for his alleged mediocrity in writing: "And do you know why I'm not a great writer? Because I was born average. Nature did not want to grant me more than a certain talent, one gift, not *genius*. ... And I placed all *my will*, from childhood to the present, to nurture it. And I am always nurturing it, without doing anything else in my life. But, as you see, will alone is not enough."[29] He acknowledged again his talent in writing as having innate origins, highlighting that no amount of "will" or desire throughout his life could raise him to genius. Further to his argument of nurture over nature, he admitted that will is not enough to reach genius level, implying that this volition can-

25 In his autobiography Xenopoulos indicated that it was only ten years since he first read Freud. The autobiography, in its final author-edited form, was published as a serial in the newspaper *Αθηναϊκά Νέα* (Athenian news) between September 26, 1938, and February 18, 1939. I assume then that Xenopoulos had not read Freud till about 1928.

26 Xenopoulos wrote on his four volume novel *Kosmakis*: "Many think that it is a Freudian work. ... But they are mistaken. Those theories were not yet known in Greece, nor were they very well known outside of Greece. My main character ... came to the conclusion that the 'center' of the universe, of humanity's, and of his own, is love and that sexual life governs and regulates the rest of it." Xenopoulos, *Η ζωή μου*, 190.

27 Ibid., 58.

28 It is likely that Xenopoulos could have been influenced very early in his works by the French thinker, literary and art critic, and historian Hippolyte Taine (1828–1893). Taine contributed to the nineteenth-century naturalist school. His method for studying literature was by means of seeing the novel "scientifically," as a "collection of experiments" which played a role in the scientific understanding of human nature and highlighted physical and psychological variables of man's behavior ("la race," "le milieu," "le moment"—that is, the inheritance, environment or background, and the historical context). Taine argued that it would provide a closer understanding of the psychology of the author. However, even within his own lifetime Taine's work was shown to have weaknesses, promising too much in terms of what his quasi-scientific method could achieve. Also his assessment of heredity was outdated by the twentieth century. See Taine, *History of English Literature*.

29 Malanos, *Δειγματολόγιο*, 98 (my italics). The letter is ca. 1925.

not undo one's destiny predetermined by nature. In a short autobiography that he wrote in 1939, he showed he was still keenly interested and believed in the concept of the power of nature over nurture.[30]

While the following passage from the autobiography does not deal with evolutionary themes explicitly, it does demonstrate Xenopoulos' views on heredity and the transmission of characteristics (acquired or not) from one generation to the next. Worth noting also in this passage are his changed views regarding his actual talent for writing: "My mother, so cultivated, certainly contributed towards my being born with a brain capable of cultivation. And of course to her I owe my love for letters, for study, for books, which certainly do not constitute talent, but which are support. *Heredity has unfathomable mysteries.* Try to find out which of my unknown forefathers had *that something uncertain, latent,* which had to reach me striding through the centuries to be revealed; and which of my descendants, in the far away future, will receive it, commencing from me, like *a natural gift,* richer, more valuable."[31] He attributed his actual writing talent only to the unfathomable powers of heredity; so strong, "that something, uncertain, latent"—that is, the gene as we know it today (word coined in 1909)—is capable of transcending centuries, showing a sense of continuity and even immortality. In his novels, Xenopoulos constantly probed the issue of nature versus nurture.

The "Athenian Letters" highlight the diversity of Xenopoulos' knowledge and reading. However, it is noteworthy that the topic of nature was the second most discussed category in his "Athenian Letters", which trailed marginally behind the topic of literature.[32] So that next to his love of literary matters, Xenopoulos revealed a love of nature—a trait he shared with the naturalist Charles Darwin. The "Athenian Letters" also show that this great interest and concern for nature spanned most of his writing career.[33]

With a conservative air, he communicated a wide range of topics to the Greek youth. His goal was to educate, often with morally didactic overtones. An examination of a selection of these "Letters" also provides an insight into how Xenopoulos transmitted ideas on science and religion, as affected by

30 Xenopoulos, "Σύντομη αυτοβιογραφία," 5–7.
31 Ibid., 6 (my italics).
32 Malafantis, *Οι "Αθηναϊκαί Επιστολαί,"* 73.
33 In his "Athenian Letters" he wrote about nature as early as 1896. See for instance Xenopoulos, "Η καλλιτέρα εποχή," 291. For one of his latest see Xenopoulos, "Και τα φυτά είναι … ζώα," 263.

Η ΔΙΑΠΛΑΣΙΣ ΤΩΝ ΠΑΙΔΩΝ

ΕΙΚΟΝΟΓΡΑΦΗΜΕΝΟΝ ΠΕΡΙΟΔΙΚΟΝ ΔΙΑ ΠΑΙΔΙΑ ΕΦΗΒΟΥΣ ΚΑΙ ΝΕΑΝΙΔΑΣ

...νισταμένον ὑπὸ τοῦ Ὑπουργείου τῆς Παιδείας ὡς τὸ κατ' ἐξοχὴν παιδικὸν περιοδικὸν σύγγραμμα, ἀληθεῖς παρασχὸν εἰς τὴν χώραν ἡμῶν ὑπηρεσίας καὶ ὑπὸ τοῦ Οἰκουμενικοῦ Πατριαρχείου Κωνσταντινουπόλεως, ὡς ἀνάγνωσμα ἄριστον καὶ χρησιμώτατον εἰς τοὺς παῖδας.

ΣΥΝΔΡΟΜΗ ΠΡΟΠΛΗΡΩΤΕΑ

Ἐσωτερικοῦ		Ἐξωτερικοῦ	
Ἐτησία	δρ. 8.—	Ἐτησία	φρ. χρ. 10.—
Ἑξάμηνος	» 4,50	Ἑξάμηνος	» 5,50
Τρίμηνος	» 2,50	Τρίμηνος	» 3,—

Αἱ συνδρομαὶ ἄρχονται τὴν 1ην ἑκάστου μηνός.

ΕΚΔΙΔΕΤΑΙ ΚΑΤΑ ΣΑΒΒΑΤΟΝ

ΙΔΡΥΘΗ ΤΩ 1879

ΙΔΡΥΤΗΣ-ΕΚΔΟΤΗΣ ΚΑΙ ΔΙΕΥΘΥΝΤΗΣ

ΝΙΚΟΛΑΟΣ Π. ΠΑΠΑΔΟΠΟΥΛΟΣ

ΤΙΜΗ ΕΚΑΣΤΟΥ ΦΥΛΛΟΥ Λ. 20

Διὰ τῶν Πρακτόρων Ἐσωτερ. λ. 15. Ἐξωτερ. λ. 20.

Φύλλα προηγουμένων ἐτῶν, Α' καὶ Β' περιόδου τιμῶνται ἕκαστον λεπ. 25

ΓΡΑΦΕΙΟΝ ΕΝ ΑΘΗΝΑΙΣ

Ὁδὸς Εὐριπίδου ἀρ. 38, παρὰ τὸ Βαρβάκειον

Περίοδος Β'.—Τόμος 25ος *Ἐν Ἀθήναις, 6 Ἰανουαρίου 1918* *Ἔτος 40όν.—Ἀριθ. 6*

Η ΜΥΣΤΗΡΙΩΔΗΣ ΝΗΣΟΣ

[Μυθιστόρημα ὑπὸ ΙΟΥΛΙΟΥ ΒΕΡΝ]

ΜΕΡΟΣ ΔΕΥΤΕΡΟΝ

ΚΕΦΑΛΑΙΟΝ ΙΔ'. (Συνέχεια)

Δὲν τὸ ἐσκέπτετο μὲ τὰ σωστά του ὁπωσδήποτε, ὁ καλὸς ναύτης προτιμοῦσε νὰ πᾶνε τώρα νὰ προγευματίσουν στὸ πλοῖο. Ὁ δρόμος ἄλλως τε ἦταν χαραγμένος κι' ὄχι πολὺ μακρύς,—μόλις ἕνα μίλι. Ξεκίνησαν λοιπὸν ἀπὸ τὸ δάσος, ὅπου γίδες καὶ χοῖροι ἐξακολουθοῦσαν νὰ φεύγουν μπροστά τους κατὰ ἑκατοντάδες. Ὕστερ' ἀπὸ εἴκοσι λεπτά, ὁ Πέγκροφ κι' οἱ σύντροφοί του ξαναεῖδαν τὴ δυτικὴ ἀκτὴ τοῦ νησιοῦ καὶ τὸν «Καλότυχο» ἀγκυροβολημένο ἐκεῖ ποῦ τὸν εἶχαν ἀφίση.

Ὁ Πέγκροφ δὲν μπόρεσε νὰ συγκρατήση ἕνα στεναγμὸ ἀνακουφίσεως... Ἐπιτέλους παιδί του ἦταν τὸ καΐκι ἐκεῖνο καί, σὰν πατέρας, εἶχε τὸ δικαίωμα νὰ ἀνησυχῆ περισσότερο κι' ἀπ' ὅ,τι ἔπρεπε...

Μπῆκαν μέσα κι' ἔφαγαν καλά, ὥστε νὰ μὴν ἔχουν ἀνάγκη νὰ ξαναφάγουν σὲ λίγο. Ὕστερα ξαναξεκίνησαν πάλι για τὴν γενικὴ ἔρευνα ποῦ εἶχαν ἀποφασίση.

Τὸ πιθανώτερον ἦταν ὅτι ὁ μονάκριβος κάτοικος τοῦ νησιοῦ ἐκεῖνου εἶχε πεθάνη. Ὥστε μᾶλλον νεκροῦ παρὰ ζωντανοῦ ἔψαχναν τώρα νὰ βροῦν τὰ ἴχνη. Ἀλλὰ τοῦ κάκου ξόδεψαν τὴ μισὴ σχεδὸν μέρα, τοῦ κάκου ἐξερεύνησαν ὅλες ἐκεῖνες τὴς συστάδες τῶν δένδρων ποῦ σκέπαζαν τὸ νησί! Ἀναγκάσθηκαν τότε νὰ παραδεχθοῦν ὅτι, κι' ἂν ὁ ναυαγὸς εἶχε πεθάνη, δὲν ἔμενε τώρα πιὰ οὔτε ἴχνος ἀπὸ τὸ πτῶμά του. Κάποιο ἀγρίμι, σαρκοβόρο, θὰ τὸ εἶχε φάη ὡς τὸ τελευταῖο κοκκαλάκι!

Ὁ Πέγκροφ κι' οἱ σύντροφοί του, στὴς δυὸ μετὰ τὸ μεσημέρι, ξαπλώθηκαν κάτω ἀπὸ πεῦκα, γιὰ νὰ ξεκουρασθοῦν λίγες στιγμές.

— Αὔριο, πρωὶ-πρωί, φεύγουμε! εἶπε ὁ καπετάνιος.

—Καὶ μοῦ φαίνεται, ἐπρόσθεσε ὁ Χάρμπερτ, ὅτι χωρὶς καμμιὰ τύψι, μποροῦμε νὰ πάρουμε μαζί μας τὰ πράγματα τοῦ μακαρίτη...

—Ναί, βέβαια, ἐπιδοκίμασε ὁ Πέγκροφ. Ἐμεῖς εἴμαστε τώρα οἱ φυσικοί του κληρονόμοι.

—Σωστά, εἶπε κι' ὁ Γεδεὼν Σπίλετ. Τὰ ὅπλα καὶ τὰ ἐργαλεῖα θὰ συμπληρώσουν τὰ ὑπάρχοντα στὸ Γρανίτινο Παλάτι. Καὶ τὸ πιὸ πολύτιμο ἀπ' ὅλα εἶνε τὸ βαρελάκι τῆς μπαρούτης καὶ τοῦ μολυβιοῦ.

—Να μὴν ξεχάσουμε ὅμως, ὑπέλαβε ὁ Πέγκροφ, να πιάσουμε καὶ κανένα ζευγάρι ἀπ' αὐτὰ τὰ γουρουνάκια, ποῦ δὲν ὑπάρχουν στὸ νησί μας.

—Καὶ νὰ μαζέψουμε ἀπ' αὐτοὺς τοὺς σπόρους, ἐπρόσθεσε ὁ Χάρμπερτ, ποῦ θὰ μᾶς δώσουν ὅλα τὰ λαχανικὰ τοῦ παλαιοῦ καὶ τοῦ νέου κόσμου.

—Τότε λοιπόν, εἶπε ὁ δημοσιογράφος, ἂς μείνωμε ἀκόμη μιὰ μέρα, για νὰ μαζέψωμε ὅ,τι χρήσιμο ἔχει ἡ νῆσος Θαβώρ.

— Ὄχι, κύριε Σπίλετ, ἀποκρίθηκε ὁ καπετάνιος. Θὰ σᾶς παρακαλοῦσα νὰ φύγουμε αὔριο τὰ χαράματα. Ὁ καιρὸς ἔχει διάθεσι νὰ γυρίσῃ καί, ἀφοῦ γυρίζουμε καὶ μεῖς, θὰ τὸν ἔχουμε πάλι πρίμα, ὅπως τὸν εἴχαμε καὶ στὸν ἐρχομό.

—Ἐμπρὸς λοιπόν! εἶπε ὁ Χάρμπερτ, σηκωνόμενος. Ἂς μὴ χάνουμε καιρό.

—Ναί, ἂς μὴ χάνουμε καιρό. Σύ, Χάρμπερτ, μάζεψε τοὺς σπόρους, ποῦ τοὺς γνωρίζεις καλλίτερά μας. Ἐν τῷ μεταξύ, ὁ κ. Σπίλετ κι' ἐγὼ θὰ πᾶμε νὰ κυνηγήσουμε γουρούνια· καί, μολονότι μᾶς λείπει ὁ Τόπ, ἐλπίζω νὰ τσακώσουμε μερικὰ ζωντανά.

Ὁ Χάρμπερτ πῆρε τότε τὸ

«Ὁ Κύρος Σμὶθ τὸν ἄγγιξε στὸν ὦμο...»

Figure 3.1. Example of a front page of *Children's Guidance*. January 6, 1918, no. 6, 45. Reproduced with permission from Anna-Maria Vlassi-Galerou (publisher) of Vlassis Brothers Publications, Athens.

6 Η ΔΙΑΠΛΑΣΙΣ ΤΩΝ ΠΑΙΔΩΝ 24 Μαρτίου

ΑΘΗΝΑΪΚΑΙ ΕΠΙΣΤΟΛΑΙ

ΣΟΒΑΡΑ ΤΑ ΠΡΑΓΜΑΤΑ

Ἀγαπητοί μου,

Η ΦΙΛΗ μας Ἑλένη Κ. Κ. — ἔχει καὶ ψευδώνυμο, ἀρσενικὸ ὅμως, ποὺ δὲν τῆς πάει, — εἶναι δεκατεσσάρων χρονῶν καὶ μαθήτρια τοῦ Γυμνασίου. Διαβάζει λοιπόν, ἀκούει καὶ μαθαίνει πολλά, ποὺ τῆς γεννοῦν τὶς μεγαλύτερες ἀπορίες. Τελευταῖα ἄκουσε ἢ διάβασε καὶ γιὰ τὸ Δαρβινισμό, τὴν περίφημη δηλαδὴ θεωρία τοῦ Δαρβίνου. Κι' ἔγεινε ἄνω-κάτω τὸ κορίτσι. Γιά, ἀκοῦστε ἐδῶ τί μοῦ γράφει:

«Εἶναι ἀλήθεια πὼς ὁ κόσμος ἔγεινε ἀπὸ τέσσερα κύτταρα μὲ τὴν ἐξέλιξή τους; Εἶναι ἀλήθεια πὼς ὅλη αὐτὴ ἡ φύσις ἡ τόσο σοφὰ κι' ἁρμονικὰ κατασκευασμένη, παρήχθη ἔτσι **ἐκ τύχης**; Τὰ διάφορα στάδια τῆς Γῆς, οἱ τέσσερες γεωλογικοὶ αἰῶνες, μαρτυροῦνται ἀπὸ τ' ἄπειρα ἀπολιθώματα ποὺ βρέθηκαν σὲ διάφορες ἀνασκαφές. Ἀλλὰ ὅτι ὁ ἄνθρωπος φάνηκε κατὰ τὴν τεταρτογενῆ περίοδο τοῦ καινοζωϊκοῦ αἰῶνος **μὲ μορφὴ πιθήκου**, πῶς θὰ τἀποδείξουμε; Εἶναι πιστευτὸ τέτοιο πρᾶγμα;

«Ὥστε ὁ ἄνθρωπος δὲν ἐπλάστηκε ἀπὸ τὸ Θεό; Δὲν ὑπῆρξε Ἀδὰμ καὶ Εὔα; Ὅλ' αὐτὰ εἶναι ψέμματα; Μά... πὼς εἶναι δυνατὸ τέσσερα **ἄψυχα** κύτταρα νὰ δώσουν ζωὴ σὲ ἄλλα καὶ νὰ σχηματίσουν ὀργανικά, ἔμψυχα ὄντα; Κι' ὅλ' αὐτά, ὅλος ὁ κόσμος,— στεριές, θάλασσες, ζῶα, φυτά, —ἔγειναν κατὰ τύχη; Μὰ ἡ Ἁγία Γραφὴ λέει πὼς τύχη δὲν σημαίνει τίποτα. Τί λοιπὸν νὰ πιστέψουμε;»

Σοβαρὰ τὰ πράγματα, βλέπετε! Κι' ὅμως ὁ Δαρβινισμὸς δὲν εἶναι παρὰ μιὰ ἐπιστημονικὴ θεωρία ὅπως ὅλες. Στὸν καιρό της,—ἐδῶ καὶ σαράντα χρόνια νὰ ποῦμε,—εἶχε μεγάλη πέραση. Τὴν πολεμοῦσαν μὲ λύσσα ἐπιστήμονες καὶ μή· ἄλλοι ὅμως, κι' οἱ καλύτεροι, τὴν ὑπερασπιζόνταν μὲ φανατισμό. Διανοούμενοι, φιλόσοφοι, τὴν ἐπίστευαν καὶ προσπαθοῦσαν νὰ τὴν ἐφαρμόσουν παντοῦ. Εἶχε τὴ μεγαλύτερη ἐπίδραση στὴ σκέψη τοῦ ΙΘ' αἰῶνα. Ἔκανε τοὺς ἀνθρώπους νὰ βλέπουν πὼς τίποτα στὸν κόσμο δὲν ἔγεινε διαμιᾶς, παρὰ πὼς ὅλα πέρασαν καὶ περνοῦν μιὰ ἐξέλιξη. Γιατὶ αὐτὴ ἡ θεωρία τῆς ἐξέλιξης, τοῦ ξετυλίγματος νὰ ποῦμε, εἶναι προπάντων ὁ Δαρβινισμός. Ἀπὸ τότε καθιερώθηκε κι' ὁ ὅρος, ἡ λέξη, ποὺ τόσο τὴν μεταχειριζόμαστε. Καὶ μαζὶ μὲ τὴ λέξη, φυσικά, ἐπικράτησε κι' ἐπικρατεῖ ἡ ἰδέα πὼς κάθε μορφὴ στὴ φύση, ὅπως στὴν τέχνη, στὴ βιοτεχνία, στὴν κοινωνία καὶ παντοῦ, δὲν εἶναι ἀρχική, καμωμένη ἐξ ἀρχῆς, παρὰ βαθμιαία ἀνάπτυξη, τελειοποίηση, μιᾶς ἄλλης ἀρχαιότερης κι' ἀτελέστερης.

Στὸν καιρό μου, ὅταν δηλαδὴ ἐσπούδαζα κι' ἐγὼ Φυσικὲς Ἐπιστήμες, ὁ Δαρβινισμὸς πιστευόταν σχεδὸν σὰν ἀξίωμα. Λέγαμε «ὁ ἄνθρωπος ἔγεινε ἀπὸ τὸν πίθηκο καὶ τὸ πουλὶ ἀπὸ τὸ ψάρι», ὅπως λέγαμε καὶ «τὰ τρίτῳ τινὶ ἴσα καὶ ἀλλήλοις». Ὅποιος εἶχε ἀντίρρηση, δὲν ἦταν... στὰ καλά του. Ἢ τουλάχιστο δὲν ἦταν «ἀληθινὸς ἐπιστήμων.» Ξέραμε πὼς ἡ θεωρία παρουσίαζε «χάσματα». Ξέραμε πὼς δὲν εἶχε ἀνακαλυφθῇ ἀκόμα ἡ διάμεση μορφὴ μεταξὺ πιθήκου καὶ ἀνθρώπου, ὁ «ἀνθρωποπίθηκος», καὶ πὼς διάφοροι παλιοὶ σκελετοί, ἀπὸ ἀνασκαφές, ποὺ στὴν ἀρχὴ εἶχαν νομισθῆ σκελετοὶ ἀνθρωποπιθήκων, βεβαιωνόταν κατόπι πὼς ἦταν ἢ γορίλλα ἢ ἀγριανθρώπου. Ἀδιάφορο! «Θὰ ἀνακαλυφθῇ» λέγαμε. Κι' ἡ πίστη μας ἦταν τέτοια, ποὺ νομίζαμε ὅτι μὲ τὸν καιρὸ ἡ θεωρία ὄχι μόνο δὲ θἄπεφτε, παρὰ θὰ τελειωνόταν μὲ τὴν πειραματική, τὴ θετικὴ πιὰ ἀπόδειξη πὼς ὅλο τὸ φυτικὸ καὶ τὸ ζωϊκὸ βασίλειο, ὅλη ἡ ζωὴ τῆς Γῆς μὲ τὶς ἄπειρές της μορφές, ἔγεινε ὄχι ἀπὸ τέσσερους-πέντε ἀρχικοὺς τύπους, ὅπως ἔλεγε ὁ Δαρβῖνος, ἀλλ' ἀπὸ ἕνα καὶ μόνο **πρωτόπλασμα**!

Ἀλλὰ τὶς Φυσικὲς Ἐπιστήμες τὶς ἄφησα γρήγορα γιὰ τὴ Φιλολογία. Κι' ἀπὸ τότε πέρασαν εἰκοσιπέντε τουλάχιστο χρόνια. Σ' αὐτὸ τὸ διάστημα, πολλὲς φορὲς ἔτυχε νἀκούσω ἢ νὰ διαβάσω πὼς ἡ Δαρβινικὴ θεωρία δὲν ἔχει πιὰ τοὺς πρώτους ὀπαδοὺς καὶ τὸ πρῶτο κῦρος. Πέρασε κι' αὐτή, ἔπεσε, ὅπως πέφτουν καὶ περνοῦν ὅλες αὐτὲς οἱ θεωρίες. Προπάντων ἡ θεωρία τῆς καταγωγῆς τοῦ ἀνθρώπου: Ἡ διάμεση μορφή, ὁ «ἀνθρωποπίθηκος» δὲν βρέθηκε! Κι' οὔτε ὑπάρχει πιὰ ἡ πεποίθηση πὼς θὰ βρεθῇ. Περισσότερα μὴ μὲ ρωτᾶτε. Ἀφοῦ ἔπαψα πιὰ νὰ παρακολουθῶ τὶς Ἐπιστήμες, δὲν εἶμαι εἰδικός, δὲν εἶμαι σὲ θέση νὰ σᾶς δώσω λεπτομέρειες, νὰ σᾶς πῶ ποῦ ἀκριβῶς βρίσκεται σήμερα ὁ Δαρβινισμός, ποιὰ ἀπ' ὅσα ἐδίδαξε μένουν σὰν ἀλήθειες,—γιατὶ μένουν κάμποσα,—καὶ ποιὰ ἔπεσαν σὰν ὑπερβολὲς κι' αὐθαιρεσίες. Στὸν καιρό μου τὰ ἤξερα. Σήμερα δὲν τὰ πολυξέρω. Τὸ μόνο ποὺ ξέρω, εἶναι αὐτὸ ποὺ σᾶς εἶπα: ἡ θεωρία ἔπαθε πολλὲς ἀβαρίες.

Ἀλλ' ἂς ὑποθέσουμε ὅτι δὲν ἔπαθε καθόλου. Ἂς ὑποθέσουμε πὼς ἡ ἀληθινὴ Ἐπιστήμη πιστεύει ἀκόμα καὶ διδάσκει πὼς τὰ εἴδη ἔγειναν τὸ ἕν' ἀπὸ τἄλλο καὶ πὼς ὁ ἄνθρωπος κατάγεται ἀπὸ τὸν πίθηκο. Πρέπει νὰ συμπεράνουμε πὼς ὁ κόσμος εἶναι «τυχαῖος» καὶ πὼς δὲν ὑπάρχει Δημιουργός, Θεός, ποὺ τὸν ἔκαμε μὲ θέληση καὶ μὲ σκοπό; Κάθε ἄλλο! Ἐγὼ τουλάχιστο, καὶ τὸν καιρὸ ἀκόμα ποὺ ἐπίστευα σὰν ἀξίωμα τὴ θεωρία τῆς Ἐξέλιξης—τὰ πάντα ἀπὸ ἕνα,—οὔτε μιὰ στιγμὴ δὲν ἔπαψα νὰ πιστεύω, νὰ θαυμάζω καὶ νὰ δοξάζω τὸν Δημιουργὸ αὐτοῦ τοῦ κόσμου. Καὶ ξέρετε γιατί; Γιατὶ δὲν ἄκουγα λόγια ἀπὸ δῶ κι' ἀπὸ κεῖ. Γιατὶ δὲν ἐμάζευα τὶς γνώσεις μου καὶ τὶς ἰδέες μου ἀπὸ τὸ δρόμο ἢ ἀπὸ τὸ καφενεῖο. Ἄκουσα τότε γιὰ τὴ θεωρία τοῦ Δαρβίνου; Θέλησα νὰ τὴ μάθω καλά. Καὶ πῆρα τὸν ἴδιο τὸ Δαρβῖνο νὰ διαβάσω. Κι' ὁ ἴδιος ὁ Δαρβῖνος μ' ἔμαθε πὼς οὔτε ἡ θεωρία του, οὔτε καμμιὰ ἄλλη κοσμοθεωρία, δὲν ἔρριχνε τὴν ἰδέα τοῦ Θεοῦ. Ἀφοῦ λοιπὸν ὁ ἴδιος ὁ Δαρβῖνος μοὔλεγε ἔτσι, θἄκουγα ἐγὼ τοὺς ἄλλους, ποὺ φλυαροῦσαν ὅ,τι ἤθελαν χωρὶς νὰ τὸν διαβάσουν ἢ τὸν παρεξηγοῦσαν γιατὶ δὲν ἦταν σὲ θέση νὰ τὸν καταλάβουν;

Ἀλλὰ σήμερα πρέπει νὰ σταματήσω ἐδῶ. Στὸ ἐρχόμενο θὰ σᾶς πῶ τί λέει ὁ Δαρβῖνος γιὰ τὴ θεωρία του καὶ θὰ σᾶς δείξω πόσο λίγο θίγεται ἡ θρησκευτικὴ πίστη καὶ ἂν ἀκόμα ἡ θεωρία ὑποτεθῇ ὅλη ἀληθινή.

Σᾶς ἀσπάζομαι ΦΑΙΔΩΝ

Ο ΒΡΑΧΟΣ ΤΩΝ ΓΛΑΡΩΝ

[Μυθιστόρημα τοῦ Ἀκαδημαϊκοῦ JULES SANDEAU].

Θ (Συνέχεια)

«Ὁ προπάππος μου, καθὼς ἦταν ξεσκολισμένος ἀπὸ Γεωγραφία κι' εἶχε διαβασμένα πολλὰ γι' ἀγριανθρώπους, κατάλαβε ἀμέσως τί τὸν περίμενε... Γιὰ φαντασθῆτε, ἀλήθεια, νὰ ἔχῃ κανεὶς νὰ φάῃ δεκατρεῖς ἡμέρες καί, πάνω στὶς δεκατέσσερις, νὰ τρώγεται ἀπὸ Κανίβαλους!.. Καὶ καλὰ ἂν ἦταν νὰ χορτάσῃ μὲ τὸ κρέας του φίλους, συντρόφους. Ἕνας ἄνθρωπος μὲ καρδιά, σὲ τέτοια περίσταση, θὰ μποροῦσε νὰ πάῃ εὐχαριστημένος. Παραδείγματος χάρη ὁ παπποῦς τοῦ Ἀγκίστρη μπορεῖ νὰ βρῆκε κάποια εὐχαρίστηση ὅταν τὸν ἔφαγαν οἱ σύντροφοί του, μὲ τὴν ἰδέα πὼς πέθαινε αὐτὸς γιὰ νὰ ζήσουν ἐκεῖνοι. Μὰ νὰ φαγωθῇ κανεὶς ἀπὸ ἀγριανθρώπους, ποὺ οὔτε τοὺς γνωρίζει, οὔτε τοὺς εἶδε ποτέ του κι' οὔτε

Figure 3.2. One of Xenopoulos' "Athenian Letters" on Darwinism, entitled "Things are serious." Reproduced with permission from Anna-Maria Vlassi-Galerou (publisher) of Vlassis Brothers Publications, Athens.

post-Darwinian thought. Undoubtedly, the physics and astronomy associated with the Copernican revolution, and the uniformitarianism[34] of geologist Charles Lyell changed traditional interpretations of the Bible; but it was Darwinism which shook the foundations of religion by questioning "man's place in nature" and with God. It is these issues of science and religion that Xenopoulos attempted to address in a series of "Athenian Letters."

Naturally, his thoughts on evolution and the sciences changed over a sixty-year span. However, what can be drawn from this section is that Xenopoulos was always a hereditarian, and this will be reflected in his writing which will be examined in the forthcoming chapters. On matters of religion and on Darwinism he showed inconsistency, although there was always an underlying support for Darwinism, as will be seen. The ideas he displayed in these "Letters" did not always coincide with those found in his more open autobiography and in other sources.

Heldreich and Gradualism

In 1902, Xenopoulos wrote a tribute to the Hellenized German botanist Theodor von Heldreich.[35] He praised the man and the scientist. It was well-known that Heldreich was a fervent supporter of Darwin's theories, having written to Darwin on a number of occasions to express his support. On such an occasion, Heldreich had indicated that at the time (1878) few Greeks had the courage to show their support for Darwinism, because of the "reign of dogmatism."[36] The tribute reflects Xenopoulos' positive association with members of the scientific community, and in particular with those who were strong supporters of Darwin. An excerpt from this tribute also presents a style of thought which Xenopoulos used in many of his "Athenian Letters," and also in his novels. In the spirit of post-Darwinian writing, Xenopoulos made reference to Heldreich's character using evolutionary concepts in his explanation: "It is observed—I have taken the opportunity to tell you

34 Uniformitarianism is the theory of pre-Darwinian geologists, in particular Charles Lyell, that all the changes associated with the history of the earth are gradual; that is, the changes do not occur in saltations (a sudden event) or in jumps. The scientific view is that because they are "gradual, these changes cannot be considered acts of special creation." See Mayr, *What Evolution Is*, 291.

35 Xenopoulos, "Θεόδωρος Χελδράιχ," 282–283.

36 Krimbas, "Ο δαρβινισμός στην Ελλάδα," 107. Krimbas' work on this is discussed in the introductory chapter, which examines the reception of Darwinism in Greece.

on another occasion—that gardeners, florists, botanists and, in general, those who deal with flowers and who love them, become *little by little* more placid than other individuals. The fine nature of Heldreich had yielded to this influence of plants and flowers. Heldreich, who, it is true, would have been a *perfect* individual in whatever science he took up, became still *more perfect* because he took up Botany."[37]

It needs to be noted here that evolution in the eighteenth century was seen as "the stages through which a living being passes in the course of its development from egg to adult"—that is, within an "account of a single life span."[38] In the nineteenth century, however, evolution was "the development of the *species* rather than of the individual," as is further emphasized by Beer: "The blurring of the distinction between ontogeny—individual development—and phylogeny—species development—in the single term "evolution" proved to be one of the most fruitful disturbances of meaning in literature ... and is a striking example of the multivalency of evolutionary concepts."[39] The point here is that Xenopoulos used this exact technique in his description of Heldreich and, as will be seen throughout the book, Xenopoulos used it often. I have already mentioned in the introduction to this study that Darwin used this type of transformation in his *OS*. In the passage quoted on Heldreich, Xenopoulos makes a point of this evolutionary observation, and also highlights that he has mentioned it before. He observes changes in the character of those individuals who are drawn to vocations affiliated to botany. According to Xenopoulos, these *gradual changes*, where these individuals are slowly developing a gentler nature as compared to those outside botany, are due to the specific conditions of the environment. This reaction to the environment indicates that he is alluding to a form of adaptive evolution.

As mentioned earlier, the concept of gradualness was a critical prerequisite for modern evolutionary thinking, specifically Darwinism. Darwin stated in the *OS*: "Natura non facit saltum" (Nature takes no leap).[40] It should be noted that the theory of gradual development (gradualness or gradualism) in terms of evolution was not new to Darwin, as it had been

37 Xenopoulos, "Θεόδωρος Χελδράιχ," 283 (my italics). In original Greek, "little by little" reads "ολίγον κατ'ολίγον."

38 Beer, *Darwin's Plots*, 11.

39 Ibid., 12.

40 Darwin, *OS*, 236.

taken up earlier by Lamarck, Lyell, and later also by Herbert Spencer. One can go as far back as philosopher Gottfried Wilhelm Leibniz (1646–1717), who also propounded the concept of gradualness in nature.[41]

In the previous passage on Heldreich, Xenopoulos did not indicate the mechanism of the influence, whether it is adaptation due to willful choice (Lamarckian influence) or due to natural selection. It is more than likely it would have been considered as Darwinian at the time, embracing Darwinian and Lamarckian thought (or called biological evolutionism).[42] The development of physical and mental attributes by humans, according to Darwin in the *DM*, appears to be due to natural selection with some contribution by inherited habit.[43] An example of this view in the *DM* is: "Man has risen, though by slow and interrupted steps, from a lowly condition to the highest standard as yet attained by him in knowledge, morals, and religion."[44]

Xenopoulos mentions Heldreich changing from a less "perfect" to a "more perfect" state, which is a key feature interpreted from Darwin's writings. When Xenopoulos refers to variations or modifications in form going from imperfect to perfect, he is referring to Darwin's process of natural selection, namely "every slight modification, which in the course of ages chanced to arise, and which in any way favoured the individuals of any of the species, by better adapting them to their altered conditions, would tend to be preserved; and natural selection would thus have free scope for the work of improvement."[45] Darwin also suggested that "as natural selection works solely by and for the good of each being, all corporeal and mental endowments will tend to progress towards perfection."[46] In evolutionary terms, the concepts of progress and perfection were highly controversial and often misunderstood in the scientific world, particularly post-*OS*. The effects of

41 Mayr, *Growth of Biological Thought*, 325.

42 This view of adaptation is found in Darwin, *DM*, 1:118. See his chapter 4 on "Manner of Development" where, for example, Darwin talks of the developed short-sightedness of watchmakers and the long-sightedness of sailors and also savages.

43 In his first edition of the *OS*, Darwin initially rejected Lamarck's theory of the inheritance of acquired characteristics through their use and disuse. Later editions embrace Lamarck's theory. See Darwin, *OS*, 459. By the time Darwin wrote the *DM*, it was clear that he attributed certain facets of evolution to not only the primary mechanism—that is, natural selection—but also to "natural selection, aided by inherited habit." See Darwin, *DM*, 1:162. So, Lamarckian elements in Darwin's later writing were utilized by creative writers and considered Darwinian. It is only where the clear mention of conscious effort or intention is made that the writing could be considered as exclusively Lamarckian.

44 Darwin, *DM*, 1:184.

45 Darwin, *OS*, 131.

46 Ibid., 459.

this flowed to the literary world. Darwin's world is one where evolution is based on random variation, and natural selection can produce progressive or regressive evolution, or even result in extinction; and the evolution is not teleological. The world for evolutionary philosopher Herbert Spencer was Lamarckian and finalistic—that is, with an intrinsic drive to perfection.[47] Although Darwin was reluctant to describe evolution in terms of progress to perfection and of low complexity to high, it is suggested that he did so in order to refute theories where species were considered to be constant, and also to refute theories which denied the possibility of improvement through natural selection.[48]

It was not unusual for literary writers of the late nineteenth and early twentieth centuries to use aspects of evolution, such as gradualness, to describe changes in the human condition within the life cycle of an individual; this was displayed as either a literary motif or a theme. Gillian Beer's study of Charles Kingsley's novel *The Water Babies* (1863) is a fine example of Darwinian analysis dealing with the transformation of individuals within the life cycle of an individual.[49] Even the French naturalist Emile Zola, who had been influenced by others on the topic of social environments,[50] in his naturalist "experimental novels" had pointed out: "Here it would be necessary ... to consider Darwin's theories." Regarding this he went on to say that the novelist needed "to show man living in the social milieu, which he himself has produced, which he modifies every day, and in the midst of which he in his turn undergoes continuous modification."[51] To add to this, in the preface to the second edition of his novel *Thérèse Raquin* (1867) he acknowledged this Darwinian approach in his novel. He wrote in the preface in retaliation against his critics who found his novels scandalous: "They would surely not be surprised by the scientific analysis that I tried to apply in *Thérèse Raquin*.

47 Spencer saw evolution as "a necessary progression toward higher level and higher complexity," unlike Darwin. Confusion between Darwinism and Spencerism was very common. Spencerism was aligned with popular misconceptions which infiltrated areas such as literature (especially in America). Spencer's theory was based on metaphysical assertions, whereas Darwin's was based on observational evidence. So although Xenopoulos had been a Spencerian early in his life, by the time he wrote most of his "Athenian Letters" and novels, he knew that Spencer's theories were outdated.

48 Mayr, *Growth of Biological Thought*, 531.

49 Beer, *Darwin's Plots*, 116–130.

50 Zola was influenced by Hippolyte Taine regarding heredity and environment in his novels. Also he was influenced by Claude Bernard but claimed that he had not read him till 1878.

51 Zola, "The Experimental Novel," 173–174. (Becker indicates his source is Zola's *Le roman experimental*, Paris, 1880. Translator is not named.)

They would recognize it as the modern method and the universal research tool that our century uses so passionately to lay bare the secrets of the future. Whatever their conclusions, they would accept my point of departure: the study of temperament and of the *profound modifications* of an organism through the influence of environment and circumstances."[52]

The notions of gradualness and adaptation were found during this period even in Greece. Xenopoulos also used them in his other works, including *Tereza Varma-Dacosta*.[53] The scholarship on the theory of gradual development in the modern Greek world has scope for further exploration.[54] Andreas Karkavitsas' formal education in medicine would have exposed him to Darwin's writings, though it cannot be verified if he had read Darwin.[55] However, he also used this view of adaptation and gradualism in his novel *The fair maid* (*Η λυγερή*) (1890). The novel is about a young Greek girl, Anthi, who is forced to marry the well-established financially, much older, and unpleasant grocer Nikolos Picopoulos (also known as Divriotis), despite her affections for the young Giorgos. Genuine feelings of love abound for her husband when she falls pregnant. This change stems from the fact that she now realizes that socially and financially she is much better off with Nikolos rather than with Giorgos, the no-hoper and dreamer. Applying Darwinian discourse metaphorically, Karkavitsas makes reference to Anthi's adaptation to the harsh treatment in her marriage by means of her transformation:

> The assimilation came on complete. Whatever was not achieved by Mrs. Panagiotena's warm advice and by Froso's continuous efforts was achieved by Nature herself. *Nature, the all-powerful goddess, who little by little* [μικρόν κατά μικρόν] *changed the body and predisposed Anthi's soul* in complete agreement with the soul of Divriotis. Just as Nature endows tropical plants, transplanted into the cold of the North, with new powers, fixes their roots, encour-

52 Zola, *Thérèse Raquin*, 7 (my italics).

53 This aspect of *Tereza* will be examined in chapter 5, which is dedicated to the novel. Other examples of this approach to adaptation will be examined in the context of the general analysis of the relevant work.

54 For a parallel in French literature during the Enlightenment period see Delon, *Ο Διαφωτισμός και η σημασία των διαβαθμίσεων*, 17–75.

55 While not dealt with here, Karkavitsas is also known for his 1896 novel *Ο ζητιάνος* (The beggar). Generally seen as a novel of the naturalist school, it is claimed to have a Darwinian message in the last few lines; that nature treats the just and the unjust equally without moral consideration. See Stavropoulou, "Ανδρέας Καρκαβίτσας," 192–193.

> ages the juices, and *little by little transforms even their species*, so that they [the plants] can be strong enough to live in their new country. ... Anthi is no more—she is not Giorgos Vranas' dreamy lover. She is the positive spouse—the wife of Nikolos Picopoulos.[56]

Jina Politi was the first to observe evolutionary elements in this passage.[57] The assimilation or adaptation referred to in the passage is the mechanism for Anthi's transformation, which involves her physical state (she is pregnant) and her emotional state (development of feelings for Nikolos). Eri Stavropoulou indicates specifically that there are Darwinian elements in this passage. She states that regarding the last section of the novel, "there is reference to it in nature ... with the example of transformation of a species due to reasons of adaptation in a new environment, that is, its assimilation."[58] However, neither commentator mentions that Karkavitsas highlights the adaptation mentioned in the passage was gradual. The gradual nature of evolution was an important aspect of adaptive evolution, specifically Darwinian. Note also the other Darwinian language, such as "the adaptation" and "the species." By adding this phrase "little by little," Karkavitsas wanted his reader to know that not only was he comparing the transformation, due to Anthi's adaptation, with that of the botanical evolution of "the tropical plants," but also with the Darwinian evolution of the species. Her name means flower or blossom in Greek, which is appropriate since her evolution is being compared to that of the "tropical plants." In the narrative, transformation or metamorphosis of a Darwinian type is characterized by its gradual nature, unlike the non-Darwinian type, such as Ovidian metamorphosis.[59] In his version of evolutionary theory, Karkavitsas highlights the idea of transformation within the life cycle. Anthi "is no more."[60] She has been transformed to the new version—that is, the New Woman. In his description of her appearance Karkavitsas mentions her tallness and "somewhat

56 Karkavitsas, *Ηλυγερή*, 155 (my italics).

57 See her analysis of the passage in Politi, "Η μυθιστορηματική κατεργασία της ιδεολογίας," 125–127.

58 Stavropoulou, "Ανδρέας Καρκαβίτσας," 192.

59 The English geologist Charles Lyell argued that geological formations were formed not over thousands of years but over millions. Darwin extrapolated that man's evolution must also have been over such a time, reflecting evolution as gradual and slow. Xenopoulos did acknowledge the symbolic nature of the Bible's version of the time span of the Creation in his "Letters." On Ovidian metamorphosis see Beer, *Darwin's Plots*, 5, 104.

60 Ibid., regarding the concept of extinction within Darwinism.

masculine height."[61] An observation of physical masculinity such as this in a woman in society and in literature was often to signal the New Woman stereotype of the late nineteenth and early twentieth centuries. This issue of masculinity is taken up in chapter 5.

To pursue gradualism as a theme, in 1923 Xenopoulos wrote an "Athenian Letter" entitled "An evolution." It is about the developmental history of the printing press. He takes the reader from the "primitive form" of the first printing press, which he viewed as a boy, through the various stages of technological development, to the latest printing press of his present day. Referring to the old printing press he states: "'Progress' had superseded it. It belonged to the past." He refers to the next development of the printer: "a new printing press, larger and more perfect than the one which I knew earlier." And he describes this latest printer:

> And nevertheless, from that came this. *Slowly, slowly* [σιγά-σιγά] over time, the small one became larger and was perfected. Each year would pass adding a small improvement to the original machine by Gutenberg. It also produced its own evolution. And from the protoplasm, let's say the less perfect, it arrived to its present form, and to amazing perfection. Nor will it stop here of course. … And so for all things—machines, systems, institutions: each of these things has its original form, *its protoplasm*, which *evolution metamorphoses bit-by-bit* [λίγο-λίγο] *to ever greater perfection*. Because even man, who does all these things himself, *evolves, becomes perfect, and progresses*.[62]

Here he has used the evolutionary tenet of gradualism to explain the technological advance of the printer, and has aligned society's technological progress with evolutionary progress. Finally, as will be seen later in this chapter, Xenopoulos used gradualism to describe Darwinian evolution in biological terms. Later it will be demonstrated how he wove it in his works to describe the social evolution of groups of people, races, and also to describe ideologies, such as socialism.

61 Karkavitsas, *Η λυγερή*, 13.
62 Xenopoulos, "Μια εξέλιξη," 220 (my italics).

The gender divide

In 1916, Xenopoulos wrote an "Athenian Letter" entitled "Boys and girls."[63] It is reminiscent of sections of Darwin's *DM*, where in his discussion Darwin utilizes the craniological studies of German naturalist Carl Vogt (1812–1895).[64] In his "Letter" he indicates that his information came from a university lecture he attended by S. Tsakonas, a professor of gynecology. He does not say when he attended this lecture—that is, whether it was when he was a student at university, or at a more recent time. Indicating that the information is only a portion of what was said in the lecture, he compares the brain sizes of the male and female, and comments on the differences in their mental powers. The reader should bear in mind that Xenopoulos wrote this as an educational message to instruct the young readers:

> [Woman] is not at all a lower being than Man, as is believed by many, but equal and in some things superior. ... So Woman, according to the latest findings of Science, has a more delicate brain than the male; natural ability in perception, memory, and imagination are stronger in Woman, and more so than Man she feels love—I mean Christian love, the "love of others"—sympathy, and compassion; in other words, she is less selfish. Woman, by cultivating her mind, can do whatever a Man does. With her heart however, she achieves more than him. If selfishness is useful for the individual, altruism however, the love of others, is indispensable for society. So Woman, with her superior mental qualities, is more social than Man.[65]

By comparison, Darwin writes in the *DM* on the "Differences between Man and Woman" and their "Differences in Mental Powers":

> [Man's] brain is absolutely larger, but whether relatively to the larger size of his body, in comparison with that of woman, has not, I believe been fully ascertained.[66]
> Woman seems to differ from man in mental disposition, chiefly in her greater tenderness and less selfishness. ... Woman, owing to her maternal instincts, displays these qualities towards her infants in an eminent degree; therefore it

63 Xenopoulos, "Αγόρια και κορίτσια," 65.
64 See for example Darwin, *DM*, 1:329.
65 Xenopoulos, "Αγόρια και κορίτσια," 65.
66 Darwin, *DM*, 2:316–317.

> is likely that she should often extend them towards her fellow-creatures. ... It is generally admitted that with woman the powers of intuition, of rapid perception, and perhaps of imitation, are more strongly marked than in man.[67] In order that woman should reach the same standard as man, she ought, when nearly adult, to be trained to energy and perseverance, and to have her reason and imagination exercised to the highest point.[68]

Darwin's text can be seen as part of the conventional nineteenth-century medical discourse on gender roles. Although the two passages are forty-five years apart, striking similarities can be seen. It would appear that Xenopoulos had absorbed Darwin's manner of observation of craniological and anthropological data, and also taken on a significant amount of the actual content. Xenopoulos' intention, as he mentions in the "Letter"—that is, to show woman's equality, if not her superiority in certain aspects—came at a time when he was supposedly an overt supporter of the Women's Movement in Greece. Between 1912 and 1921, he wrote a series of "Athenian Letters" that by the standards of that period supported gender equality. However, his fiction presents women in a much more sexist manner, sometimes with misogynistic undertones, even by the standards of that period.

Darwin's detailed study, of which the above passage is only a small excerpt, comes to some further significant conclusions; perhaps Xenopoulos would have been aware of these, but he did not include them in his "Letter." After observing that "with woman the powers of intuition, of rapid perception, and perhaps of imitation, are more strongly marked than in man," Darwin goes on to say "but some, at least, of these faculties are characteristic of the lower races, and therefore of a past and lower state of civilization."[69] He extends his observations to state that in terms of mental faculties "man has ultimately become superior to woman."[70] He also argues that even after cultivation of women's minds it would take many generations of evolution for women to reach the same level of mental power as men. He concludes that men's "severe struggle in order to maintain themselves and their families ... will tend to keep up or even increase their mental powers, and, as a

67 Ibid., 326.
68 Ibid., 329.
69 Ibid., 326–327.
70 Ibid., 328.

consequence, the present inequality of the sexes."[71] Assuming that Xenopoulos did use Darwin's observations, it is likely he only selected a few aspects because of the comments on inequality. It would suffice to say that by 1916, whether one believed in gender equality or not (in terms of ability and opportunity), it was common practice in countries such as Greece to publicly promote it.[72]

Darwin's comments associating women with a lower state of evolution had enormous repercussions for women in areas such as medicine and psychology. Although others, such as Herbert Spencer, had propounded the biological inferiority of women, it was Darwin in this statement who was interpreted as providing the universal scientific validation. This view on women was also translated into the literary world, and became a theme in novels for at least up until 1940.[73]

Although the idea was too controversial for Xenopoulos to add the "inferior" evolutionary nature of women to his "Athenian Letter," he was well aware of it. This is reflected in some of his novels, such as *Tereza Varma-Dacosta*, as will be seen in chapters 4 and 5.

Science and Religion

Darwinism's great impact on religion also became an issue for Xenopoulos. A series of the "Athenian Letters" reveal Xenopoulos' preoccupation with the status of religion versus science, with Darwin's theory and its effects on the traditional concepts of God and creation, and also on "man's place in nature." These "Letters" begin in the early 1900s and continue through to the 1940s.

Despite his life-long positivist views, which he discusses in his autobiography, Xenopoulos declared his support for religion in these "Letters." At various points they reflect his attempts to reconcile religion with science and, in particular, religion with Darwin's theory of evolution. This trend was established by the 1880s, when Western clerics had realized that Darwinism was not going to go away and that completely condemning it was dangerous

71 Ibid., 329.

72 Malafantis, *Οι "Αθηναϊκαί Επιστολαί,"* 184.

73 Greenslade, *Degeneration*, 290. Greenslade highlights literary writers' use of scientific material on the gender differences which "adopted a standard post-Darwinian position on women's inferiority to men."

for the Church—a repeat of the Galileo case was not to occur. So, Western churches sought to utilize those same evolutionary concepts of Darwinism to substantiate the fundamentals of Christianity. This trend permeated the works of literary writers who attempted to reconcile religion and evolutionary thought.[74]

In 1914, Xenopoulos wrote an "Athenian Letter" entitled "The atheists."[75] According to Malafantis, this is an example of how in his "Letters" "Xenopoulos often turns wholly against atheism and its representatives."[76] It is as if Xenopoulos wants it to be publicly known that he is averse to atheism. He wrote to his readers that he did not actually know exactly what the details were in relation to the famous case of the "atheists" from Volos. Basically, due to pressure from the Church, the progressive girls' school in Volos, the Higher Primary Girls' School, was closed down in 1911 for its ideas, such as the teaching of demotic Greek and the elimination of daily morning prayers.[77] The school's principal, liberal educationalist and demoticist Alexandros Delmouzos, and his associates were tried for their alleged immoral and atheist teachings. It was supposed that they were circulating material on Darwinism.[78]

As outlined in chapter 1, Delmouzos was one of the intellectuals who propounded that the teaching of youth in the demotic (or spoken) language would free their minds from the difficult and synthetic katharevousa, thus allowing the youth to learn and think freely in the language that they were accustomed to. Along with this would also come the learning of progressive ideas, including in science. Xenopoulos believed Delmouzos and his followers had misconstrued Darwin's theory of evolution as atheistic. Xenopoulos claimed in his "Letter": "These are people with cracked brains who have either read booklets of other madmen, or they couldn't understand and assimilate some scientific theories—e.g., the Darwinian theory of evolution—they arrived at the most irrational conclusions, e.g., that God does not exist!"[79] The trial was not necessarily targeted at any Darwinian views, but

74 For evidence of this general trend see Paul, "Religion and Darwinism: Varieties of Catholic Reaction," 406; and for its presence in literary works see Gibbons, *Rooms in the Darwin Hotel*, 6; Henkin, *Darwinism in the English Novel*, 141–167.

75 Xenopoulos, "Οι αθεϊσταί," 167.

76 Malafantis, *Οι "Αθηναϊκαί Επιστολαί,"* 222.

77 For further discussion on this famous school see Delmouzos, *Το κρυφό σκολειο.*

78 Delmouzos was a follower of the work of Ernst Haeckel, strong German advocate of Darwin's theories. See Krimbas, *Δαρβινισμός και η ιστορία του*, 128.

79 Xenopoulos, "Οι αθεϊσταί," 167.

Xenopoulos here appears to use the trial as a basis for arguing that Darwinism is not atheistic. Whether he did so to protect himself (he also detached himself from having had any real connections with those tried) or whether he was attempting to promote Darwinism as theistic for his own purposes is difficult to know. Note that Xenopoulos will go on to discuss Darwinism in theistic terms here and in "Letters" to come, up until 1939.

Further pursuing the "Letter," Xenopoulos maintained that Darwinism could coexist in harmony with religion. In doing so, though, he highlighted his support for Darwinism, despite the sentiment of most at the time, who understood the theory to discount the existence of God. The following quote shows how Xenopoulos first conveyed Darwinism to his young readers:

> Darwin himself ... was anything but an atheist. On the contrary, he believed that his theory on evolution showed even more the power and grandeur of a God-Creator. As you would have heard, Darwin, an English naturalist, one of the greatest, through his observations arrived at the conclusion that animals were not created all at once, but that from two or three original *forms*, all this variety of the animal species was produced *by degrees* [βαθμήδον] through *evolution*. And the great scientist Darwin says at the end of one of his books: how does my theory harm the idea of God? Whether all the animals were created independently, or whether there were at the beginning two or three, or only one protozoan, from which the other animals were formed—is it not the same? And indeed, does not the creative power of God appear still more wondrous and more magnificent when we accept that only one protozoan was created by Him in such a way as to produce the whole of the animal kingdom? You see, my dear readers, that one can accept Darwin's theory and still believe in God the creator and lord of everything.[80]

Here he tries to gloss over any difference between the significance of independently created species—fundamental to the theory of creationism—and the Darwinian creation of species from few protozoa. He then tactfully pushes for the acceptance of the latter version, but also promotes a theistic approach to evolution—that is, with a purposeful designer or God—an approach he pursues in all his "Letters."[81]

---

80 Xenopoulos, "Οι αθεϊσταί," 167 (italics in original).

81 There appear to be various forms of creationism. The original form was the belief in the literal truth of Creation as in the Book of Genesis. Creationists may disagree as to how literally this should be taken.

In the aforementioned passage he refers to another important Darwinian tenet, which identifies biological evolution as occurring gradually. Xenopoulos' use of evolutionary gradualness implies a belief that the highest living form could only have been reached from the lowest living form through many small intermediary stages, which could only have occurred over millions of years. This reflects Xenopoulos' anti-creationist views against the biblical version of creation, which maintained the appearance of immutable individual perfect species in a more modest period of time.

Further on in the 1914 "Athenian Letter," Xenopoulos argued that the supposed atheists should not be punished because of their alleged atheism, and described the trial in terms of "medievalism and barbarism."[82] In addition, he pointed out that: "Today each of us is free to have an opinion, whether it is right or wrong. Our civilization demands it. If I was a judge I would not have convicted them."[83] He will take this same view regarding freedom of speech when in 1923 a biology teacher in Tennessee in the United States was convicted and charged for teaching Darwin's evolutionary theory. This will be discussed later in this chapter, as part of the Scopes trial.

In a 1915 "Athenian Letter" entitled "Religion," Xenopoulos wrote again on religion and science, believing that they are issues to be examined separately: "Science is one thing and Religion is another; that is, the matters, the problems, which Science solves, and those which Religion solves. ... For example, the existence of God, his relationship with man, the immortality of the soul, the afterlife, are matters of Religion. ... Religion starts from where Science's authority ends and it [Science] declares itself incompetent. That some teachings of Religion, cosmogonic or other, do not agree to the letter with today's Science, is not important. Religions are much older than Science, and they explain phenomena of the universe according to the understanding of the people of that time."[84] Xenopoulos believed that the concepts of the immortality of the soul, the afterlife, and humanity's posi-

82 Delmouzos and his associates were eventually declared innocent due to the evidence of the witnesses who supported them. See also the following article published in 2007, which also maintains that Delmouzos and his associates were tried for circulating information on Darwinian theory: Giannakidis, "Ο πίθηκος έδωσε το μήλο στον Αδάμ και στην Εύα," 40.

83 Xenopoulos, "Οι αθεϊσταί," 167.

84 Xenopoulos, "Θρησκεία," 79. In another letter, "Αποκάλυψις" (Revelation), he argues further on the differing natures of science and religion. He indicates that science is based on proof, experiments, and reasoning, whereas religion is based on the belief in revelations. For this letter see Xenopoulos, "Αποκάλυψις," 105.

tion relative to God are issues only to be answered by religion. They are themes which he repeated in this series of "Letters" relevant to the study. He also emphasized to his readers that alleged scientific books attempting to answer issues of religion, such as those of the German materialists, should be avoided. Perhaps he was attempting to shield the youth from the shattering experience of losing their traditional orthodox faith, which is what happened to him. In 1917, he wrote a "Letter" entitled "The sophism of atheism," in which he discusses the concept of the soul, the nature of God, the creation, and the punishment of sinners after death, from various religious perspectives.[85] Again he will argue against atheism, emphasizing that questions querying these ideas cannot be answered using reasoning only, but also require one's calling on instinct and feelings.

In 1923, Xenopoulos further developed his Darwinian ideas in a "Letter" entitled "Things are serious," in which he discussed the rise and fall in popularity of Darwinism and of other evolutionary theories.[86] The following passage from the "Letter" gives his historical perspective on Darwinism at the time.

> Darwinism is but a scientific theory like others. In its time—forty years ago—it was very popular. It was fought fiercely by scientists and nonscientists; others however, and indeed the best, supported it with fanaticism. Intellectuals and philosophers believed in it and tried to apply it everywhere. It had the greatest influence in thought in the nineteenth century. It made man see that nothing in the world was made at once, instead that everything has undergone and undergoes an evolution. Because this *theory of evolution, that is of unfolding*, is the essence of Darwinism. And from then the term, the word which we use so much was established. And together with the word, naturally, the idea prevailed that each form in nature, as in art, in industry, in society, and everywhere, was not original—that is, created from the beginning—but that it was a *gradual* development, a *perfection* of another more ancient and less perfect form. In my time, that is when I was also studying the natural sciences, Darwinism was believed almost as an axiom. We said "man came from the ape and the bird from the fish," just as we would say "two things equal to a third are equal to each other." Whoever had an objection, was not ... all there; or at least he was not a "real scientist."[87]

85 Xenopoulos, "Τα σοφίσματα της αθεΐας," 87.
86 Xenopoulos, "Σοβαρά τα πράγματα," 116.
87 Ibid. (my italics; suspension points in original).

This "Letter" is important for the development of Xenopoulos' own ideas. Historically, he acknowledged the Darwinian revolution in Greece, dating it from around 1880, later than it occurred in other European countries.[88] He observed firsthand how the gradualist theory "of unfolding" (του ξετυλίγματος) was pervasively applied to culture and society. Like many others, he saw it as an evolution from the imperfect to a perfect state. Gradualism and evolution to perfection are concepts that Xenopoulos would draw on in his other writings. He noted that the popularity of Darwin's biological theory of evolution would continue throughout the decade of the 1880s, which was also the period when he was studying at the University of Athens. He asserted that the validity of the evolutionary doctrine was held by most intellectuals in Greece, by aligning it to a revered Euclidean axiom. This would have been at the same time when mathematicians and natural scientists, along with the academics of the medical and philosophy schools were fighting for Darwinism against the theologians. Xenopoulos continued in his "Letter":

> We knew that the theory exhibited "gaps." We knew that the intermediate-form between ape and man, "anthropopithecus" and various old skeletons from excavations, which initially were thought to be anthropopithecus skeletons, were confirmed later on as either those of a gorilla or a savage. It didn't matter! "It will be discovered," we would say. And our belief was such that we thought that with time the theory not only would not have fallen, instead it would have been completed by the experimental and, by now, positive proof that all the plant and animal kingdoms, all life on Earth with its infinite forms, was created not from four or five original types, as Darwin said, but from one single protoplasm![89]

Xenopoulos' reaction of hope in relation to proof of Darwin's theory reflects the same sentiment of many intellectuals, not only in Greece but also in other countries that had been exposed to the doctrine. At this point there is a change in the direction of his life and an apparent modification in his views on Darwinism:

88 See the introductory chapter further on this.

89 Xenopoulos, "Σοβαρά τα πράγματα," 116.

> But I quickly left the natural sciences for Literature. And since then at least twenty-five years have passed. In that period, it has happened many times that I heard or read that Darwinian theory does not have its former followers and its former authority anymore. It too has had its day, it has collapsed. It fell just like all those theories which pass and fall; above all, the theory of the origin of man. The intermediate-form, "anthropopithecus," was not found! And neither does the belief exist that it will be found. Don't ask me to tell you more. Since I have ceased to follow the Sciences, I am not in a position to give details, to tell you exactly where Darwinism is today, which of his teachings remain as truths—because quite a few remain—and which have collapsed as exaggerations and arbitrariness. In my time I knew these things. Today I don't know them very well. The only thing that I know is what I have told you: the theory suffered many blows. [90]

In contrast to the 1914 "Letter" "The atheists," there is a deliberate negative tone here, in his 1923 "Letter." It is estimated that Xenopoulos had withdrawn from his science degree by the late 1890s. At that time, the scientific arguments and any evidence were not sufficient to convince many of Darwin's theory, reflecting a period where Darwinism was at an all-time low. Nevertheless, in the aforementioned passage he does not wish to give up on Darwinism entirely, highlighting that "quite a few" truths remain. In addition, while Xenopoulos mentions that he "quickly left the natural sciences to study literature," he does say otherwise in his autobiography.[91] Indeed he would invoke Darwinian and other post-Darwinian scientific ideas in his writings. In his "Letter" of 1923, Xenopoulos imparted an unexpected message to his readers:

> But let us assume that it did not suffer anything. Let us suppose that true Science still believes and teaches that the species were created one from the other, and that man descended from the ape. Should we conclude that the universe happened "by chance," and that there is no Creator God who made it with purpose and design? Far from it! Even the time when I still believed

90 Ibid.

91 He described how after leaving the sciences he still involved himself with the positivist Giorgakis Hairetis: "Because I was a mathematician, I would regularly speak with Hairetis about the 'spirit' of that science, essential for every philosophy, as he would say, but also for every art form. He would show me his original mathematical works and they would get me so excited that I nearly returned to my Mathematics." Xenopoulos, *Η ζωή μου*, 261.

> the theory of evolution as an axiom—that is, everything from one—I, at least, did not for one second stop believing, admiring, and glorifying the Creator of this universe.
> And do you know why? Because I did not listen to hearsay; because I did not collect my knowledge and my ideas from the street or from the coffee-house. I heard then about Darwin's theory and I wanted to learn it well. *And so I read Darwin for myself.* And Darwin himself taught me that neither his theory nor any other cosmic theory could bring down the idea of God.[92]

What is quite significant here is that despite wanting to distance himself from Darwin's theory, at the end of his "Letter" Xenopoulos is indirectly hinting to his readers to read Darwin for themselves, just as he had done. One would suppose that if Xenopoulos really believed that Darwinism was outdated, he would not have alluded to this, let alone written this "Letter." Xenopoulos knew of the destructive effect on one's beliefs that the reading of Darwin could have. At the very least, the direct reading would have aroused a course of questioning and doubt of one's religion among the readers. Hence, Xenopoulos thought Darwin's theory important enough that the Greek youth should be somehow told to read it; this at a time when it was not taught at schools. He would have been quite aware of the fact that he could have ruined his own writing career had he been more open about his views.

Further to this, in this "Letter" Xenopoulos would contribute to the argument on the "designer." He emphatically professed that he did not accept the naturalistic element of "chance" in the evolution of humanity and the universe. He spoke of a God who created with "purpose and design," and so opted for traditional monotheistic religion that attributes all creation to a God of design and teleology. His point here was that Darwinism is not incompatible with these beliefs. It is possible that Xenopoulos is representative of those individuals who had found an atheistic idea of Darwin's evolution difficult to accept. For instance, many Christians that accepted Darwinism incorporated the theory within a revised "theism."[93] It is worth noting that Darwin's removal of a "designer" in his theory of evolution would do

92 Xenopoulos, "Σοβαρά τα πράγματα," 116 (my italics).
93 Oldroyd, *Darwinian Impacts*, 247.

away with God's crucial role. This would attribute less power to God and have negative repercussions for the Church.[94]

Overall, his "Letter" suggests a certain ironical tone, communicating one message to the Greek youth—that is, that they *must* read Darwin despite the negativity—and another one to the establishment. To the establishment he was conveying that he was not interested in Darwinism, that it was a superseded theory, and that his faith in God had not been affected. This approach could have been necessitated by the circumstances surrounding the publication of the "Letter." Three days after it was published in *Children's Guidance,* the Athenian newspaper *Ethnos* commenced the publication in serial form of the first volume of Xenopoulos' novel *Kosmakis 1: The first awakening.* As mentioned earlier, the first volume relates the story of his turning away from religion because of the influence of Büchner's book *Force and Matter.* It is not until the fourth volume *Kosmakis 4: The return* that the main character Dionysis Alibrande returns to Orthodox Christianity. In the novel Xenopoulos includes Darwin as one of the materialists.[95] He would not have wanted to display any overt support of Darwinism that could have jeopardized the publishing of his serial. This may have necessitated a negative tone in the 1923 "Athenian Letter."

A week later, Xenopoulos wrote a follow-up "Letter" entitled "Science and God" ("Η επιστήμη κι ο Θεός"). His attitude towards Darwinism appears to have changed from the previous week; it seems more positive. Referring to Darwin's "epilogue" in the *OS,* he maintains again, as he did in his "Letter" of 1914, that all life forms are derived from a few original forms and that the theory does not affect the idea of a creator. He declares that this results in "all the variety of species that we see ... with so many different forms, from the imperfect to the more perfect!" adding: "Don't you find that the protoplasm, which was created with the capacity to produce from within itself a whole world, is something infinitely more wondrous compared to any one species which continues *to multiply eternally the same and unchanged?*"[96] Although Xenopoulos had repeatedly referred to the concept of one or a few original forms evolving into all life, it is not till this "Letter" that he actually talked about the creationist version, where the original life forms are con-

94 Ibid., 247–248.

95 Xenopoulos, *Ο Κοσμάκης Α',* 65.

96 Xenopoulos, "Η επιστήμη κι ο Θεός," 124 (my italics).

sidered to multiply without change. In the same "Letter," Xenopoulos continues to call on what he apparently read in the "epilogue" to the *OS* regarding the mutability of species. He interprets, as many have done, that Darwin in his writing admitted to admiration for "this new God": "I remember very well, however, that this was its meaning. And I remember it because it left then the greatest impression on me, because I recognized the great truth in it, and because much more than the God which I knew—the God which created the animals and plants one by one—I admired, like Darwin, this new God, the great Creator, who had the power to create everything from one!"[97]

It can be seen that in his writings Xenopoulos constantly referred to Darwin's acknowledgement of a God. Darwin was more guarded about his faith publicly, and declared he was an agnostic. In his private letters, however, he was more frank and tended towards atheism. Darwin was well aware that his theory of evolution in the *OS* would extend to the evolution of the human species from lower life forms; and that this would cause a backlash against traditional religion. It is argued that to counteract in some way this negative response and to appease the conservatives, he linked his evolutionary theory with God and nature in the last pages of his concluding chapter of the *OS*. Here, Darwin speaks of a Creator who rules by law through natural selection, implying that natural selection is progressive towards perfection and that humanity is the endpoint of a process formed by the Creator. Darwin does not specifically include a supernatural Creator in the last paragraph of the first *OS* edition.[98] It is only in the subsequent editions that Darwin was pressured by the Church to add the Creator.

Throughout his "Athenian Letters," Xenopoulos constantly pursued the idea that one should continue to believe in God and that no scientific theory can touch on the concept of God. In 1923 he referred in the "Letter" "Science and God" to the previous century's fall of materialism and to the German materialists, such as Büchner, Moleschott, Vogt, and Haeckel, who denied the existence of God, pointing out that "they endeavored to explain the world only by matter, by power, by evolution, and much later by autogenesis."[99] To this he added: "And those who still believe in Darwinism

97 Ibid.

98 Darwin, *OS*, 459–460.

99 Autogenesis, also known as abiogenesis, is the process whereby life is considered to have arisen from nonliving matter.

don't misunderstand him like those much earlier, the ones who thought that God could be replaced with 'evolution.' Today, true scientists believe in God. And those who happen to think materialistically, it means that they have not followed or understood the new science. So—don't listen to rumors!"[100] By mentioning the "new" or latest science, Xenopoulos contradicts the "Letter" "Things are serious," which he had only written the week before, where he mentioned that he was not up to date with the latest on Darwinism. This is an indication that he probably downplayed his support for Darwinism in the "Letter" from the previous week.

In our times the issue of creationism has resurfaced in the United States, the United Kingdom, and Australia, and has emerged in the form of Intelligent Design (ID). Supporters of ID assert that some or all features of living things are best explained as the work of a designer rather than as the result of a random process of genetic variation together with natural selection.[101] Xenopoulos appears to take on the attitude close to that held by supporters of ID, although his approach is more like that of Darwin's after his first edition, where he includes an initial creator with the subsequent random process of evolution.

Out of step with the conservative tone of Xenopoulos' other "Athenian Letters" is a more daring "Letter" that he wrote in 1925, entitled "The trial of ... Darwin."[102] It supports freedom of speech and many aspects of Darwinism.[103] It confirms that Xenopoulos was still very interested in issues associated with Darwin, despite the suggestion of the opposite in earlier "Letters." Here Xenopoulos discusses his views on the famous and much publicized Scopes trial of March 23, 1925. The state of Tennessee passed an antievolution bill, also known as the Butler Act, which noted "that it shall be unlawful for any teacher in any of the universities, normals and all other public schools of the state which are supported in whole or in part by the public school funds of the state, to teach any theory that denies the story of the divine creation of man as taught in the Bible, and to teach instead that

100 Xenopoulos, "Η επιστήμη κι ο Θεός," 124.

101 They may use the language of science to validate their views and may or may not accept that creation arose from a single or a few living things.

102 Xenopoulos, "Η δίκη του ... Δαρβίνου," 292 (suspension points in original).

103 Malafantis also notes that Xenopoulos' stance on this was daring, given the period in which the letter was written. Malafantis, *Οι "Αθηναϊκαί Επιστολαί,"* 240.

man has descended from a lower order of animals."[104] Violation of the law incurred a $100 to $500 fine. The science teacher John Thomas Scopes of the Rhea High School in Dayton, Tennessee, admitted to violating the law when teaching biology, and so was tried and convicted.

In his "Athenian Letter," Xenopoulos pondered the two laws which he believed clashed—that is, the law forbidding the teaching of Darwin's theory and the law that he interpreted as "no one should be hindered from exercising their profession." He believed the essence of the case was the following: "Does a scientist have or not have the right, in a free country, to teach whatever he knows and whatever he thinks, without being concerned about what the Bible says about the origin of man, for example, or the creation of the world? And one more thing: Is Darwin's theory true, and if it is, does it raise the idea of religion, of God, of the immortal spirit and so on? That is why the trial which is happening today in a small American city is stirring the attention of the civilized world."[105]

On reading this "Letter" about the Scopes trial, also known as the Darwinian trial or the Monkey trial, one can see Xenopoulos' intense interest and concern regarding what he believed to be true. Xenopoulos' observations were based on newspaper clippings sent to him by an American friend, which contained minutes of the case and included the opinions on it by prominent men. He also read about the case in other foreign papers and in Greek papers: "I was pleased to see that the top-ranking intellectuals everywhere are saying exactly those things which I have said to you at other times on the topic of Darwin's theory and, in general, on the relationship between Religion and Science."[106]

Xenopoulos thus suggests that science does not have anything to do with religion; that science demonstrates its ideas experimentally, whereas religion works through the revelation of God. He pointed out, once more, that science is unable to solve the great problems of God and of the soul: what we are, from where we originate, and where we are going. He argued that these questions could only be answered by the teachings of religion, intuition, and conscience, which, he advised his readers, have no relation to sci-

104 For the antievolution act, transcripts of the Scopes trial, and other information associated with the case see the website of the University of Missouri–Kansas City, School of Law, 2002, last accessed January 2014: http://www.law.umkc.edu/faculty/projects/ftrials/scopes/scopes.htm.

105 Xenopoulos, "Η δίκη του ... Δαρβίνου," 292.

106 Ibid., 292.

ence. He again indicated that Darwinism had fallen in popularity due to the nonacceptance of all life arising from one or several original forms, and also because the skeleton of the "missing link" or intermediate form of man had not been found. He pondered what the outcome would be if Darwin's theory was proven to be true and man did evolve from the ape: "What does it mean? Maybe it will be shown that God does not exist? Not so! Again a God would have created man from the ape and again a God would have created the protoplasm. Poor Darwin said it: 'My theory does not overthrow God. On the contrary, I would have admired more the magnificence of a God who had the power to create, from only one protoplasm, all that multitude of plants and animals.' Just imagine, really, what life this microscopic little thing had within itself!"[107]

Finally, in this "Letter" Xenopoulos concludes that Scopes should be allowed to freely teach Darwinism.[108] Xenopoulos also appears to have changed somewhat regarding his negative views on Darwinism: "True or false, this theory belongs now to Science. It may not have been proven yet as a whole; it has however many excellent details proven and truly amazing. Those at least, a scientist, and indeed a biologist, is obligated to know."[109]

Xenopoulos sends a final double message regarding religion and science, although given the times, a rather bold message to the Church: "Neither is it right that Religion should oppress Science anymore, nor that Science should ever consider Religion as superfluous." This "Athenian Letter" clearly removes any doubt which may have been observed in the earlier "Letters" mentioned; Xenopoulos had not lost interest in Darwinism. His intense interest is also voiced in his strong support for freedom of speech. The cited passages also indicate that despite the lack of evidence for the missing link, Xenopoulos still strongly supported aspects of Darwinism. On the one hand, he maintained a harmonious coexistence between God the creator and Dar-

107 Ibid.

108 Prosecutor in the Scopes trial was the fundamentalist politician William Jennings Bryan, and the defense lawyer was the agnostic Clarence Darrow. The trial was followed closely by millions of readers in the United States and Europe, taking several months. The judge ruled that the civil liberties component and the validity of the doctrine of Darwinism were not to be tested. According to the judge, the only issue to be dealt with was whether or not Scopes had taught Darwinism. He admitted that he had. Although widely considered a win for the evolutionists, according to Bowler, "the fundamentalists succeeded in keeping evolution out of the school curriculum for decades." Bowler, *Evolution* (2009), 376. Similar laws occurred in other American states. The Butler Act was abolished in 1967.

109 Xenopoulos, "Η δίκη του … Δαρβίνου," 292.

win's theory, yet on the other hand, he assumed that religion and science are totally unassociated, and should be treated as separate entities.

So although Scopes was finally convicted and charged one hundred dollars for violating the bill, the enormous publicity only managed to expose the fundamentalists to international ridicule. Therefore, the trial became not a test of freedom of speech, but a display of the "battle" between science and religion.

The subject of science versus religion that Xenopoulos discussed in this series of "Letters" was, as it still is, a highly controversial issue. Xenopoulos' preoccupation with the subject is reflected here particularly through his repetition of certain points. He constantly claimed that science cannot broach the subject of religion, and that Darwin himself had indicated that his theory did not reject God. What is pertinent here is that Xenopoulos would have been well aware that Darwin's theory was not taught at schools in Greece due to religious bias. So he was able to use the Scopes trial as a way to express his views on freedom of speech and on evolution.

In 1927, Xenopoulos wrote a "Letter" entitled "A new cosmic theory," which seems to be at odds with the more progressive scientific views that he presented in the previous "Letters."[110] Here he discussed an alleged new cosmological theory of the universe. This theory was supposed to have refuted Copernicus' well-established heliocentric theory of 1543, which was later validated by Galileo's refracting telescope of 1610; and although the "Letter" is not directly related to Darwinism, it is relevant as Copernicus' theory and Darwinism have been viewed hand in hand when investigating the relation of science to religion.[111] By dislodging earth as the center of the universe and questioning biblical teachings associated with the sun, Copernicus' theory challenged traditional Christianity and humanity's place in the universe, as did Darwinism. After explaining the pre-Copernican theory, Xenopoulos comments that "the people of early times were very proud of their place in creation: the great Earth, the center of the universe, and they, kings on the great Earth!"[112] The following passage encompasses his views

110 Xenopoulos, "Μια νέα κοσμοθεωρία," 140.

111 Quoting from Freud's "A difficulty in the path of psycho-analysis" (1917), Beer refers to the "three severe blows" to humanity inflicted by science. She states that Freud names them as "the *cosmological*, associated with Copernican theory, the *biological*, associated with Darwinian theory, and the *psychological*, associated with psychoanalytic theory" (italics in original). See Beer, *Darwin's Plots*, 8–9.

112 Xenopoulos, "Μια νέα κοσμοθεωρία," 140.

of Copernicus' theory: "Naturally, with this cosmic theory, the Earth fell miserably from its throne, it was reduced to a particle of dust in Infinity and Man, the king, had now only the value of a 'being' which is what they used to call microbes! … To tell you the truth, each time I would ponder these things, I too would feel something like doubt, disillusionment, and sadness. I don't know, but I would have preferred that the Earth was still the center of creation, or at least that we could believe it like the people of older times."[113] Xenopoulos confesses that he always hoped that Copernicus' theory was wrong. In a newspaper he reads about a renowned German astronomer's new cosmic theory, which is almost identical to the pre-Copernican theory. Oddly, he does not recall the name of the astronomer and, to date, there appears to be no record of his identity. Xenopoulos relates the theory according to the astronomer:

> The Earth returns, he says, to the center. The Universe, whatever that is, we can see with our eyes and with telescopes, is a sphere where we are in the center. This sphere with the Sun, the Moon, the planets, and fixed stars, revolves around us. The furthest star is no further than a few thousand kilometers from us. The Sun, for example, takes once more the dimensions of the Peloponnese—thereabouts—and the Moon that of Aegina. The other planets are small also in proportion. The fixed stars are even smaller. None of them are inhabited. So all things were created and exist for the Earth, and the Earth was created for Man the king.[114]

He appears to be optimistic that the unnamed astronomer's theory will prevail over that of Copernicus, and he concludes: "Something tells me that all those boundless and breathtaking things that the astronomers reeled off were … 'exaggerations' because of miscalculation or optical illusion. This, sooner or later, will be proven. And the world which we had lost for so many centuries—from the time of Galileo—will be all ours again."[115]

This "Letter" can only be assessed at face value, as there is no other information currently available which could throw light on the name of the astronomer. Assuming the information which Xenopoulos supplied is correct, then one can only say that the support for the theory was so short

113 Ibid.
114 Ibid.
115 Ibid. (suspension points in original).

lived that it does not appear to have registered a place in general cosmology. The "Letter" may seem odd at a glance, but it nonetheless provides an understanding of Xenopoulos' sentiment associated with Copernicus' theory, the consequences of which demoted humanity from the central position in the universe, robbing it of its superiority among living beings. It is clear that he, personally, was very affected and disillusioned when he was exposed to Copernicus' theory, and that he was still not able to come to terms with its repercussions for humanity. Further mystifying the situation, Xenopoulos highlighted that the unnamed astronomer "has very significant arguments and the scholars of the world are always discussing them." However, the "Letter" seems to have an ironical message. The new theory that he described elaborately is absurd, even by the standards of early twentieth-century astronomy. The subtext discloses the message that if Copernicus' theory was true (and which his readers would have known to be true), then every human would be doomed to a useless "being."

Xenopoulos' views on the relationship between science and religion would change when he wrote in 1931 an "Athenian Letter" entitled "Religion and science."[116] Inspired by a talk given by the academic and theologian Dimitris Balanos of the University of Athens, he decided to pass on the views of the speaker, with whom he agreed. The speaker, according to Xenopoulos, indicated that science and religion are "the closest of friends and that one helps the other for the prosperity and progress of Humanity."[117] Xenopoulos' past "Letters" examining the relationship between science and religion have displayed contradicting opinions; for instance, within one "Letter" he emphatically maintained that science and religion do not mix, and then, to make his case for Darwinism, in the same "Letter" he indicated that science does recognize religion as a different kind of "knowing." Nevertheless, Xenopoulos admitted: "True Science—not superficial knowledge—recognizes that Religion has its say and that it is something necessary and indispensable for man, who is tormented by the great and worrying problems: Who created me? Where did I come from? Where am I going? What is my destiny on earth?—on the other hand, Religion—today's Religion, free from fanaticism, superstitions, and the fears of the Middle Ages—rec-

116 Xenopoulos, "Θρησκεία και επιστήμη," 8.
117 Ibid.

ognizes the need and the value of Science, and instead of persecuting and hindering it as in the past, it helps it. Science and Religion are friends; the world pushes on with progress, civilization, and happiness."[118]

In 1939, Xenopoulos wrote a "Letter," "Darwin's theory," celebrating one hundred and thirty years from Darwin's birth.[119] He reiterated the same observations on Darwinism and God as he did in his past "Letters," acknowledging that paleontologists still had not found evidence of the missing link of the ape-man. However, he mentioned that there was fossil proof exhibiting the evolution of lower forms. The "Letter" differs from the previous "Letters" because, despite his acknowledgement of the lack of evidence of the missing link, he fully accepts Darwinian theory. The following passage exhibits his overwhelmingly supportive stance towards Darwinism at that time: "All scientists today accept Evolution, at least up to the ape. Some believe that all the plant and animal kingdom came from only one original form, and others that the original forms were four or five. Darwinism has been, up to the present, the grandest, most enterprising scientific theory. Ever since it appeared, we can say that it changed human thought. From the plant and animal 'species,' Evolution spread to all things, material and moral. Everywhere we see Evolution."[120] It is only in this "Letter" of 1939, well into the twentieth century, that he wrote to the young readers of the children's magazine on Darwinian evolution as a ubiquitous phenomenon that has changed the way that humans think.

The "Letters" examined in this study, though not always fully supportive of Darwinism, display one individual's significant post-Darwinian response associated with science and religion. Interestingly, he had been an advocate of Herbert Spencer's theories before they became outdated. Interviewed a few years before his death, he was well informed on the demise of Spencerism, acknowledging that "Spencer did not verify Darwinism, because there are many deviations from the road that the author of the *Origin of Species* mapped out."[121] His comment, further to this, reflects his awareness of the ongoing changes in Darwinism: "But, as it happened, didn't Darwinism become neo-Darwinism, with variations substantially in research?" Accept-

118 Ibid.
119 Xenopoulos, "Η θεωρία του Δαρβίνου," 139.
120 Ibid.
121 Kouhtsoglou, "Ο 'σπενσεριστής' Ξενόπουλος," 1016.

ing that Spencerism was now past history, he made it clear in the interview that he did not consider himself a person with antiquated ideas. On the contrary, he believed he was progressive, indicating that had one particular work of his on Spencer been saved from his house fire, he would have published it, but with numerous revisions.[122]

In 1942, Xenopoulos in the "Athenian Letter" entitled "Theology of the idle" appears again to be at odds with his past views. He advised his young readers not to attempt to use logic or reasoning to ponder the nature of God or the existence of free will in man, as this could lead to their becoming "skeptics, pessimists, [and] atheists."[123] He touched on God's immortality, also asking himself whether the human soul is immortal, and if the souls of those that commit suicide die also. Immortality of the soul is a theme he pursued in several of the "Athenian Letters."[124] Discouraging his young readers from enquiry, and in accordance with an old school teacher of his, he advised: "Let's leave God alone, or better still, let's remain unperturbed and believe in him without any doubt."[125]

Nevertheless, Xenopoulos' advice to youth is not consistent with his own past searchings for the meaning of life and death. In 1904, Xenopoulos wrote an article in the periodical *Panathinaia* (*Παναθήναια*) on "The instinct of death," which is based on the studies of Russian zoologist and microbiologist Elie Metchnikoff (1845–1916).[126] In the article, Xenopoulos noted that science does not recognize the immortality of the human soul.

Physiognomies and expressions

I have included in this chapter an article spanning three consecutive issues (June 7, 14, 21, 1941) of the *Children's Guidance* magazine.[127] The first issue consists of several paragraphs and two sets of photographs, and the second and third issues only have photographs. The segments are not derived from

122 Ibid., 1017.

123 Xenopoulos, "Θεολογία αργόσχολων," 235.

124 He mentions the immortality of the human soul and God's immortality in the following "Athenian Letters": "Religion" (1915); "Revelation" (1917); and "The trial of ... Darwin" (1925).

125 Xenopoulos, "Θεολογία αργόσχολων," 235.

126 Xenopoulos, "Το ένστικτον του θανάτου," 8–11, 41–46. Xenopoulos indicates that he has taken Metchnikoff's work from his *Études sur la nature humaine. Essais de philosophie optimiste* (1903).

127 M.D.S. [pseud.], "Πέντε διαφορετικές παιδικές φυσιογνωμίες," 209–210, 220, 224.

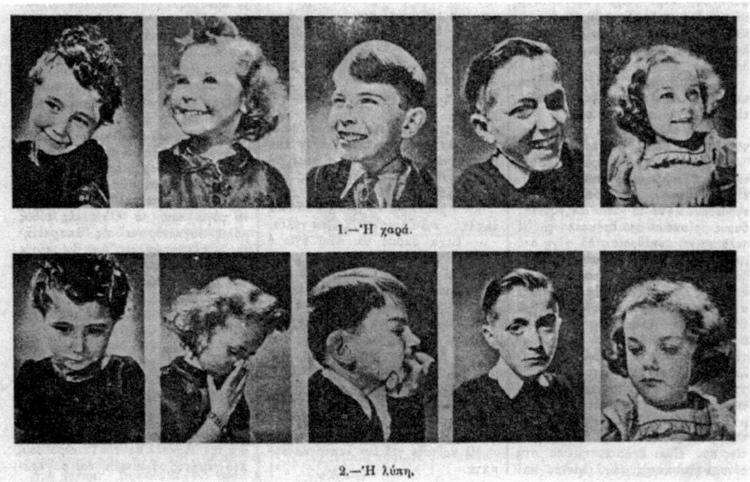

Figure 3.3. The expression of the emotions of happiness and sadness. Reproduced with permission from Anna-Maria Vlassi-Galerou (publisher) of Vlassis Brothers Publications, Athens.

the "Athenian Letters," neither is the identity of the author known (the first segment is signed with the pseudonym "M.Δ.Σ.," that is, "M.D.S.").

The three segments that I refer to are a series of five sets of photographs of the facial expressions of five children. Each set of photos shows a particular expression in each of the five children. These five expressions are of happiness, sadness, fear, sneering or contempt, and pain. The first article, entitled "Five different children's physiognomies," is taken from the front page of the June 7 issue (see fig. 3.3) and points out that "depending on the way in which the children show happiness or pain, their physiognomy changes."[128] The writer asks his readers to compare their own physiognomies with the one most likely to be similar to their own expression when they are happy, sad, or another emotion. He ends the article, curiously, by saying that his readers should try to correct their expressions. The first issue shows the photos of happiness and sadness, in the June 14 issue are those showing fear (see fig. 3.4), and in the June 21 issue are photos showing negative feelings of sneering, contempt, and mockery (see fig. 3.5).

The point of all these observations is that the person who wrote this article in *Children's Guidance* is very likely to have read Darwin's *EE* and been

128 Ibid., 209.

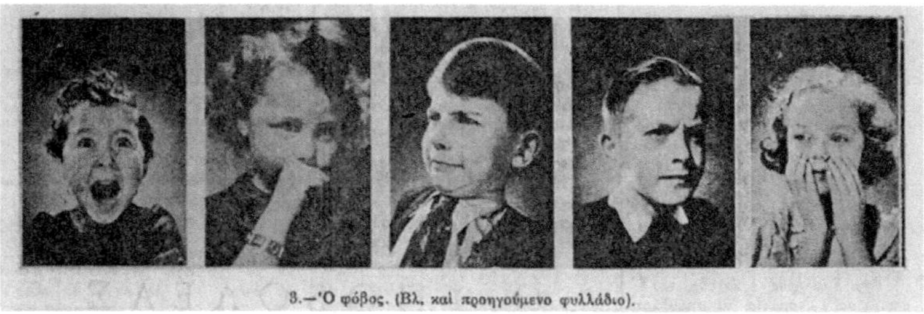

Figure 3.4. The expression of the emotions of fear. Reproduced with permission from Anna-Maria Vlassi-Galerou (publisher) of Vlassis Brothers Publications, Athens.

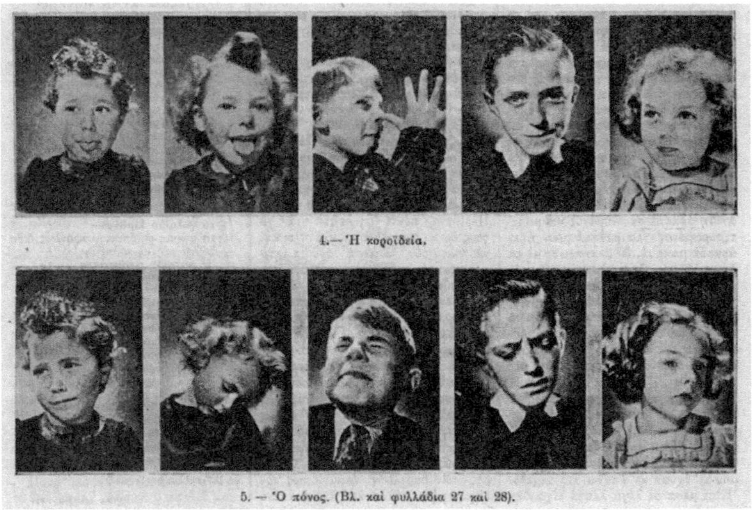

Figure 3.5. The expression of negative emotions such as sneering, contempt, and pain. Reproduced with permission from Anna-Maria Vlassi-Galerou (publisher) of Vlassis Brothers Publications, Athens.

strongly influenced by it. These photos are reminiscent of those in Charles Darwin's *EE*, where he used a very similar approach to displaying the expression of the emotions of humanity. Darwin demonstrated this with a series of up to six people for each emotion, which included happiness, sadness, negative expressions of sneering and contempt, and pain.[129] As will be seen in this next chapter, Darwin's photos were meant to show the universality of expressions among all people, and also to align the human species' expressions with those of the lower primates.

129 See Darwin's chapters 6–12 on these emotions in humans. Darwin, *EE*, 146–308.

As to who might have actually written the article, one could make some suggestions. Many Greek writers had pseudonyms often consisting of only one, two, or three initials. Delopoulos records thirty-eight different pseudonyms for Xenopoulos, and it is likely that he had more which have not yet been identified.[130] Delopoulos does not record M.D.S. as one of them; nor is this pseudonym anywhere in his book. However, he notes that the literary critic and writer Mihail D. Stasinopoulos (1903–2002) wrote in the *Children's Guidance* magazine under the pseudonym Kimon Alkidis.[131] Stasinopoulos also had several pseudonyms, though M. D. S. has not been recorded as one of them.[132] Delopoulos also notes that Stasinopoulos shared the pseudonym Kimon Alkidis with Xenopoulos and others, which is all rather intriguing.[133] So, Stasinopoulos may have written the article in collaboration with Xenopoulos, remembering that Xenopoulos contributed to the whole magazine, not just the "Athenian Letters."

What is certain is that Xenopoulos, as editor-in-chief, would have read this article on physiognomies, particularly because of the nature of the topic, which would have interested him. Two of the three segments were considered significant enough to be placed on the front pages of the magazine. What this article confirms is that by 1941 those individuals writing in the magazine—which includes, of course, Xenopoulos—were well aware of or had read Darwin's *EE,* even though it had not been translated into Greek.

Xenopoulos' "Athenian Letters" and other articles and essays have served to show a Xenopoulos who was interested in science and Darwinism on a biological level, and on the impact of Darwinism in other fields as well. Evolutionary gradualism as a theme or motif features significantly in his "Athenian Letters" in various forms, reflecting his extensive knowledge of it scientifically and in its application to literature. It influenced his novels, as will be seen in the coming chapters. In his "Athenian Letters" Xenopoulos discussed Darwinism and other evolutionary ideas in a pedagogical manner, as well as how they created an impact on matters of religion, science, and also the gender issue. His overall intention seems to have been to educate the youth, who in Greece would not have been taught about Dar-

130 Delopoulos, *Νεοελληνικά φιλολογικά ψευδώνυμα,* 309–310.
131 Ibid., 371.
132 Ibid.
133 Ibid.

win specifically till the 1930s. Although he was able to explain the mechanical side of evolution, he was unable to take a consistent approach to the real issues, such as humanity's place in nature and the relationship between science and religion. It is very likely that the conservative ideology of the magazine—which Xenopoulos had helped shape—and that of the general Greek society played a major role in formulating Xenopoulos' often contradictory writing. His reliance on his writing for a living deterred him from writing against the Church.

Xenopoulos presented a different message to young readers, compared to what was going on in his own mind, on various aspects of science and religion. Although his intellectual background provided him with the ability to rationalize, he tended not to convey this to the readers of the children's magazine. Instead, for answers to questions associated with humanity's place in nature and with God, he attempted to promote faith and the belief in revelation within the traditional church. Although Xenopoulos presented to the readers an essentially anti-creationist view, in terms of the presence of only one or two original forms from which all living things arose, he did not exclude God as the Designer of creation.

Perhaps Xenopoulos' philosophy can be summed up from an interview he gave at some time between 1949 and 1951, just before he died. In his statement he encompassed a theory which embraces the essence of evolutionism from Spencer to Darwinism to neo-Darwinism: "Here on Earth everything is modified in the long-term: geological formations, plant and animal species, complex societies, nations, races, ways of life, civilizations, administrative and political systems. ... Using these natural phenomena as a basis, man tries philosophically to give some explanations, and to achieve a makeshift solution for a normal way of life. Social life is a direct reflection of natural oppositions."[134]

This affiliation between biological evolution and social evolution transcends his literary works, as has been seen with his "Athenian Letters" and will be seen in the chapters to come. In terms of evolution, Xenopoulos acknowledged Darwin, among others, as an important intellectual influence. In his highly acclaimed novel *Rich and poor*, Xenopoulos mentions and discusses Darwin's theory of natural selection, which plays an important

134 Kouhtsoglou, "Ο 'σπενσεριστής' Ξενόπουλος," 1017.

role in the story through the experiences of his main character Popos. Also in the novel, Xenopoulos provides a perspective on how Darwinism permeated various fields—such as science and politics, in particular socialism—and played an important role in the making of the intellectual discourse of that period. Certain matters which Xenopoulos has covered in his "Athenian Letters" will be also covered in his novel *Rich and poor,* discussed in the next chapter and also in chapters 5 and 6.

CHAPTER IV

A REREADING OF *RICH AND POOR*: IT'S IN THE EYES

> Excuse the bad writing—I have a pen which if natural selection influenced pens, would have been cast into the fire long ago: but the disturbing moral element makes me too lazy to cast it thereinto—& to find a new one.
>
> CHARLES KINGSLEY[1]

Xenopoulos' novel *Rich and poor* (*Πλούσιοι και φτωχοί*) (1919), which has been considered his finest, is also the most reviewed of all his work.[2] The Darwinian aspects of the novel cannot be ignored, as the novel does not hold together without the evolutionary framework. Evolutionary theory and the sciences of man provided a framework for the literary representation of man in relation to body and mind, social concepts (such as class, race, and gender), and political ideologies (such as socialism and capitalism). Themes of class and class mobility were common in novels in the nineteenth and early twentieth centuries, and after Darwin these themes were often closely interwoven with Darwinian theory. This chapter aims to specifically demonstrate some of the biological ideas that were derived from Darwin's theories and absorbed in Xenopoulos' novel. These ideas may also be

1 From a letter to Charles Darwin, dated December 11, 1867. See Burkhardt et al., *Correspondence of Charles Darwin, 1867*, 15:478.

2 The notion of the Darwinian perspective found in this novel was first published in 2006 in my conference paper Zarimis, "Darwinism in Modern Greek Literature and a Rereading of Grigorios Xenopoulos' *Πλούσιοι και φτωχοί*," 393–402. The research was extended in my 2007 PhD thesis. The chapter here is a revised and further extended version of that found in the thesis. This notion in the novel appears, to a degree, to have also been taken up by Athanasios Blesios in 2008, in his book *Κράτος, κοινωνία και έθνος στη νεοελληνική λογοτεχνία*. While he acknowledges a social Darwinian and Spencerian sociobiology in the novel, his analysis is primarily on the sociopolitical aspect in Greece rather than the scientific aspect.

found more generally in European literary culture. Importantly, this novel embraces what Greek intellectuals with specific knowledge of Darwin's theory and associated literary trends were contemplating at the time.

The title of the novel denotes a theme of poverty, which was utilized extensively by writers in the late nineteenth and early twentieth centuries in Western countries. The plight of the poor becomes a real issue in most countries, and Greece was no different.[3] The bipolarity of the classes creates a literary discourse which sees the poor as a separate breed or race. The plight of the poor, a theme of class, receives a biological literary treatment. This approach is seen by other commentators in such works as H. G. Wells' *The Time Machine* (1895) and *A Modern Utopia* (1905), Jack London's *People of the Abyss* (1903), and Francis Galton's eugenic utopia "Kantsaywhere" (1910).[4] As highlighted by Angelique Richardson, the biologization of the poor would indicate that poverty was "immune to the environmental changes that could be brought about by social reform."[5] The literary, political, and social discourses became a debate between the hereditarians (based on neo-Darwinism) and the environmentalists (based on Lamarckism).

In this chapter I argue that *Rich and poor* is Xenopoulos' representation of the 1880s in Greece, a period that exhibited a pervasive Darwinian influence among intellectuals in the sciences (including biological), politics (including socialism), and in literature. I also regard it as a parody in which Xenopoulos was able to comment, with the historical and scientific hindsight of forty years, about those Darwinian influences which preoccupied Greek intellectuals around the 1880s. The novel has been shown to be semi-autobiographical,[6] but as will be seen with the novel's analysis, there is the

---

3 On the extreme poverty in Greece see Korasidou, *Οι άθλιοι των Αθηνών και οι θεραπευτές* (The miserable of Athens). Note that Ioannis Kondylakis also wrote a novel of the same name, *Οι άθλιοι των Αθηνών*, published in 1894. The title was inspired by Victor Hugo's novel *Les misérables*, published in 1862.

4 For their treatment along these lines see Henkin, *Darwinism in the English Novel*, 242–246; Richardson, *Love and Eugenics*, 21–22. Galton's "Kantsaywhere" is found in Pearson, *Life, Letters, and Labours of Francis Galton*, 3a:411–425.

5 Richardson, *Love and Eugenics*, 24.

6 Popos, the main character of *Rich and poor*, shows similarities to Xenopoulos, who abhorred the rigidity of the classes and had in his youth (in the late nineteenth century) taken up socialism; his enthusiasm for socialism was short lived, as he became disillusioned with it. Furthermore, Popos, like Xenopoulos, had read widely, also having read Darwin. See Xenopoulos, *Η ζωή μου*. Other creative writers of the late nineteenth and early twentieth centuries who had also read Darwin, who were scientifically minded and had read foreign works, who realized that this approach was very popular, and felt strongly about the inequality of the classes, used a Darwinian approach to validate the inequality of the classes. These would include Thomas Hardy, D. H. Lawrence, and George Gissing.

recurring problem of trying to separate the author's views from those of the narrator, main character, and others.

Xenopoulos' distance from the ideas of the various characters signifies that the characters represent views which Xenopoulos had once considered but now rejected; or it may be that Xenopoulos is engaged in an inner debate with himself, or that he merely wants to maintain ambiguity. I would speculate that many of the views that his main character expresses are to some degree those which Xenopoulos also had held in the 1880s but now had modified or rejected. It can thus be argued that this novel is a culmination of Xenopoulos' reassessed views on Darwinism and its application in society, at a time when natural selection was probably the only concept which he retained, while he rejected the applications of Darwinian thought to physiognomy and to socialism.

The novel is generally known as one of a social trilogy;[7] hence, commentary has tended to focus on a social or sociopolitical perspective and on the theme of socialism. The novel, set in Zakynthos and Athens in the 1880s, is about the young, lower middle-class Popos Dagatoras, who appears initially in his life to have it all: nurturing parents and a university education in mathematics. In spite of this, he struggles to establish himself financially and turns to the socialism of his mentor, the philosopher Leon Harisis. Finally, Popos' alleged political views and his own human failings see him in jail, where he dies.[8] During his life Popos develops a "scientific" theory of the eyes, derived from his physiognomical observations and Darwin's theory of natural selection. He uses this combination to highlight the inequality of the social classes, which he refers to as "the race of the rich" and "the race of the poor."[9] So, the basis of this chapter is the examination of Popos'

---

7 *Rich and poor* was published in serial form from January 2, 1919, to October 5, 1919, and as a book in 1926. The other novels in this trilogy are: *Τίμιοι και άτιμοι* (Honorable and dishonorable), published in serial form from October 10, 1921, to February 5, 1922, and as a book in 1926; and *Τυχεροί και άτυχοι* (Fortunate and unfortunate), published in serial form from April 29, 1924, to August 30, 1924, and as a book in 1927. In serial form they were published in the Athens *Ethnos* newspaper and as books by the Athens publisher Kollaros. Each novel is independent of the others, and *Rich and poor,* which is the first of the three, is the most significant for the purposes of this study. My references in this study are to these novels in the *Collected works* published by Biris in Athens in 1972. See Xenopoulos "Πλούσιοι και φτωχοί," *Άπαντα*, vol. 2; "Τίμιοι και άτιμοι," *Άπαντα*, vol. 3; "Τυχεροί και άτυχοι," *Άπαντα*, vol. 4.

8 See Xenopoulos, *Η ζωή μου*. Like his character Popos, Xenopoulos was fluent in at least French (162); his father and uncle shared a timber business which burned down causing the family's economic downfall (182–183); he was born around the 1860s, he grew up in Zakynthos, and would later study mathematics and the natural sciences at the University in Athens (159–162).

9 See the discussion on the meanings of the word "ράτσα" further in this chapter.

theory, its effect on him, his response to those around him, and his turning to socialism.

Further to this, I will examine firstly Popos' Darwinian approach to the human mind and behavior and to the inequality of the classes; and secondly, though to a lesser degree, I will demonstrate how at that time Darwinism was manipulated to validate socialism internationally. In all, the analysis aims to highlight the pervasive nature of what was considered to be Darwinian thought at the time, extending to the disciplines of psychology and sociopolitical theory.

Although nineteenth-century physiognomy in itself has been regarded as a pseudoscience, renewed interest into reading people's faces, whether subconscious or not, can reveal aspects about their personality and life.[10] Xenopoulos' masterly use of the physiognomy of his characters becomes not only a literary tool but also a scientific one in the probing of his characters' minds. Physiognomy, in association with Darwin's theories on physical and mental traits, is an example of the tentative beginnings of a science of the mind, which prevailed in the late nineteenth and early twentieth centuries. To date, partial readings of *Rich and poor* have omitted the subtleties of Xenopoulos' work, hence *Rich and poor* has not been placed internationally in its literary context.

Literary Criticism, the "note," and prologue

This section highlights the gap in the scholarship associated with Xenopoulos' Darwinian approach by exploring briefly the past trends of literary commentary. These only view the novel from a social or sociopolitical perspective. A "Note" (written after 1919 and before 1926) is included at the end of the novel. After much commentary (negative and positive) on his novel, Xenopoulos wrote this "Note" that provides clues to how he wanted it to be interpreted. The "Note" mentions Greek socialism and also alludes to

10 The last decades have shown enormous research interest in the correlation between facial expression and emotion by researchers such as the authoritative Paul Eckman and his associates. His work spans from the 1970s to the present. See Eckman, "Universals and Cultural Differences," 207–283; and "Facial Expression and Emotion," 384–392. More recently, significant work by psychologists has homed in on a "physiognomical" approach using photos of people to reflect their true feelings and also predict their relationships with others. See for instance Harker and Keltner, "Expressions of Positive Emotion," 112–124.

Popos' ideas of Darwinism and to those of an unnamed American sociologist (possibly William Graham Sumner). Xenopoulos reveals: "The psychological portrayal of this man is the primary aim of the book, and so is the representation of the environment with its ideology. It is therefore a psychological and social novel, which does not support any ideology. So it will not please the socialists or the conservatives. It will please however, I hope, the reader, as an impartial work of Art, as well as a picture of the Athenian urban society of that period, when socialism—and it certainly is a curious story—first appeared."[11] If we are to take this "Note" as a reasonable insight into the thematics of the novel, then Xenopoulos' main concern was the "psychological portrayal" of Popos, which involves Popos' detailed physiognomic observations of the eyes. I would interpret "the representation of the environment with its ideology" as a Darwinian representation, which reflects the Zeitgeist of the 1880s.

As indicated at the beginning of this chapter, many commentators have viewed the theme of class inequality and the introduction of socialism in a stereotypical manner. They have also shown concern to analyze Xenopoulos' stance on these issues. In his "Note" above, Xenopoulos makes a point of detaching himself from the views of his characters, which is typical of his middle-of-the-road attitude to many issues.

Consequently, the fact that he denies taking a position on any application of Darwinism reveals his inability, at the time, to understand that Darwinism could never be completely devoid of ideology. Of course, the social application of Darwinism has long been used and abused by social scientists, politicians, and creative writers among others, so as to sanction their ideological theories. So in fact Xenopoulos is stating an ideological position even if he does not realize it. While he believes that he provides "a picture of the Athenian urban society of that period," I also maintain that by parodying the "Darwinian" physiognomical theory propounded by Popos and the "scientific socialism" in the story, Xenopoulos is actually taking a stance against them.

In his prologue Xenopoulos believes that the cause of class immobility lies in the eternal inherent inequalities between humans produced by nature: "Each person is born into the world predestined by his nature, which

11 Xenopoulos, *Πλούσιοι και φτωχοί*, 317.

is his Fate, to either remain poor or become rich. There exists a race of the Poor and a race of the Rich. The vast injustice which this natural and fateful distinction creates in today's society will be modified, as much as is possible, by the society of the future."[12]

The prologue generates the idea of the eternal laws of nature. The human is born "predestined by his nature." The concept here is purely naturalistic. According to one's character traits, one is assigned to either the rich class or the poor class. The innate injustice among people that Xenopoulos talks about here has nothing to do with sociological factors. Christian doctrine may have indicated that we are all born equal, but the prologue reflects another story: Darwin's theories of variation and heredity attested that we are not all born the same, since physical and mental abilities, which are heritable, vary from person to person. The last sentence of the prologue reflects a pessimistic representation of a society which may modify, to a degree, the huge injustice one day only in the future.

The earlier commentators of *Rich and poor* seem to have been more concerned with forming an opinion as to whether Xenopoulos had a proclivity for socialism or conservatism, and with damning or praising him for it, or even protecting him from his alleged views. They seem to not have been concerned with taking the novel to another level.[13]

Gianis Kordatos is one of the critics who viewed the novel as pro-socialist due to its emphasis on class inequality.[14] But he failed to see the undermining satirical treatment of the socialist group in the novel; also he did not understand the workings of Popos' mind that led him to his destruction.[15] Although Petros Charis saw the theme of the novel as the "first appearance of socialism in Greece," he believed that Xenopoulos had taken an apolitical stance.[16] By contrast, Apostolos Sahinis also believed that Xenopoulos showed pro-socialist views in the novel, and maintained that Popos' demise

12 Ibid., 11.

13 See Doulaveras' social perspective on *Rich and poor* as part of the social trilogy: Doulaveras, "Κοινωνικές δομές και ιδεολογία," 131–141. She acknowledges the social Darwinian views of Antonis Roukalis (friend of the protagonist Popos in *Rich and poor*) and so states it is not a coincidence that in the novel Popos is examining Darwin's work (137).

14 Kordatos, *Ιστορία της νεοελληνικής λογοτεχνίας*, 515.

15 Another example of a critic with a similar view is Takis Adamos, who saw it as one of the social trilogy, having "purely a social context with class conflict, and with which Xenopoulos takes the side of the poor and the exploited masses." See Adamos, *Η λογοτεχνική κληρονομιά μας*, 86.

16 Charis, *Έλληνες πεζογράφοι*, 6:42–43.

was due to sociological reasons.[17] More recently, Georgia Farinou-Malamatari, who also saw the novel as reflecting the beginnings of socialism in Greece, indicated that Popos was one of a number of Xenopoulos' heroes who "is defeated socially and biologically"; however, she does not pursue the biological aspects.[18] Essentially, these critics did not really view the novel's overall intellectual orientation, which was associated with Darwinian and post-Darwinian thought. Instead, I will argue in this chapter that Xenopoulos' intention was to display a representation of Darwinian influence in the period around the 1880s in Greece.

Further to my argument that the novel is not purely on the beginnings of socialism in Greece, Giorgos Fragkoglou argues that by 1919, the time Xenopoulos had completed the serial version of *Rich and poor*, he had become apolitical and saw socialism as a pipe dream.[19] Consequently, it was a far cry from the days of his zeal for socialism seen in his *Margarita Stefa* of 1893.[20] In addition, Fragkoglou remarks on the fact that the first mention of socialism is more than halfway into the book.[21] He argues that the novel is not really about a socialist approach because of the flimsy representation of its socialism and its characters.[22] Moreover, earlier in 1915 Xenopoulos explained why he did not know of the newspaper articles of a journalist named Stephanos Granitsas: "because I hardly ever read the main articles in the newspaper—politics holds little interest for me!"[23]

Xenopoulos, who had been the enthusiastic supporter of socialist youth around the 1890s, had distanced himself from socialism by the time he wrote *Rich and poor*.[24] He certainly did not want to be affiliated with the Greek socialist group, which in 1920–1921 was renamed as communist. For Xenopoulos as a commercial writer, any association with the communists could have destroyed his writing career.

Literary historian Roderick Beaton sees Popos following the same pattern as Alkis Sozomenos in Theotokis' *Slaves in their chains* (*Οι σκλάβοι*

17 Sahinis, *Το νεοελληνικό μυθιστόρημα*, 255–257.
18 Farinou-Malamatari, "Γρηγόριος Ξενόπουλος," 322.
19 Fragkoglou, "Το λαϊκό αισθηματικό μυθιστόρημα," 79–80.
20 Ibid., 81.
21 Ibid., 80.
22 Ibid.
23 Xenopoulos, "Στέφανος Γρανίτσας," 150. See also Xenopoulos, *Η ζωή μου σαν μυθιστόρημα*, 240
24 Xenopoulos' views on socialism through his "Athenian Letters" will be further examined later in this chapter.

στα δεσμά τους) (1922); that is, as "a character-role which would be taken up again and again in Greek fiction in the twentieth century: that of the young socialist whose beliefs find no outlet in action and who becomes finally the victim of society and his own weaknesses."[25] Further to this, as one of Xenopoulos' Zakynthian novels, Beaton perceives *Rich and poor* historically as "the belated fictional expression of a society that had undergone rapid change since the islands became part of Greece in 1864." He highlights that "the genre of the novel ... was slow to emerge in the Ionian islands" and its "rise is traditionally associated with the rise of the middle-class in Europe."[26] While Beaton's commentary is historically correct, he does not register the significant Darwinian influence among Greek intellectuals in literature, the sciences, and politics in the 1880s and 1890s, a period captured creatively by Xenopoulos in *Rich and poor.*

Although Kostis Palamas is an earlier critic, his comments have been dealt with last in this section because he appears to have treated *Rich and poor* differently, steering clearly away from a viewpoint on the treatment of socialism in the novel. In the novel's prologue he comprehends the distinction between the rich and poor as follows: "This law is no more than a more vivid adaptation of the old renowned doctrine of the free and enslaved, by birth, or still more simply, of the strong and the weak."[27] Referring to Popos he remarks: "He acts each time, in all his tragicomic adventures, as if under the influence of inexpiable Fate. And it [Fate] is nothing but the creation of the two dominant powers which contribute more than anything else to human affairs, which are heredity and environment."[28] Palamas refers to Xenopoulos' prologue, alluding to the social Darwinism of the nineteenth century when he mentions the doctrine of "the free and enslaved by birth"; and hints at Darwinism when he refers to the doctrine of "the strong and the weak." In spite of this, Palamas wants to stress that the book is not directly ideological. Rather he takes the view that Xenopoulos acquired abstract ideas and transformed them by using art.[29] Xenopoulos, in his prologue, implies a certain sense of timelessness associated with the class problem,

25 Beaton, *Introduction to Modern Greek Literature,* 107.

26 Ibid., 106.

27 Palamas, "Εισηγητική έκθεσις," 448.

28 Ibid.

29 Ibid.

which does not seem to be subject to any improvement. According to Xenopoulos' prologue, the "huge injustice" of the inequality of the classes will not change. It might only be modified, to a certain extent, in the future.

As noted with Palamas, Xenopoulos does not mention Darwinism by name and dismisses any obvious ideology in *Rich and poor*. Presumably, Palamas knew only too well from personal experience that one could be criticized for allegedly flaunting one's knowledge of the intellectual trends of the time in his writing.[30] Palamas avoids mentioning any social implications of biological or evolutionary theory, perhaps so as to protect Xenopoulos from any negative consequences. After all, Popos' theory of the eyes, implying a certain biological determinism from which one could not escape if one was poor, evoked pessimism and helplessness, and would have been quite confronting for the young and poor modern Greek nation.

Also, such an analysis after the publication of *Rich and poor* in Greece was unlikely at a time when biological determinism was not overtly considered in the studies on the fabric of the modern Greek nation.[31] This to a degree reflected the trend outside Greece as well. From the late 1920s till the 1960s, evolutionary theory, embracing "hard" inheritance, was not a desirable idea to openly communicate in society, and of course this trend would spill over to literary theory.[32] This could also explain why Palamas omitted to actually mention Darwinism or social Darwinism by name, and also why Xenopoulos did not mention them by name in his "Note."

Despite Palamas' comments, it is difficult to ignore the pervasive presence of Darwinism throughout the novel. Firstly, Popos' application of his evolutionary theory to the inequality of the classes played an influential role in his life, from its early beginnings to the very end on his death bed. Secondly, a social Darwinian or more aptly, a social Spencerian style is applied to a capitalistic milieu within which the story is told. This is achieved by exhibiting a climate which shows the elements of the struggle for survival between the strong, or rich, and the weak, or poor, with no government aid to the poor. Thirdly, the presence of the Darwinian environment is high-

30 Ambatzopoulou, "Ο Κωστής Παλαμάς," 374. Her study shows that Palamas was criticized for the name dropping of European scientists, which she maintains was perceived "as a flaunting of knowledge or an attempt to follow the spirit of the age."

31 As I have indicated in the introduction, basically most Greek writers were more concerned with Greece's internal needs than with a "foreign" ideological trend such as Darwinism or its social applications.

32 Carroll, *Evolution and Literary Theory*, 24.

lighted when Harisis finally introduces his form of scientific socialism (the Darwinian environment and this socialism were, at the time, considered to go hand in hand).[33]

This, I believe, was the perspective that Xenopoulos had when he wrote *Rich and poor*. By the time Xenopoulos wrote the novel, he had no interest in politics. He was disillusioned with all sides of politics, presenting them in an unfavorable manner in the novel. The critics cannot ignore Xenopoulos' assertion that the novel's main goal is "the psychological portrait of this man ... and the portrayal of the environment with its ideology," and nor can they ignore "the theory of the American sociologist [which] had such an influence on the life and shaping of Popos Dagatoras." While Xenopoulos maintained that he did not view social Darwinism as necessarily true, or agree with its concepts, he did use Darwinism to expound his personal views of socialism and the inequality of the classes.

Hence, the psychological aspect of *Rich and poor*, with the representation of a particular group of individuals and their ideology, will be explored in this chapter. Literary writers commonly utilized the tools of psychology in their work to observe human behavior. This is found in the work of writers such as Hardy who, according to Peter Allen Dale, also had read "contemporary epistemology and psychology" when preparing for the writing of *Tess of the D'Urbervilles* (1891).[34] Dale highlights that in the second half of the nineteenth century scholars were "all most concerned ... with the extent to which our subjective consciousness determines the form of external things."[35]

This intensified after Sigmund Freud introduced his psychoanalytic theories in the early 1900s.[36] Not all took up his theories, including Xenopoulos, who revealed that he had not read Freudian theory till the late 1920s. He chose to apply his own methods independently which, he claimed, were analogous to Freud's.[37] The methods that Xenopoulos has used in *Rich and*

33 Marx and Engels had attempted to sanction their socialism by claiming it was a necessary consequence from Darwinism that showed that all men are not born the same, due to inherent variations in capabilities; this they said caused the inequalities of the classes.

34 Dale, *In Pursuit of a Scientific Culture*, 249–250.

35 Ibid.

36 Unlike Darwinism, generally speaking Freudianism tended to rely on situations in a person's life history, rather than the role of heredity.

37 See his autobiography for further reading on his theories on the sexual instinct. Xenopoulos, *Η ζωή μου*, 345.

poor involve the examination of the outer person as a reflection of the inner, utilizing visual images in the form of physiognomy, as well as Darwin's theory of natural selection. Further to this, Xenopoulos used this biological representation of Popos and those around him to reflect a social representation of them.

In *Rich and poor,* Popos observes the physiognomic traits (in particular the eyes) of himself and those around him. With these observations he goes to great lengths to formulate a biological theory based on Darwin's theory of natural selection. He uses this theory to explain why people behave in certain ways, differentiating those who succeed financially (the rich race) and those who fail (the poor race), in a society perceived as laissez-faire. In evolutionary terms, this translates as the "struggle for survival" between those who are "fit" or "strong" and those who are "unfit" or "weak," in a competitive environment; in other words, the "survival of the fittest" or natural selection. Popos' association with the philosopher Leon Harisis and his followers introduces him to socialism, which only serves to reinforce his evolutionary theory.[38] The following section will provide a background to scientific and literary physiognomy and its relevance here to Darwinian theories, which will facilitate the analysis of *Rich and poor.*

Physiognomy, expressions, and natural selection

Physiognomy, in the context of this work, is the art or "science" of allegedly judging a person's nature by his or her external characteristics, such as facial features, expression, and body. In various studies of the face, the permanent features act as a baseline for expression, so these are often inextricably connected.[39] That expression is dependent on eye physiognomy will be clearly demonstrated in the next section on the analysis of *Rich and poor.* Physiognomy was also used as a literary tool in Europe and was at its peak in popularity in the nineteenth century, when it was adopted by writers such as Balzac, Flaubert, Dickens, and Hardy in their work to reflect the nature of a person or other relevant information about that person.[40]

38 Further on this and on Leon Harisis, see the section on scientific socialism.

39 Darwin said that he considered physiognomy as comprising only permanent features of the face, and not the expressions. However, it has since been argued that he did rely on both facial forms and expressions to propound his theories in *EE.*

40 See Tytler, *Physiognomy in the European Novel*; Rivers, *Face Value*; Greenslade, *Degeneration.*

Xenopoulos used physiognomy in many of his novels to reflect the inner nature of his characters. In chapter 5 physiognomy is demonstrated in Xenopoulos' *Tereza Varma-Dacosta: The Middle Ages today* (1926). Also, in the prolegomena of one of his plays, *The secret of Contessa Valerena* (1915), Xenopoulos provides an elaborate description of eye physiognomy to reflect, in his words, the "stigma" of illness.[41] His fascination with the eyes is also extended to *Rich and poor*, where the physiognomy is used both as a literary device or motif and as a "scientific" theory applied by Popos. Although the origins of physiognomy predate Darwin's work by many centuries, it is Xenopoulos' use of it in combination with Darwin's theory of natural selection which makes it significant and relevant to this work. He combines a modified biological theory with social theory to provide a framework for his representation of social and political ideologies.

The other important aspect is that Xenopoulos held that mental attributes evolve and adapt just as physical attributes do, subject to natural selection. I argue that Xenopoulos, through his character Popos, used a primitive form of psychology based on the natural selection of mental traits in combination with physiognomy to propound his views on the classes.

From a scientific perspective, Xenopoulos—who, as already mentioned, had read most of Darwin—would have read in Darwin's introduction to the *EE* his lengthy preliminary discussion of the various works of physiognomists and anatomists and of their drawings.[42] These included Charles Le Brun, Johann Caspar Lavater, Dr. Guillaume Duchenne, and Petrus Camper. Indeed Xenopoulos would have been aware that, following on from the centuries-old tradition of studying the resemblances of humans and animals, Darwin in his *EE* had emphasized that the only way possible to investigate "the causes of Expression" was not to view "man and all other animals" as "independent creations."[43] In an evolutionary manner, Xenopoulos would analyze the physiognomies and expressions of such resemblances in his *Rich and poor.*

41 Xenopoulos, *Το μυστικό της Κοντέσσας Βαλέραινας*, 8. The use of the word "stigma" implies an element of degeneration as seen in Nordau's *Degeneration*. Xenopoulos also uses the word in *Rich and poor* in a similar manner, as a possible reason for the emergence of the poor race.

42 Darwin, *EE*, 1–19.

43 Ibid., 12. Darwin refers here also to common expressions between humans and animals, such as the "bristling of the hair under the influence of extreme terror, or the uncovering of the teeth under that of furious rage ... [and] the movement of the same facial muscles during laughter by man and various monkeys."

From the literary aspect, as already mentioned in chapter 3, Xenopoulos was well read regarding foreign works and would have also received his ideas on physiognomy from writers like Flaubert and Dickens. As noted earlier, Xenopoulos used physiognomy in various forms throughout his novels. In addition to this, creative writers of the nineteenth century applied Lavaterian physiognomy to other aspects of an individual's outward appearance, including one's clothing and other belongings, such as their house. The appearance of a person's house, for example, reflected their social standing and inner being.[44] Chapter 5 looks at Lavaterian physiognomy in *Tereza Varma-Dacosta*.

Herbert Spencer wrote the pre-Darwinian *Principles of Psychology* (1855), which proposed a pattern of mental evolution via his development theory. He argued that the human was slowly and gradually adapting to the conditions of social life and gradually adapting to psychological change; he claimed these were requirements for humanity's movement toward order and completeness. His work is a starting point for the psychology of adaptation.[45] However, his theories were of a philosophical nature rather than scientific, and were Lamarckian. On the other hand, Darwin in *EE* quotes favorably from Spencer's *Principles of Psychology* in his description of the facial expression and emotion associated with fear: "Fear, when strong, expresses itself in cries, in efforts to hide or escape, in palpitations and tremblings; and these are just the manifestations that would accompany an actual experience of the evil feared. The destructive passions are shown in a general tension of the muscular system, in gnashing of the teeth and protrusion of the claws, in dilated eyes and nostrils, in growls; and these are weaker forms of the actions that accompany the killing of prey."[46]

The use of the word "physiognomy" in its loose sense—that is, meaning "the appearance of the face and at times the body"—was strongly related to psychological works on the description of emotions. In his chapter on "The Emotions" in his book *The Principles of Psychology* (1890), American psychologist William James (1842–1910) quoted Danish physiologist Carl Georg Lange's lengthy and detailed description of grief, from which the following is an excerpt: "The chief feature in the physiognomy of grief is per-

44 Tytler, *Physiognomy in the European Novel*, 252.

45 Leahey, *History of Psychology*, 326.

46 Darwin, *EE*, 8–9.

haps its paralyzing effect on the voluntary movements. ... The neck is bent, the head hangs ('bowed down' with grief), the relaxation of the cheek and jaw muscles makes the face look narrow, the jaw may even hang open. The eyes appear large, as is always the case where the *orbicularis* muscle is paralyzed, but they may often be partly covered by the upper lid which droops in consequence of its own levator."[47]

Darwin's cousin Francis Galton was the first of his followers to apply his principles of variation, selection, and adaptation to humans in the field of psychology. By various measurements, Galton studied the effect of heredity on the physical and mental characteristics of humans. In one of his enterprises he used photographic techniques (composite portraits) to observe facial similarities among different members of the same family, observing such groups as criminals and those with chronic illness, with variable results. This is undoubtedly a form of physiognomy applied to science, though much of it is now discredited.[48] Galton collected data on an individual's inherent tendency for talents, the results of which he documented in his book *Hereditary Genius: An Inquiry into Its Laws and Consequences*.[49] He assumed the all-importance of heredity with the total absence of the environmental component; that is, individuals were born not only with differences in facial features but also with variability in talents and intellect.[50] Xenopoulos utilized these concepts of hard inheritance in *Rich and poor*, which will be mentioned further at the end of this chapter, exemplifying a period in Greece where intellectuals were actively debating the issues of nature-nurture, the features of a genius, particularly that of poets, and an association with physiognomy. Out of all those quoted only Spencer, James, Galton, and Darwin himself believed in the mutability of the species. Despite Darwin's claims of not using physiognomy, it has been thoroughly argued that both he and Bell used it and even carefully examined numerous photographs and works of

47 James, *Principles of Psychology*, 1058 (italics in original). He quotes from Lange's *Über Gemüthsbewegungen. Eine psycho-physiologische Studie* (1887). William James may very well have inspired Xenopoulos in many of his works, including *Rich and poor*. This is due to the fact that as one of Darwin's followers, in his book James asserted that the basis of the human mind and behavior had animal origins, and that instinct in both humans and animals obeyed the Darwinian law of natural selection.

48 A certain amount of Galton's other work is still considered valid today. See Murphy, *Historical Introduction to Modern Psychology*, 117–122.

49 Darwin quotes him on this in the *DM*, 1:111.

50 Galton also invented the word "eugenics," a science through which he believed humans could direct the course of human evolution.

art so as to study external facial anatomy in relation to expression.[51] As well as Bell's detailed knowledge of muscle and nerve anatomy and of the physiology of the face, Darwin drew in the *EE* from Bell's knowledge of "emotional representation in art and literature."[52]

In the late nineteenth and early twentieth centuries, those novelists interested in the psychology of the mind and in Darwin's theories of the expression of emotions would develop their own Darwinian literary response to human behavior. Bert Bender identifies the novels of Harold Frederic, Edith Wharton, Kate Chopin, Henry James, and William Dean Howells as examples of novelists who "made what they could of the new theory."[53] Such writers, explains Bender, "developed psychological insights from Darwin's theory of expression, concealment, and repression of emotions."[54] According to J. B. Bullen, Hardy used physiognomy to provide an insight to the characters in his novels, and he also used the visual arts to facilitate his writing.[55]

In relation to Xenopoulos, as a supplementary note, one cannot dismiss the fact that he used picture postcards, particularly of women, as an aid to his literary descriptions.[56] In 1922 in a letter to his long-time friend Katina Papa, he asks her to send him not drawings, but actual photographs of young women's faces, "faces only" and "as nature would have them."[57] He admitted: "The strange thing is that although it will be an 'artistic' work the cards must be ... vulgar."[58] Papa explained why he wanted these pictures: "His ideas were exhausted by now by the numerous images of girls and young women which he described in his books, all his stockpile of representations, and so he wanted new faces for new inspirations. And naturally the further they were from the faces of the painters and from art, the closer they would come to nature and to reality, which inspired him."[59]

51 See for instance Hamilton, "Policing the Face," 76–77; Montagu, *Expression of the Passions*, 101; and R. Richards, *Darwin and the Emergence of Evolutionary Theories*, 231.

52 R. Richards, *Darwin and the Emergence of Evolutionary Theories*, 231.

53 Bender, *Descent of Love*, 278. See, for example, Bender's analysis of Frederic's *The Market-Place* (1899).

54 Ibid., 275. He also cites William James, Granville Stanley Hall (1844–1924), and Sigmund Freud as "constructing their own theories of instinct and emotion from Darwin's evolutionary texts" (278).

55 See Bullen's physiognomic analysis on Hardy's *The Return of the Native* (1878). Bullen, *Expressive Eye*, 96.

56 See further on this chapter 5, in relation to physiognomy and *Tereza Varma-Dacosta*.

57 Papa, "Η αλληλογραφία μου," 167.

58 Ibid. (suspension points in original). Céline Surprenant has discussed the influence of Darwin's *EE* in the work of French novelist Marcel Proust (1871–1922). Similarly, she points out that Proust had others collecting photographs of people's faces and their expressions, so as to create the characters in his novel *In Search of Lost Time* (1913-1927). Surprenant, "Darwin and Proust," 449.

59 Papa, "Η αλληλογραφία μου," 167–168 n5.

Xenopoulos was criticized by Greek critics such as Alkis Thrylos, who did not understand his use of literary physiognomy. These critics considered Xenopoulos' characters to be superficial, and could not appreciate the rationale behind this technique. Xenopoulos responded to a critique by Alkis Thrylos who maintained that Xenopoulos tended to "paint the surface and not below"[60] in his works. Xenopoulos argued: "The best works, of both my own and everyone else's, are those which have the least mistakes; those, that is, which copy the surface in a more faithful and lifelike manner; because they are the more profound. Knowing this, I always paint with a model. That is how I painted Nitsa Gazelis in *The Three-Sided Woman*."[61]

The following section analyzes in Darwinian terms Xenopoulos' use of physiognomy in *Rich and poor* in relation to Popos' theory of the eyes, discussing issues of mind and body, class, and race. Furthermore, an exploration of the "scientific" socialism, the nature-nurture question, and the Galtonian issue of genius and hard inheritance will also be raised in *Rich and poor*. The aim, once again, is to demonstrate that Xenopoulos exemplified a period in Greece that actively embraced intellectually Darwinian and post-Darwinian ideas in a number of fields.

The eyes—mirror of the soul?

Early in the novel *Rich and poor*, the narrator alerts the reader to Popos' knowledge of French rationalist philosopher and mathematician René Descartes' *Discourse on the Method of Rightly Conducting One's Reason and Searching for Truth in the Sciences* (*Discours de la méthode pour bien conduire sa raison, et chercher la vérité dans les sciences*) (1637). Also, the narrator notes that Popos had begun to read Darwin's *OS*: "So that you can get an idea—at that time he had finished the *Discourse on Method* by Descartes and he had started *On the Origin of Species* by Darwin."[62] The mentioning of Descartes and Darwin together provides a definite association and sets the perspective he utilizes in his novel. Both compared human and animal abilities. Descartes believed there was a separation of mind and body in humans who,

60 Here Xenopoulos is indicating what Alkis has said about his writing. See Xenopoulos, "Η *Τρίμορφη γυναίκα*, ο φεμινισμός και ο ανθρωπισμός," 3.

61 Ibid.

62 Xenopoulos, *Πλούσιοι κα φτωχοί*, 35.

he said, acted out of their own free will, and a separation between humans and animals that acted only out of instinct in a mechanical manner. From this followed his Cartesian dualistic approach to the human mind (which included the soul) and body (which included the brain); hence the mind was considered autonomous from the body. That said though, Descartes did realize that mind and body do not function entirely independently but do interact with each other. He indicated, for instance, that the mind is able to send messages to the brain, which controls muscular movements in the face.

Darwin closed the gap between human and animal by situating them in the same evolutionary tree. Variation occurred in the mental attributes of each species. Those with advantageous attributes would survive the struggle for existence via natural selection. In terms of mental traits, Darwin pointed out that there was no real separation between humans and animals. In other words, all human traits were present in animals, particularly the primates, in some rudimentary or complex form. It is worth noting also that Descartes believed that all humans through volition could apply themselves to achieve high intelligence. This went counter to Darwin's work, which asserted variations in mental ability between humans.

Popos associated his lower class origins with his bloodlines ("το σόι"/"η ράτσα") and, while he aspired to become upwardly mobile, he was unable to beat the class system. What results is angst and frustration, especially when he wants to marry the upper-class Klementina and, due to his lowly origins, her brother Antonis and father, old Mr. Roukalis, disapprove. Popos' obsession with his bloodlines and class penetrates the novel, triggering a life-long search for answers to these perceived injustices. Even outside Greece, this storyline was common. Xenopoulos' novel alludes to the French novel with the same name, *Riche et pauvre* (1836) by Emile Souvestre.[63] Thomas Hardy's novel *A Pair of Blue Eyes* (1873 in book form) follows a similar theme.

Although the common literal meaning of the word "ράτσα" is race, Xenopoulos will use it in reference to humans or lower animals, to mean a "species" or "race" or a specific group such as a "stock," depending on the context in which it is used. As we know, Darwin makes no specific reference to humans in the *OS*. He spoke collectively of all the animal species, implying

63 Farinou-Malamatari, "Το παρα-κείμενο της πεζογραφίας," 392–393. For further on *Riche et pauvre* see also Hemmings and Carsaniga, *Age of Realism*, 144–145. Here it is viewed as a "moralistic social novel." It is about the lives of two lawyers, one from an aristocratic and the other from a humble background.

that the human species was included. And so, the complete title of his book would imply that species and race are interchangeable for both humans and lower animals—that is, *The Origin of Species by Means of Natural Selection or the Preservation of Favoured Races in the Struggle for Life*. Further to this, Darwin in his *DM* used the terms "race" and "species" and "sub-species" according to the topic of his discussion, often interchangeably. For example, when he discussed the "variability of body and mind in man" he stated: "the present discussion ... bears on the origin of the different races or species of mankind, whichever term may be preferred."[64]

The problem of speciation resulted in observations that placed humans' evolutionary ancestors under the same umbrella as certain contemporary ethnic races. Darwin suggested that classification did not matter. He pointed out that "it is almost a matter of indifference whether the so-called races of man are thus designated, or are ranked as species or sub-species; but the latter term appears the most appropriate."[65] Darwin's writing on the races of man would later be misconstrued to sanction racism. Darwin did however specifically contribute to the discourse against racism in his *EE*. He asserted that he had shown the universality of expressions throughout humanity; and so from this one could deduce that "the several races [of man] descended from a single parent-stock, which must have been almost completely human in structure, and to a large extent in mind, before the period at which the races diverged from each other."[66]

The social/biological variations of meaning of "race" here used are a common feature in Darwinian creative writing and in the corresponding critical discourse. It reflects a biological approach that was and is still taken when examining a society or nation. Its use by Xenopoulos in this way should alert us to a sociobiological reading of the novel. Xenopoulos uses the word "ράτσα" for "race" loosely. He uses it interchangeably in the following with "σόι" (which means clan, family tree, family, descent, lineage, and even race): "So the family tree, the race, with its eyes, the Roukalis eyes ..."[67] Further to this, when Popos says "as there are races [breeds] of animals, there could be races of man," he reveals a biological/anthropo-

64 Darwin, *DM*, 1:108.

65 Ibid., 235.

66 Darwin, *EE*, 359.

67 Xenopoulos, *Πλούσιοι και φτωχοί*, 42.

logical emphasis on the word, maybe even a zoological meaning, by placing humans under the same umbrella as animals. The word "race" in the context used by Xenopoulos in *Rich and poor* has a racial connotation when Popos uses it on several occasions in relation to the saying "Jews have no dealings with the Samaritans."[68] This is taken from John 4:9. Generally speaking, in biblical times the Samaritans were discriminated against by the Jews. With respect to anthropology in Greece, Sevasti Trubeta has outlined the polysemy of the word *fili* ("φυλή")—that is, "race"—and its various meanings in religious, national, racial, biopolitical, or eugenic discourses;[69] in fact, the word "φυλή" also has links to "ράτσα" and "σόι," the words used by Xenopoulos in this novel.

Xenopoulos, via the narrator, takes the reader on an exploratory journey through Popos' life. It is reminiscent of Darwin's investigations in the *OS*. Here Darwin confirmed his hypothesis on natural selection with the observation of the animal species, which he later extended to humans in the *DM*. In what appears to be an attempt to imitate Darwin, Popos' life observations are carried out with (by his criteria) scientific rigor, and interpreted according to the theory of natural selection.

Popos' observations are elaborate and also often imaginative and subjective. From the start, the narrator is skeptical of Popos' observations, as he connects eye physiognomy with heritable traits of those around him. The narrator prejudices the reader's thinking by emphasizing that Popos knew something of Darwin's theory of natural selection. In other words, Popos' reading of Darwin was partial, which would lead him to misinterpretations. However, the narrator is not really impressed with Popos' level of such knowledge, or with his intellectual ability: "We saw that the 'bush philosopher' had read quite a bit of Darwin. Of the theory of 'natural selection' he knew bits and pieces."[70]

With each new observation of the eyes, Popos accumulates information for his theory. The journey begins when Popos, while still a student, describes Klementina's appearance. It is the description of the eyes that dominates: "She showed a splendid chestnut little head, a face with rosy cheeks, with vivid red laughing lips, and with two eyes, jet-black and bril-

68 Ibid.
69 Trubeta, *Physical Anthropology*, 147–286.
70 Xenopoulos, *Πλούσιοι κα φτωχοί*, 43.

liant—indistinguishable from her brother's—which looked at you as if to devour you ..."[71]

In *A Pair of Blue Eyes,* Hardy, the physiognomist, describes the aristocratic Elfride Swancourt, but again it is the eyes that dominate and communicate her inner being:

> One point in her, however, you did notice: that was her eyes. In them was seen a sublimation of all of her: it was not necessary to look further: there she lived.
> These eyes were blue; blue as autumn distance—blue as the blue we see between the retreating mouldings of hills and woody slopes on a sunny September morning. A misty and shady blue that had no beginning or surface, and was looked *into* rather than *at.*[72]

Each time Popos makes a new observation, the reader is invited to a scientific world of observation followed by questioning, then speculation, often in the form of theorizing. So it is with Antonis' eyes:

> And he saw two jet-black eyes, which although they were fixed forward in space, would turn suddenly towards the sea and shoot lightning bolts.
> "Strange eyes! ..." Popos thought. "Klementina's eyes ... They frighten you but you still love them ..."[73]

In the following passage Popos admits his preoccupation. The reader must be reminded that Popos at this stage has not read Darwin, as he is still at school.

> "Just now I was looking at your eyes ... indistinguishable from Klementina's eyes ... Tell me, where did you get them from? Where did you inherit them from? They are neither your mother's nor your father's."
> "You're mistaken!" answered Antonis. "They are Roukalis eyes. My old man has them the same. Didn't you ever notice?"
> Popos was intrigued. Did old Roukalis have the same eyes as Antonis and Klementina? But his were small, half-closed, wrinkled, dull, red, maybe even sticky ... He never managed to see them properly as they were so deeply sunken![74]

---

71 Ibid., 20.
72 Hardy, *Pair of Blue Eyes,* 1–2 (italics in original).
73 Xenopoulos, *Πλούσιοι και φτωχοί,* 38 (suspension points in original).
74 Ibid., 39 (suspension points in original).

In the same conversation, Antonis asks Popos what he thinks about the Roukalis eyes. The effect that Antonis' eyes have on Popos starts to intensify. Popos' involuntary lowering of the eyes and his whispering voice is enough to show that Antonis has achieved a psychological advantage over Popos:

> "I don't know," said Popos. "Nice eyes. Lively, very lively ..."
> "They emit fire, don't they?" said Antonis goggling his eyes in a showy manner.
> "Yes," whispered Popos and, unintentionally, he lowered his own.
> Truly, what was exceptional about those Roukalis eyes? Why did they leave an impression on him? And there were many others which were black, large, luminous, and lively. They weren't, however, the same! He didn't feel, as with those, a vague fear and an unbeaten fascination ...[75]

A few days later Popos bumps into Antonis' father, whose eyes appear to be different to the way they usually look. The following describes Popos' reaction to this: "In the beginning he felt something like fear. The sensation didn't last of course so much as a second, but it was the same as the sensation that he had once when he suddenly found himself in front of the eyes of a hawk. Equally wild, equally strong; those eyes also looked in the same way as if to devour you. And the first sensation of fear gave way to a big surprise when he did this—in the duration of the same second—and he saw that those eyes belonged to old Roukalis!"[76]

At this point, Popos compares old Roukalis' eyes to those of a hawk. The literary physiognomy relays an instant visual message to the reader. The analogy between human and predatory bird functions to reveal the personality of Mr. Roukalis as "wild," "strong," and "predatory." This is typical of a comparative examination of human and eagle physiognomies, and in Darwinian terms it also hints at humanity's animal link to nature. When old Roukalis sees Popos, his eyes revert to the aged expression that Popos was familiar with: "But—a strange thing! As soon as the old man set eyes on the son of Dagatoras, he recovered from his preoccupied state; he put aside his thoughts and then his eyes moved, became smaller, sank, became tamer,

75 Ibid. (suspension points in original)
76 Ibid., 40.

and were extinguished. And they again became the old, the wretched, the rheumy eyes of the old man; those which Popos would not have found possible to imagine as the same as Klementina's and Antonis' eyes!"[77] The passage shows that Xenopoulos was not just dealing with a fixed or permanent physiognomy, but was also exploring expressions which are the central idea of Darwin's *EE*. The significance of Roukalis' "reversion" to a hawk-like expression and of his return to a human one draws attention to what Darwin argued as "man's bestial nature," due to evolution. An example of this is well described in the *EE*: "The expression here considered, whether that of a playful sneer or ferocious snarl, is one of the most curious which occurs in man. It reveals his animal descent; for no one, even if rolling on the ground in a deadly grapple with the enemy, and attempting to bite him, would try to use his canine teeth, more than his other teeth."[78]

Is Popos seeing things?

> But no, impossible! Indeed, truly, he had seen for a few moments, in place of the eyes of old Roukalis, the eyes of Klementina and Antonis; which means that they were the same, the same strange Roukalis eyes, to use his friend's words. His wretched old age had changed them for the old man; but here were some moments where they reverted like those of the young, that is, large, illuminated, strong, wild, greedy, and threatening.
>
> But even sweet?—Why not? The moment when he saw old Roukalis, his rejuvenated eyes did not of course have any sweetness about them. And the youth remained with the fear which was caused by their ferocity and their greediness. On odd occasions, however, there could have been moments when the old man's eyes had even taken on the sweetness—the same sweetness as Klementina's and Antonis' eyes, which had bewitched him so many times ...[79]

Old Roukalis' expression appears to have an "adaptive" quality, dependent on the situation with which he is confronted. His reversion to the human expression reflects that he did not see Popos as a threat. Popos realizes that the Roukalis family, in particular old Roukalis, his son Antonis, and daughter Klementina, has the same eyes:

77 Ibid.

78 Darwin, *EE*, 251.

79 Xenopoulos, *Πλούσιοι και φτωχοί*, 40.

> So these were the Roukalis eyes! This was their peculiar characteristic: the gaze which looked at you as if to devour you—the sweetness which later removed the fear and filled you with trust. And for Popos to understand this, he had to see for a moment old Roukalis' eyes as he saw them without their sweetness, that is, only with that ferocity of the hawk, the same however and indistinguishable—in color, in the way they opened, the shape, in everything—as his children's eyes! [...]
> He would try to remember if Aunt Eugenia Farena's eyes were like that. Wasn't she related to the Roukalis family? ... He would visualize them and—how strange!—he found them to be the same as they were siblings. [...] Oh, poor Eugenia was always placid, always reserved and sweet. You would never see her looking at you as if to devour you. But still, how do you know? She too may have had her moments ...
> That is the clan then, as the Roukalis family would say. The race, as his father had said ... could it be, really, that its hallmark, its stamp was those strange eyes?[80]

In this passage another eye trait shared by the Roukalis family is the enticing and hypnotic glance, which manages to remove any fear from its trusting victim before it pounces. Aunt Eugenia could also have a "dark side," a reversionary state that could reveal the beast in her. Popos' observations lead him to constant questioning and theorizing. Were these eyes the "family" characteristic? He applies his theory to other members of the extended family, for confirmation of its validity. Popos contemplates the eyes of the relatives of the Roukalis family: "He didn't know the extent of the kinship which connected each of those people with the Roukalis and Faraos families; nor which were related through his father and which were related through his mother. ... And yet, among the multitude of relatives Popos thought now that he saw, here and there, shining eyes identical to those of the Roukalis family."[81]

He elaborates on one particular relative of the Roukalis family named Marketis, a banker: "He had eyes the same as Antonis'; even more lively, more intense, more fierce. Many times when they would encounter each other on the road and would look at each other for a moment, Popos could

80 Ibid., 40–41 (suspension points in original).
81 Ibid., 41.

not tolerate his gaze. It humiliated him; it caused him to feel weak, small, almost to be ashamed. Later, when Marketis would go past though, his sense of pride would be provoked and it made him think: 'Don't you worry, I will be better than you! ...'"[82]

The above passage reveals the effect that the Marketis' eyes have on Popos. Popos thinks of other relatives with the Roukalis eyes, and though the narrator shows some ambivalence towards Popos' ideas, he does relay Popos' hypotheses:

> It could now have even been his imagination; but nobody could deny that all those eyes, even if they didn't look perfectly alike, had many common and idiosyncratic characteristics. And which was the most outstanding one, or better still the most common one?
> Popos thought about that too—having always a tendency to generalize; and he found that the common characteristic was *the ferocity*. All those eyes had something which reminded one of a hawk, the bird of prey, the predatory bird. In others—like the eyes of the banker Marketis—this *expression* prevailed; in others—as with Klementina's eyes—this appeared less. All of them though had it. Above all, it distinguished them. And Popos remembered very well all the times when one such *expression* of his friend Antonis' eyes would upset him. So the clan, the race, with its eyes, the Roukalis eyes ... [83]

In the above passage one can start to see how the eye physiognomy is enmeshed with the expression created. This will be seen repeatedly throughout the analysis. The eye expression of "ferocity," possessed by the Roukalis family, also alludes to a quality of savageness, which again the narrator qualifies as belonging to predatory birds such as the hawk. He clearly aligns the eyes, expression, and nature of humans with that of the bird. The last line in the above quotation expresses a reflective Popos, who is attempting to understand what might be the connection between the family and the eyes: "There were moments when Popos wanted to laugh. Race, so be it. The way there are races [breeds] of animals, maybe there are races of man. And it certainly must be so. But what significance can a race have? How can some characteristics—eyes, nose, hair—so radically separate the Roukalis fam-

82 Ibid.
83 Ibid., 42 (my italics).

ily, for instance, from the Dagatoras family, so that the first would say 'Oh no! Jews have no dealings with Samaritans.' But that is laughable! Popos would laugh ..."[84]

Popos' aligns humans with the animal species, giving a biological perspective to races of humanity. Both Popos and the narrator realize in disbelief that a theory is developing, according to which not only do the rules of nature apply to species of animals, but these same rules apply to races/species of humans. As discussed earlier, Darwin commonly used the words species and races interchangeably. At the time this concept prevailed due to the absence of genetic science which today defines humanity as one species. The implication that different ethnic groups or races of humans could also be different species created a biological hierarchy. Little did Darwin realize that this hierarchy would be misconstrued for at least up to the 1940s, and aid in the perpetuation of racial discrimination. This appears to be the case in the above passage, where Popos does connect familial physical traits with ethnic traits, and declares the biblical "Jews have no dealings with Samaritans," which I mentioned earlier. Popos extends the theory to other facial features. As noted already, during biblical times Jews discriminated against the Samaritans, and it is possible that what Xenopoulos is saying is that this occurred despite the well-known parable of the "Good Samaritan"[85] who helped a man in distress. It is likely that Xenopoulos viewed the Jews as symbolizing the Roukalis family, and the Samaritans as the Dagatoras family.

Xenopoulos drew on quotes or ideas from books and included them in his novels as a basis for Popos' theorizing. Although he has not named his source, this phenomenon is exhibited in the following passage. The narrator here is continuing to refer to Popos and his application of his physiognomical theory of the eyes: "There were moments when he would take it very, very seriously; because he remembered something which he had read in a book, that '*the soul makes its garment,*' and he concluded then that mental differences or different mental constitutions could actually be insurmountable obstacles among the races of humanity. And in such moments, he contemplated the eyes of the Dagatoras family and he would look at his own in the mirror."[86]

84 Ibid.

85 The parable is found in Luke 10:30–37.

86 Xenopoulos, *Πλούσιοι και φτωχοί*, 42 (my italics).

In his novels Xenopoulos applied the approach of reading the outer being to obtain an understanding of the inner, the soul, and for this he was criticized for being superficial.[87] Also he appears to have believed that variations in mental traits could be highly disadvantageous in certain races of humans. His physiognomical approach was not uncommon in the literary world in the late 1890s and early 1900s. Joseph Conrad in his novel *Victory* (1915) referring to a character indicated that "the body is the unalterable mask of the soul."[88] Daniel Pick writes that Conrad used this type of physiognomy as a "powerful reach of heredity in the determination of the individual and the race" in novels such as *Victory*.[89] The year of publication of *Victory* is around the period that Xenopoulos would have been writing his *Rich and poor*. Pick indicates that around the decades of 1900 there was "a deep concern with the mysterious relation between the body and the determination of character," found in many novels.[90]

In the following passage Xenopoulos compares Popos' eyes with those of his family:

> What gentle, sweet, weak eyes! ... Light chestnut, not small, but not large, with short eyelashes and unemphatic eyebrows. Regarding their *expression* he himself could not of course judge well: he imagined though that they would never look at someone as if to devour them ... He imagined also that on many occasions they would have taken on an *expression* which would have moved people to feel pity. [...] Such were Popos' eyes. But these, more or less, were also the eyes of his father, and his uncle Dionysakis. Gentle, sweet, weak. The eyes of good people, quiet, patient, submissive—Submissive to what? To a sense of duty, Popos would say. And with this he covered everything ... He tried also to remember his relatives, all those little, humble, poor folk who were proud of their relationship with Stathakis and Dionysakis. He didn't see them much, he didn't know them well, but he remembered their eyes—all the same: lambs' eyes![91]

87 It is difficult to find the whereabouts of the saying that he quotes in the above passage "the soul makes its garment," which he read in some book. There are numerous sayings which relate to this issue in many languages, and it is likely that Xenopoulos utilized sayings from other languages, which he paraphrased. For instance, as will be seen further in this section, Popos ponders on some Latin saying, though he does not mention what it is.

88 Conrad, *Victory: An Island Tale*, 86, quoted in Pick, *Faces of Degeneration*, 162.

89 Ibid., 162.

90 Ibid. He refers to Gissing's *The Nether World* (1889) and *Demos* (1886).

91 Xenopoulos, *Πλούσιοι και φτωχοί*, 42–43 (my italics; suspension points in original).

Popos is surprised by this and makes his own deductions:

> But that is unbelievable!
> So the lambs, God's sheep, are they not a separate race from the hawks, the birds of prey? And couldn't he [Popos] suppose that some instinct made the Roukalis family reject this mixing, which would not have been an advantage, not of course to the race of lambs, but to the race of birds of prey?[92]

The physiognomy and expression of the lambs' eyes can be contrasted to those of the hawks'. As noted earlier, at this point the narrator ironically comments on Popos' knowledge of Darwin's theory of natural selection, which he infers is the explanation for the advantage birds of prey have over the lambs and, by analogy, the social advantage the Roukalis family has over the Dagatoras family. The narrator refers to Popos' knowledge of natural selection and then comments that Popos knew that "the emotion love is a 'lure of Nature,' which looks wholly to its own purposes independent of the will of humanity" (43). It is likely this is in reference to sexual selection, a concept Darwin propounded in the *DM*. These theories, though creating doubt regarding his relationship with Klementina, were often dispelled in his mind because he was sure of her love for him: "Klementina loved him. He could see it clearly in her Roukalis eyes. And then his theories would appear laughable to him; the similarities and the differences in the eyes were imaginary. The clans and races were just words and hot air."[93]

Surprisingly, Popos sees the characteristics of the Roukalis eyes as a feature in others, such as the law student and wealthy friend Kleomenis (or Menis) Manias: "And his eyes—oh! His eyes were something strange: Even though they were chestnut and opened differently, they reminded me very much of Klementina's."[94] The Roukalis trait, though clearly an inherited trait, is no longer exclusive to the Roukalis kin. The previous passage highlights that the characteristics associated with these eyes can belong to others—that is, those who are rich and successful. (Though Menis is not of aristocratic descent, he is wealthy and belongs to the "new upper class.") When Menis and Antonis first meet, much to Popos' shock they get on very well, and Popos hypothesizes that it is because they share the same characteristic expression found in the

92 Ibid., 43.
93 Ibid., 44.
94 Ibid., 54.

Roukalis eyes. Xenopoulos will make a clear association between the heritable trait of the "Roukalis" eyes and the psychological connection which occurs between Menis and Antonis: "The eyes of them both glistened with the same light. They had the same *expression* and, more or less, the same opening, the same curve. ... It appears that it was the first time that he saw the Athenian's eyes as having such a striking version of the 'Roukalis' *expression*."[95] Menis was in no way related to Antonis, but Popos extrapolates a different commonality: "'So there is a kinship of another type,' Popos thought, 'let's call it a *mental kinship*.' And the fact here was not only the resemblance of their eyes; it was this mutual eagerness and willingness that they showed as soon as they met—that connection between the two of them so like those which is made by the always inexplicable, mysterious *chemical kinship*."[96]

The kinship was psychological, indicating that the Roukalis expression was aligned with a certain internal characteristic; that is, an eager and willing nature. The "always inexplicable, mysterious chemical kinship" refers to the genetic component common to members in a kinship. At this point, the narrator expresses his view of Popos, referring to his "fiery imagination" as Popos applies his hypothesis of the eyes to the remainder of Menis' family and relatives (104). The narrator presents Popos' ideas on Darwinism in a manner which reduces them to inaccurate, imaginative generalizations. He diminishes Popos' credibility by stating that Popos was not fully informed on Darwin's theories; or that he had forgotten them; or, as in the following passage, by signaling that he was heavily intoxicated while thinking about them. These unreliable states of Popos mitigate the impact of any Darwinian message on the reader:

> He pondered over ... Darwin's epilogue in his book *On the Origin of Species*. He was trying to remember it all. But there was one part which he could not remember at all. The train of thought was cut suddenly and it stopped in the dark.
> Then he pondered on other things, varying and different. His twenty-five drachmas, Mr. John's daughter Aglaia, an equation, a Latin proverb, uncle Dionysakis, and the clothes which he would take to the laundry. And he would return to Darwin's epilogue.[97]

95 Ibid., 103 (my italics).
96 Ibid., 103–104 (my italics).
97 Ibid., 128–129 (suspension points in original).

The contrast of the miscellaneous "other things" against the twice-mentioned final section of the *OS* manages to highlight the latter. What is the narrator trying to tell the reader? Why is Popos so obsessed with Darwin's *OS*? Does this allude to an obsession Xenopoulos may have had for Darwinism? Are these thoughts which Xenopoulos had when he was a university student studying mathematics and the sciences? As shown in chapter 3, Xenopoulos was well read on Darwin and had published significantly on him in the *Children's Guidance* magazine, writing on evolutionary theory and various applications of it.[98] Perhaps Xenopoulos wanted the reader to consult the "concluding remarks" of Darwin's last chapter, "Recapitulation and Conclusion," which refer to natural selection: "I have now recapitulated the chief facts and considerations which have thoroughly convinced me that species have changed and are still slowly changing by the preservation and accumulation of successive slight favourable variations."[99] "And as natural selection works solely by and for the good of each being, all corporeal and mental endowments will tend to progress towards perfection."[100]

Throughout the novel, it seems that biological evolution of humans (Darwinian) is aligned with a social evolution. Antonis maintains: "I don't categorize humanity except as human and nonhuman. There are poor people who deserve as much as rich people."[101] Antonis considers the Manias family as human because they have money and social standing, and the Venetis family as nonhuman because, though well-off, they are of a lower class. Antonis implies an evolutionary argument for this, emphasizing that the Venetis family needs time "to evolve" in social standing. The evolutionary states seem to be dependent on one's social class:[102] "Well, that is what the difference is! What did you say, after a few years?—which means that the Manias family are people who are already made, whereas the Venetis family are people who are being made constantly. They are in the making, as you bush philosophers would say."[103]

98 In his 1923 "Athenian Letter" entitled "Science and God" ("Η επιστήμη κι ο Θεός"), he discussed Darwin's epilogue as having had an enormous influence on him. In the "Letter" Xenopoulos wrote: "It left me with the greatest impression because I realized immediately its great truth, and because much more than the God which I knew, the God which created animals and plants one by one, I admired, like Darwin; the new God-Creator, great, who had the power to create everything from one!" See Xenopoulos, "Η επιστήμη κι ο Θεός",124.

99 Darwin, *OS*, 452.

100 Ibid., 459.

101 Xenopoulos, *Πλούσιοι και φτωχοί*, 146.

102 This social Darwinian idea is also demonstrated by Tereza in Xenopoulos' *Tereza Varma-Dacosta*.

103 Xenopoulos, *Πλούσιοι και φτωχοί*, 147.

Unlike Antonis, Popos was spendthrift and was exploited by others when it came to money. He observed the differences between himself and Antonis in terms of mental traits, and noted a similar contrast with the eyes:

> How hopelessly he managed things, and how well, how skillfully did Antonis handle things! Strange how two people can differ so much! What a huge radical difference!
> And once again the Roukalis eyes beamed in front of him. How different from his own and from those of his family! With time passing, *the occurrences accumulated* so much so that he became possessed with the old idea.[104]

With each observation of the eyes, Popos accumulated evidence to support his theory. He then imagined the faces of Klementina and Menis: "And—something unbelievable—although everything changed—nose, mouth, cheeks, hair—the eyes in both pictures remained the same, the same! But the most incredible thing was this: Later, with the same eyes, indistinguishable, he visualized the face of Mr. John Venetis' daughter Aglaia!"[105]

Popos asked Antonis for Klementina's hand in marriage with the following reaction: "Antonis remained open-mouthed for a few moments; and his eyes, the Roukalis eyes, were riveted on Popos Dagatoras' eyes. Then he laughed enigmatically and he replied slowly, emphasizing each word: 'Yes! When you ask me for her ... I will give her to you ... Yes ... you have my word!' And he said it so softly that his voice died away and the last phrase could just be heard."[106] Antonis is furious when Popos suggests that he should marry Aglaia, the daughter of Mr. Venetis, the wealthy milk vendor. Antonis considered the Venetis family socially lower, even though they had money. Popos pondered on why Antonis, despite his general unattractiveness and sloppy dress, was so attractive to females. He will conclude again that it was because of the "Roukalis" eyes. Potentially, Darwin's theory of sexual selection is observed here. Antonis' eyes also have an advantage over Popos' because they attract all the females. The narrator sees that Antonis' "Roukalis" eyes are favored, and queries:

104 Ibid., 130–131 (my italics).
105 Ibid., 131.
106 Ibid., 141–142 (suspension points in original).

> So what did he have? What was it about him that they envied? What was it about him that they loved? Because really good looking he wasn't. Popos was much better-looking.
> Only his eyes which were very lively, very expressive, very intelligent could attract. Maybe his color, deep, dark, and warm. ... With all his wealth, his dress continued to be the same. You might say it was due to his constitution to have the same clothes made by all the tailors in the world! Oh, and what about those ties, the narrow ones, the wretched ones with *the loud and vulgar colors*?[107]

Could this be a "primal" attention-drawing mechanism caused by Antonis' bright and vulgar ties? This could be likened to the courtship habits, particularly of birds, which Darwin refers to in the *DM*, and which are based on sexual selection. Popos concludes that the attraction these girls have for Antonis is instinctive and involuntary: "a real liking, without a vested interest, without any other calculation but only that most wise calculation which instinct does unconsciously."[108]

The mysterious law in the following passage has its hereditary roots in the origins of the objectified mechanism of Darwin's evolutionary process—that is, the law of natural selection. Popos wonders: "Was it a coincidence or some mysterious Law, relentless, inescapable, which hindered him from prospering, from finding relief, from living? ..."[109]

Antonis will explain in a practical manner all Popos' reservations in relation to his unwieldy and obscure theories and laws. He claimed that Popos had a characteristic of complacency, which rendered him incapable of progressing (184). While Xenopoulos utilized the concept of physiognomy as a literary technique, it was also a "diagnostic" one. He made his character Popos perceive that the eyes were an indicator of genetic difference, reflecting the psychology of the two social classes. Popos would reflect on certain physical differences in people's eyes which he saw as mirroring certain types of personalities and behavior; consequently, he supposed the eye differences determined what made these individuals rich or poor. Concurrently, Xenopoulos utilized the theory of natural selection to indicate how the Roukalis family (the strong) prevailed economically over the Dagato-

107 Ibid., 153 (my italics).
108 Ibid.
109 Ibid., 182.

ras family (the weak). The power game was psychological, though it was superficially manifested through the eyes and facial expression—that is, the physiognomy. The power game was the evolutionary struggle, with the strong surviving due to some heritable favorable trait, which was reflected in the "Roukalis" eyes.

Popos lacked survival traits because he could not adapt to the social and natural laws of competition. He was passive, impractical, and always theorizing.[110] Antonis, on the other hand, was practical and hardworking, traits which are conducive to survival. Ultimately, Popos would envisage his fate as biologically predetermined. Despite his early favorable portrayal, he also possessed an "injurious" trait of inherent morality. Nothing could save him, not even socialism. Antonis would also recognize Popos' fate.

Popos' biological theory of the eyes runs parallel with the narrator's social theory regarding the immobility of the social classes. This is exemplified in the case of Mathios Adamis, a wealthy industrialist who Popos thought was related to him. On realizing the relationship was not clear and most likely nonexistent, the narrator and Popos develop their theories. Interestingly, it is apparently the narrator who draws attention to Mathios' eyes that allegedly "reveal" his Chios background. Till now it has been Popos who has made such connections between the eye physiognomy and one's personality. So that while in the past the narrator has distanced himself from Popos' theories, he is now making the same type of associations. This tends to present the narrator as unreliable, which is the case, as noted by Farinou-Malamatari, with a number of Xenopoulos' novels.[111] By doing this, she argues, Xenopoulos is able to exploit any idea or view without fully adopting it.[112] The narrator examines the situation, observing Adamis: "In reality, even Mathios Adamis' eyes appeared different to those of Popos. They were the eyes of people who are or are becoming rich. They had that particular expressionless quality that even the Roukalis, Manias, and Venetis eyes had, and so many other well-heeled people had who he had met in his life and had watched. That, you see, had become a true psychosis."[113] When

110 Ibid., 118, 124, 130, 157, 159, 184, 192, 201, 211, 299.

111 Farinou-Malamatari, "Γρηγόριος Ξενόπουλος," 310. She adds that this phenomenon occurs with first person narrators in some of Xenopoulos' novels. In *Rich and poor* the narrator is speaking in the third person.

112 Ibid.

113 Xenopoulos, *Πλούσιοι και φτωχοί*, 190–191.

Popos decided to ask Harisis if one can tell whether someone belongs to the rich class or the poor by looking at their eyes, Harisis remarked:

> "From the eyes?" he whispered. "I don't believe it. Eyes are deceiving."
> But Popos did not believe that eyes can be deceiving! He did not dare insist and he asked something else.[114]

Popos is attracted to the socialist theory, hoping it will help him financially. However, regarding the social Darwinian theory of the American (who is possibly William Sumner, as I mentioned earlier) which, according to Harisis, destines those born poor to stay poor, Popos will feel that he is unable to escape this fate. Out of work and poverty stricken, he loses interest in his theories and socialism (297). He becomes ill and dies after ending up in jail for publishing an article against the King. On his death bed his hallucinations include the following scenario, one that had also plagued him in the past (41):

> He found himself completely alone, and all he saw were the heavens, which instead of stars contained eyes. How alike they looked! And he recognized them straight away. They were all the Roukalis eyes which he had come across in his life. He concentrated one-by-one on each pair and he tried to remember whose they were. And he worked it out.
> There are Antonis' eyes! There are old Roukalis'! There are Klementina's! There are also Meni Manias', indistinguishable! There are Mr. John Venetis'! There are Mathios Adamis'! ... How they are looking at him! But what did they want from him? To devour him? Oh, but a Popos Dagatoras cannot be destroyed so easily!
> Nevertheless he became frightened. He lowered his head so that he would not see them, and again the eyes would draw him like magnets ... That became like torture. Popos started to cry out: "But what did I do to you? What did I do to you? ..."[115]

On the surface, this hallucination is a culmination of Popos' obsession associated with his eye theory. At a deeper level, it reflects an individual who even on his deathbed was unable to come to terms with the "injustices" of

114 Ibid., 205.
115 Ibid., 315 (suspension points in original).

the inequality of the classes. The hallucination implies that he saw himself finally defeated by the strong upper class—that is, by those with the "Roukalis" eyes. The Darwinian element of natural selection has come into play with Popos, the weak being eliminated.[116]

On Harisis' quasi-scientific socialism

In his autobiography Xenopoulos mentions that his encounter with positivist philosopher and mathematician Giorgakis Hairetis and his followers during his university years had a great influence on the making of *Rich and poor.* According to Xenopoulos, Hairetis preached an academic type of utopian socialism, based on the ideas of Platon Drakoulis (1858–1934), who was one of the men dominating the socialist scene in Greece in the 1880s and 1890s. Hairetis and his "philosophical group" were the inspiration for the characters Leon Harisis and his followers in the novel.[117]

In reality, this early period of Greek socialism that Xenopoulos wrote about was a period of confusion in terms of its definition.[118] Although there was no element of revolutionary socialism, George Leon signifies that Drakoulis "ideologically ... failed to transcend his own eclectic brand of utopian socialism, which at times reflected an incoherent combination of Christian socialism, Fabianism, utopian socialism, and even theosophism with traces of Kropotkin and Tolstoi."[119] Furthermore, he argues that Drakoulis "reflected the eclectic confusionism characteristic of many intellectuals of his generation, yet he was able to escape the negativistic Nietzschean infatuation of many of his compatriots."[120]

In the novel, philosopher Leon Harisis comes into Popos' life at a time when he is vulnerable and searching for answers to his financial problems. Popos hears Harisis' version of the American sociologist's theory, describing what was perceived as a capitalist society at the time (197–199, 203–207). It

116 Around the turn of the twentieth century some activists were imprisoned for their socialist views and publications. These included Rokkos Choidas, who died in jail in 1890. Popos' death, claims Farinou-Malamatari, is reminiscent of Choidas'. See Farinou-Malamatari, "Γρηγόριος Ξενόπουλος," 320–321.

117 Xenopoulos, *Η ζωή μου*, 259–262.

118 Note that Enrico Ferri's book on socialism and evolutionary thinking was circulating in Greece at the time. See Ferri, *Socialism and Positive Science.* It was first published in Italian in 1894, in Greek in 1895, and in English in 1901.

119 Leon, *The Greek Socialist Movement*, 3.

120 Ibid., 4.

was presented as social Darwinism, which in its extreme form assumed that the lower classes were biologically inferior and that the unfavorable traits were biologically transmitted; also, it advocated a policy of no government aid to those who could not survive the severe evolutionary competition of society. This type of social Darwinism, really a form of Spencerism rather than Darwinism, was used to explain the economic competition and intervention by the state.[121] In the novel, this is the milieu that Harisis and his followers believed existed in Greece, convincing Popos of this as well.

American sociologist and economist William Graham Sumner spread the theory of social Darwinism to other American intellectuals in the late nineteenth and early twentieth centuries, including John Fiske and Benjamin Kidd.[122] Sumner played the most significant role in applying evolutionary theory to American economics, an example of a phenomenon which showed how ideologies such as capitalism used Darwinism to sanction their ideas. In the novel, in conversation with Popos, Harisis expounded the American's sociologist's theory:

> "There are the poor and the rich from birth. ... Rich and poor do not become that way, instead they are born that way"
> "Like the poets—"
> "Exactly! In other words, that humanity is divided into two races: the race of the rich and the race of the poor. If you are born in the second, whatever you do in your life, you will never be able to get into the first. And if you suddenly become rich because of some blind accident, it will be temporary; your wealth will be lost by your descendants if, of course, it is your own descendants that inherit your wealth. Also, if you are born in the first you will never cross to the second. And if you become poor, at some time, this will be temporary; you will become rich either you yourself or your descendants."[123]

Harisis' panacea for this social Darwinism is his quasi-socialism, which parodies the real and confused socialism at the time in Greece. The reader is first introduced to Harisis' version of scientific socialism that he became acquainted with in England:

121 Spengler, "Evolutionism in American Economics," 218.
122 Faris, "Evolution and American Sociology," 162–163.
123 Xenopoulos, *Πλούσιοι και φτωχοί*, 204.

> It was however a scientific socialism, as he called it, which was proven necessary like a mathematical theorem, and which had no connection with the charlatanism, according to Harisis, of some other ignorant and unread people who had appeared then in Greece.
> ... Leon Harisis prophesized only for a peaceful revolution. *Gradually*, with time, with teaching, the people would become enlightened, they would understand the advantage to them; and the rich, like the poor, would come to an understanding. They would decide on a change. They would legislate it, and in this way automatically, peacefully, the new society would emerge.[124]

Here, this gradual evolution is associated with the concept of socialism and is also a tenet of Fabianism. Xenopoulos continues to mock this socialism when Harisis adds his own theories to that of the American. Harisis' misguided views on being able to tell who will be rich and who will be poor just by looking at them lends little to the credibility of his socialist theory. He predicts that Popos belongs to the poor class, the reason being that he was drawn to Harisis and his followers who were poor (205). Harisis' followers are also parodied, portrayed as a group of "misfits" or "degenerates"; their association with this socialism only serves to undermine it (196). The depiction has physiognomic qualities—that is, they are depicted as odd-looking and eccentric, which reflects their odd and eccentric views. Popos questions Harisis about how socialism could bring equality to society if inequality was a permanent phenomenon. Harisis declares: "After I read the American's study, I relied more on my socialism. It is the only salvation. Equality in society will never happen. Always the rich by birth will have more than the poor by birth. Now though, the way society is, the strong race, boundless, unrestrained, oppresses the other, the weak. ... The strong one will always have something more. ... In other words, if this natural inequality did not exist, [that is] these two races of people, socialism would be redundant."[125]

Subsequently, Harisis did not actually provide a solution as to how this socialism will work. Popos tried to merge his theory about the eyes with Harisis' theory. The narrator discloses the following in relation to Popos: "It was a theory which suited his intellectual disposition and his current eco-

124 Ibid., 197 (my italics).
125 Ibid., 206–207.

nomic position. He knew that he would draw great comfort from the idea, from the hope, from the dream that in the future there would not be poor and rich; and that all those people with the Roukalis eyes and the predatory instincts would not be achieving any more in the world than the average people and those content with little, that is, those with the Dagatoras eyes ..."[126] Harisis refers to the poor: "It concerns us only that there are actually people like that whose mental constitution is hereditary, as there are others, the opposite, people, that is, who were born to be rich, even when they are ... poor!"[127] Popos was convinced of the American's theory. The narrator explains: "He was completely immersed in the theory. It seemed to him that with it, all the social phenomena, all the mysteries of the world could be explained" (228). The American's theory played on his mind constantly until his death (209).

The socialism that Xenopoulos refers to in *Rich and poor* appears to have had its origins in Britain at the end of the nineteenth century. It consisted of both a revolutionary type and a reformist type; and on certain points with regards to Darwinism, these two types were in agreement[128]—the latter being the more influential and the one relevant to the novel. This aside, when Darwin published his *OS*, Marx and Engels welcomed it because it provided a scientific validation for the concept of class struggle. In his speech at Marx's graveside, Engels drew an analogy between the history of life and human history by stating: "Just as Darwin discovered the law of development of organic nature, so Marx discovered the law of development of human history."[129]

Marx and Engels used these "connections" so that Darwinism would appear to scientifically sanction socialism. Although initially they recognized that natural selection could be an analogy in a capitalist society for individual competition, they did not actually utilize the concept in their works. Consequently, they did not see the struggle for existence as the product of some universal law of nature; instead, they saw it as a sociohistorical condition caused by the difficult class relationships produced by capitalism. Bowler argues that the relationship between Marxism and Darwinism "has

126 Ibid., 199.
127 Ibid., 203 (suspension points in original).
128 Jones, *Social Darwinism and English Thought*, 63.
129 Marx and Engels, *Selected Works in One Volume*, 428.

been one of suspicion arising inevitably out of their roots in the two opposing branches of nineteenth-century thought."[130] In *Rich and poor*, Xenopoulos portrayed socialism in this Darwinian manner. In the late 1800s this practice was very popular in the social sciences, and consequently was taken up by the literary world.

In *Rich and poor*, Xenopoulos implied a need for the "evolution" of the lower classes, so that they could be integrated into the higher classes. Is this biological evolution or is it a social evolution in the form of socialism? He approached the theory of scientific socialism in an evolutionary manner which, relevant to Darwin's theory of selection, utilized the phenomenon of gradualness.[131] This approach to scientific socialism, which Xenopoulos later believed to be a pipe dream anyway, is evidenced in his "Athenian Letters." Instances of these are the first passage below, from his 1909 "Letter" entitled "A response," where he refers to two types of socialism; and the second passage, from his 1921 "Letter" entitled "A little water in the wine,"[132] where he prescribes a slow, gradual change in society in order for it to be transformed, which alludes to evolution in nature:

> The scientific and the fake—the sensible and the insane. The first accepts that one day society will change its structure—as it has changed so many times—and has considered the method and the means of *gradual variation*. The second thinks that all these [changes] can happen at once, suddenly, even tomorrow by force.[133]
>
> It is predestined and unavoidable that society will change its structure. But the change should not be by force and a premature overthrow. *Slowly, little by little, gradually* ... Socialism is a nice theory. But its implementation is difficult and dangerous; and we must be careful so that we don't support with our innocent enthusiasm the schemes of the cunning and not contribute to a bad implementation of the theory.[134]

---

130 Bowler, *Evolution* (1989), 108.

131 As mentioned earlier, gradualness is an evolutionary concept also applied to Fabianism. Note that Henkin refers to gradualism philosophically: "the concept of gradual change, gradual progress, especially suited the British temper." Henkin, *Darwinism in the English Novel*, 143.

132 The title is translated literally but it would be best explained in English as "To pour oil on troubled waters."

133 Xenopoulos, "Εις απάντησιν," 263 (my italics).

134 Xenopoulos, "Λίγο νερό στο κρασί," 220 (my italics; suspension points in original). Note that he writes this "Letter" four years after the Russian Revolution.

In a 1916 "Letter" entitled "The Socialist"[135] he reveals he was no longer a socialist, which he later also confirmed in a 1929 "Letter" entitled "Optimism," where he writes: "Once ... I was a socialist, almost a Bolshevist. And with every opportunity which appeared I would speak to the children in that spirit. But with time, I saw that the dream of equality of humans is also chimerical, unrealistic; I have watered down the claims and now when the opportunity arises sometimes, I urge the children not to be led by empty talk."[136]

In *Rich and poor,* Popos perceived that this socialism would gradually and slowly produce a utopian state where the two classes or races of humans, as he saw them, would be the same. Xenopoulos represents this idea as a parody throughout the novel because in his own mind this utopian idea had become a pipe dream.

Popos' hard inheritance and Galton

Xenopoulos was fully informed regarding Darwin's views on inheritance and also those of Galton. Galton's views were extreme, in that he attributed nearly all traits to heredity and nothing to environment. Darwin did allow some environmental input, but believed that criminality, intelligence, and morality were hereditary traits. Throughout *Rich and poor,* Popos examines the nature versus nurture argument in relation to various traits in individuals. Of course he is told by Harisis that, in social Darwinian terms, social class is determined by heredity. Popos discusses with his mother about traits such as beauty being hereditary also. But his observation of the two boys Zeppos (Popos' landlady's son) and Kostakis (the son of the milkman Mr. Venetis) draws attention to the fact that the two boys come from the same social background (139). Instead, a slow and steady evolution of a Darwinian type is implied, with one boy taking an evolutionary progressive path and the other one a regressive path. Popos asks himself: "I am curious ... I would have liked to have known, after ten years, what would that Zeppos be in society and what would Mr. Venetis' son Kostakis be. The prototype of morality, the one, and the prototype of corruption, the other. And yet now they are two children nearly the same. Two students, equally well dressed,

135 Xenopoulos, "Ο Σοσιαλιστής," 175.
136 Xenopoulos, "Αισιοδοξία," 572.

who go to the same school and who appear to belong to the *same social class*. How will they gradually grow apart? What will become of them? Where will the one with his morality and the other one with his corruption end up?"[137]

As mentioned earlier, Popos maintained that all poets are born and not made; environment, it seems, has nothing to do with it (204). This raises the argument over the role played by heredity and environment in the development of abilities; in particular with gifted individuals, in this case, poets. Xenopoulos would have known of Galton's work on inheritance in geniuses.[138] Throughout his famous book *Hereditary Genius* Galton maintained that genius was a quality endowed by nature (inherited) and that it could not be acquired in one's lifetime. In his field studies he dedicated a chapter to poets.[139] Later, Darwin himself in his *DM* concurred with Galton's view, writing: "We now know through the admirable labours of Mr Galton that genius, which implies a wonderfully complex combination of high faculties, tends to be inherited."[140]

When Xenopoulos portrayed Popos with this attitude on poets, he was highlighting a view of hereditarianism among many intellectuals in Greece also. An example of this is seen in chapter 1, in the analysis of the views expounded by those intellectuals who wrote in the periodical *Grammata* between 1913 and 1915. In addition, someone like the poet Kostis Palamas, who was also an enthusiast of the sciences, would have had an appreciation of this view.[141] In her 2003 article, Ambatzopoulou points out Palamas' desire "to track down 'scientifically' the mystery of poetry and the genius of poets, which clearly is connected to pathology."[142] The scientific investigations on these matters were associated with post-Darwinian influences, such as Galton's emphasis on heredity as the primary factor involved in the genius of poets. In addition, intellectuals such as Xenopoulos and Palamas were also aware of the alleged "dark" side of genius in poets, which writers such as Max Nordau in his *Degeneration* considered as a hereditary stigma of degeneration. Xenopoulos does deal with this in his novel *The night of degeneration* (1926), as will be discussed in chapter 6.

137 Xenopoulos, *Πλούσιοι και φτωχοί*, 138 (my italics; suspension points in original).

138 Popos also believed that like poverty, beauty, ugliness, talent, and the proclivity to crime were all inherited characteristics (252).

139 Galton, *Hereditary Genius*, 225–236.

140 Darwin, *DM*, 1:110–111.

141 As has been mentioned before in this work, Xenopoulos wanted to be read by the general public but also the intellectual elite like Palamas.

142 Ambatzopoulou, "Το ψιθύρισμα της Επιστήμης," 42.

Further to this, as mentioned in chapter 2, in 1877 literary writers Angelos Vlachos and Emmanuel Roidis were the key players in a controversy about whether "a poet is born or made." Roidis believed that only with the appropriate environment could a poet's talents flourish, focusing on what he believed was the poor quality of poetry in Greece. This he claimed was due to a lack in an ambient environment in the Greek poetic circles, drawing on Hippolyte Taine's "milieu" as the important factor. Vlachos argued that a poet "is born and not made," that talent was God-given, and that no environment could affect it. His intention with this approach was to support the standard of poetry in Greece. The dispute between Roidis and Vlachos would include a dialogue on the importance of the arts in contributing to the progress of a nation and its society—an issue discussed by many intellectuals in Greece and the wider European community. The Skliros letters of 1913–1915 that were examined in chapter 1 include such a discussion. While the Roidis–Vlachos controversy has been largely forgotten today, for many decades afterwards it exerted a significant influence in the Greek literary world.[143] So it is very likely that when Xenopoulos in *Rich and poor* made reference to poets being made and not born, he was taking into consideration the spirit of the age, one of hard-wired inheritance and neo-Darwinism.

On Harisis, again Xenopoulos used the eye physiognomy to reflect the characteristic of genius. The ideas appear to be those of Popos, but it seems that the narrator is in agreement. Generally, the narrator does not agree with Popos' physiognomic judgments, but on occasion he does: "The eyes gathered still a lot of life, and even though they were tired, they beamed with something like that of genius."[144]

Rich and poor is about a period in Greece, the decade of the 1880s, when Darwinism permeated the Greek intellectual communities. The novel depicts Xenopoulos' representation of the debates on heredity and environment, in relation to humans, the mind and body, the classes, and socialism during this period. Although he does this in a parodic manner, he highlights the relevance of these key issues as part of the intellectual Darwinian discourse of the period. Writing this novel in the second decade of the twentieth century, he views the historical and scientific issues with forty years

143 Further on this see Dimiroulis, "Διαμαχή Ροΐδη-Βλάχου: ο ποιητής γεννιέται ή γίνεται;" 31–37; Dimiroulis, *Εμμανουήλ Ροΐδης*.

144 Xenopoulos, *Πλούσιοι και φτωχοί*, 196.

of hindsight. For instance, Xenopoulos superficially depicts the beginnings of socialism in Greece, having the hindsight of one who does not perceive socialism to be the answer to the eternal inequalities of society. He presents the class divide of rich and poor as an eternal problem which political ideologies, such as socialism, would probably not be able to eradicate quickly, if at all. He exaggerates the use of the physiognomical sciences in relation to natural selection and hard inheritance, so as to address the issues of the mind and body, class, heredity, and genius that were circulating at the time. The view of natural selection itself is not parodied, and also Popos' philosophical crisis regarding Darwinian evolution could reflect that of Xenopoulos.

The theme of the eyes in *Rich and poor* is not merely an aesthetic device but central to the book's intellectual background. In this study it has been related to theories of physiognomy which were linked to Darwinian theory; hence, the eyes theme reflects Xenopoulos' strong interest in Darwinism.

What Popos' theory of the eyes, the American's theory of social Darwinism (presumably Sumner), and Harisis' scientific socialism have in common is that they all base their ideas (whether rightly or wrongly) at some point on Darwin's natural selection. Xenopoulos' ability to write with hindsight reflects his ongoing interest in Darwinian ideas in that forty-year period—which, as shown in his "Athenian Letters," was to continue for decades to come. Whatever Xenopoulos' political or philosophical aspirations were, the novel demonstrates his ability to absorb the Darwinian and other evolutionary ideas of the late nineteenth and early twentieth centuries.

Xenopoulos reworks Darwin's theories but at the same time downplays them. The narrator refers to Popos as having incomplete knowledge of Darwin's theories, a strong imagination, and a tendency to generalize. In other words, as he says in his "Athenian Letters," it could be that Xenopoulos points out indirectly how dangerous ideas can be if one does not study them fully (see chapter 3). Also, the narrator himself is guilty of inconsistencies, where at times he appears to distance himself from Popos' views and at other times he is saying the same thing as Popos.

In reality, for all its ambiguities and contradictions, *Rich and* poor reveals a writer who was adept at the social sciences, the natural sciences, evolutionary ideas, in particular Darwinism, and especially at heredity and the transmission of characteristics. Daniel Pick also refers to writers such as Joseph Conrad who wrote "in contradictory and elusive ways" on the "possible implications for human meaning of current biological and physical theories," particularly when linking his work with Darwin, regarding their "vast shared problem of reading nature, evolution, language, the human face, [and] evil."[145] The ability to play with such concepts and to satirize some of them requires a confident writer who is fully informed and has the ability to view all aspects of the arguments.

Rich and poor has hitherto been viewed as referring to, and often reflecting, socialism in Greece in the late nineteenth to the twentieth century. There is a need to see the novel beyond the political framework, and it needs to be considered as a literary exemplar of Greece's development in the spirit of contemporary Western thought. Furthermore, commentary on the novel has tended to be associated with the socialist aspect. No one has ventured to address Popos' theory of the eyes, possibly because of its racist and racial connotations; but also because Greek commentators have tended not to embrace evolutionary approaches to literature.

In the novel, Xenopoulos portrays Popos as gullible and too ready to take on Harisis' socialism—which Xenopoulos satirizes—as the solution to the injustices of laissez-faire capitalism. The novel could also represent Xenopoulos' covert belief that society was unable to understand humanity as part of nature and directed by Darwinian laws. This may reflect the social and political climate of his period.

Popos was seriously trying to find an answer for his life: whether it was his application of his eye theory that he used to try to understand the workings of the mind and its application to the classes, or whether he was searching for a utopia. Popos' constant observing, questioning, and theorizing are reflections of this. Early in the novel, the reader is told that Popos was reading Descartes, whose work is deliberately mentioned fully by title—*Discourse on the Method of Rightly Conducting One's Reason and Searching for Truth in the Sciences*. One could speculate that if the meaning of this title was also to be taken at face value, then its message is self-explanatory in terms of Popos' search, and potentially also for that of Xenopoulos. Finally, the novel also shows Popos as an individual who was not able to resolve his issues due to his biology and consequently his position in society. In other words, he is a victim first of nature and then of society.

145 Fick, *Faces of Degeneration*, 61.

CHAPTER V

METAMORPHOSES OF WOMAN: DANGEROUS FANTASIES

On Woman: "[Her qualities are] her deep-rooted, primitive nature, with its infinite resources and caprices—qualities that are today as alive as in the prehistoric past, and that make our modern woman a true daughter of the primal Eve."
WILLIAM J. FIELDING [1]

Echoing the social evolution in relation to class and race, and discourses on inheritance and the mind-body relationship, Darwinism also provided a new "scientific" framework for the thinking of sexual differences between, and within, the genders. Darwin's statements in his *DM* on woman's lower state of evolution and on her biological inferiority fuelled a medical, psychological, artistic and, of course, a literary gender discourse that was pervasive at least up until 1940.[2]

Ann Heilmann documents that as early as 1865 the term New Woman was used in the *Westminster Review*.[3] As well as emerging as a social phenomenon, the New Woman appeared in literature. In 1894, the British feminist writer Sarah Grand used the term New Woman to describe fictional characters and actual women who asserted their independence and equal-

1 Fielding, *Woman: The Eternal Primitive*, 64.

2 Darwin, *DM*, 2:316–329. As mentioned in chapter 3, in the *DM* Darwin discusses the differences between the "secondary sexual characters" in man and woman and differences in their mental powers. Note also that Darwin's investigation in the unexplored field of sexual psychology was followed by his disciple George John Romanes, who further utilized a Darwinian framework. At the time, Herbert Spencer, Francis Galton, British psychiatrist Harry Campbell, and American psychologist G. Stanley Hall also contributed to the "scientific consideration of sex differences in psychology." See further Russett, *Sexual Science*, 44–45.

3 Heilmann, *New Woman Fiction*, 22.

ity, and who wanted to have a role outside the domestic sphere. Grand and antifeminist writer Ouida (pseudonym of Marie Louise de la Ramée) would battle over the concept of the New Woman; a defining moment in the concept played out in the *North American Review*.[4] Further to this, the reality of the New Woman and her literary representation would be hotly debated in prose and in parlors. Those who saw the New Woman as a cause of social decline represented her in a negative manner, and often as a degenerate.

Intellectuals and the mainstream scientific world "validated" the notion of the degenerate New Woman. Her emergence in Western countries coincided with the controversial Woman Question, which had spread by the fin de siècle's decadent period.[5] Some of the factors which were perceived by antifeminists to have caused social decline in this period were the changes in attitudes towards women's emancipation, changes in sexual hierarchy, and the growing participation of women in the public sphere.

In Greece around 1900, the feminist movement led by Kalliroi Parren was making its mark within the Athenian *Ladies' Newspaper* (*Εφημερίς των Κυριών*) (1887–1917), and so provided an agenda for writers to express their representations of the New Woman. Parren wrote a number of feminist serials in the newspaper. Noteworthy is her trilogy *The books of dawn* (*Τα βιβλία της αυγής*) (1899–1903), consisting of *The emancipated woman* (*Η χειραφετημένη*), *The witch* (*Η μάγισσα*), and *The new contract* (*Το νέον συμβόλαιον*), which depicts the morally upright New Woman and her post-Darwinian nature.[6]

Literary representations of this "degenerate" version of the New Woman often depicted her as a revisionary form of the *femme fatale* (or fatal woman) stereotype, a powerful and sexually assertive woman who destroyed her lovers physically or emotionally. One could say that the *femme fatale* has always existed and is observed in the literature and the visual arts of almost every period from classical times to the present, taking on numerous forms, such as the Lamia, Medusa, Sirens, Scylla, Salomé, and Lucrezia Borgia.[7] How-

4 Grand, "The New Aspect of the Woman Question," 270–276; Ouida, "The New Woman," 610–619.

5 In the late nineteenth and early twentieth centuries, physical and mental degeneration was aligned culturally with a perceived decline in morals and values in Europe. This was the "decadent" period, where literary writers, artists, philosophers, sociologists, theorists, and political figures described this trend.

6 For the feminist movement in Greece see Varika, *Η εξέγερση των κυριών*.

7 Praz, *Romantic Agony*, 199–300.

ever, at the end of the nineteenth and into the twentieth century, the *femme fatale* was very often aligned with the degenerate New Woman.

Few, if any, commentators have viewed the novel *Tereza Varma-Dacosta: The Middle Ages today* (*Τερέζα Βάρμα-Δακόστα: ένας σύγχρονος Μεσαίωνας*) (1926) in its true light— that is, as a post-Darwinian reading.[8] The story, set mainly on the Ionian island of Zakynthos, around 1885, is about Stephanos Angelikopoulos, an up-and-coming lawyer, and Tereza Varma-Dacosta, the beautiful young daughter of an aristocratic family. Stephanos falls in love with Tereza, but she will not reciprocate his love because of their different social classes.

In this chapter, with particular reference to *Tereza*, it must be mentioned that Xenopoulos, like many male writers (and some female), conflated the ideas of woman (the fatal woman and versions of the New Woman) in order to attack female independence and to express alarm at the assertion of female sexuality, which was often perceived as uncontrolled. Xenopoulos' overt support for gender equality, expressed to the youth in his "Athenian Letters," did not filter through to his presentations of women in many of his novels.[9]

In the novel, Xenopoulos' main character Tereza is a representation of the degenerate New Woman, who is described in evolutionary terms. Throughout the novel, there is a discourse, influenced by Darwinian writings and other Darwinian creative writers, which portrays Tereza's sexuality and mental characteristics in terms of physical and moral degeneration. Stephanos narrates these observations, which he perceives as a physical and mental transformation from an intelligent and pleasant female to an oversexed predatory beast and *femme fatale*.

I will also argue, in this chapter and the next, that the New Woman was simultaneously representative of the "degeneration of society and ... that

8 Here I use the following Greek edition: Xenopoulos, "Τερέζα Βάρμα-Δακόστα: ένας σύγχρονος Μεσαίωνας," *Άπαντα*, vol. 9 (Athens: Biris, 1972), 12–147. The novel was published in serial form in the *Ethnos* newspaper from April 24, 1925, to July 24, 1925, and in book form by the Athenian publisher Estia in 1926. As regards "the Middle Ages" (Μεσαίωνας), in Western European history this is widely accepted as the period between the 5th and 15th centuries (perhaps more narrowly ca. 1000–1453); "Μεσαίωνας" is also synonymous with a period of retrogression and obscurantism, of intellectual stagnation, of strict theocracy forbidding questioning and doubt, of magic, the supernatural, and superstition. See Babiniotis, *Λεξικό της νέας ελληνικής γλώσσας*, 1081.

9 For Xenopoulos' representation of women in his urban novels see Doulaveras, "Τα δυο φύλα."

society's moral regeneration."[10] As stated by Heilmann, the New Woman could be any number of women, ranging from a writer to a sexual libertarian.[11] Pykett sees the New Woman (like the physically and morally degenerate male and the homosexual) represented in contradictory ways: "She was figured as both a figure in a counter-discourse of renovation, often utopian in form, which represented a brave new world which would come into being through redefinitions of gender and of relations between the sexes."[12] Those writing (factual or fiction) about the term New Woman were "apt to shift and contest the parameters" of the term.[13] Heilmann sees it in terms of a "semantic instability" which "derives in part from the multiplicity of agents who had an ideological stake in constructing her"; indeed, she sees the New Woman as a "complex historical phenomenon which operated as both cultural (textual and visual) and sociopolitical levels."[14] In chapter 6, relevant to Xenopoulos' work and other works, I will also be discussing a masculinized "degenerate" version and the regenerative variant of the New Woman.

The ultimate aim of this chapter is to examine those intellectual trends and modes of thought in *Tereza* which were fuelled by the advent of Darwinism; and in doing so, to probe the contemporary ideologies and theories underlying Xenopoulos' thoughts. Also, I will endeavor to replace *Tereza* in the context of similar literary works by Xenopoulos' contemporaries (non-Greek and Greek).

Furthermore, in the sciences and in the literary world this inherently atavistic nature of the newly portrayed woman would be, in evolutionary terms, perceived as driving humanity down the path of degeneration and extinction. Unlike his degenerate counterpart, the male was at the same time striving to create and move forward. This notion of woman proliferated in the sciences and literature up to at least the 1930s, and was readily absorbed in the Western literary and art worlds; bear in mind though that the idea in various forms had been and continues to be utilized.[15]

10 Heilmann, *New Woman Fiction*, 1.
11 Ibid., 2.
12 Pykett, *Engendering Fictions*, 37.
13 Heilmann, *New Woman Fiction*, 2.
14 Ibid.
15 Kazantzakis, in some of his earlier works, takes up this theme where women were seen to impede progress. These will be mentioned later in this chapter.

The turn of the twentieth century brought about major political and economic upheaval. Issues such as the levelling of the socioeconomic classes and socialism filtered into Western novels. Greece was no exception. Literary representations of the allegedly degenerate Ionian aristocracy[16] were aligned with that of the degenerate version of the New Woman in the form of an aristocratic protagonist.[17]

Academic scholarship in relation to such literary works has abounded throughout the twentieth century. However, despite Xenopoulos' obsession with female representations in his literary work (from the 1880s), his apparent support for women's rights, and his clear association with evolutionary theories, as far as I am aware there has been no study which examines his female representations in a post-Darwinian spirit. More specifically, there is no evidence of academic scholarship in relation to the evolutionary ideas associated with the *femme fatale* or the New Woman in *Tereza Varma-Dacosta*. The following section outlines the reasons Xenopoulos gave for writing the novel. A letter to a critic also provides an insight into the essence of the novel and its reception in the Greek literary world.

Xenopoulos' comments

Xenopoulos dedicated *Tereza* to Kostas Oikonomidis, editor-in-chief of the Athenian newspaper *Ethnos* (Έθνος).[18] In the dedication he includes a comment made by Oikonomidis: "Well and good with all these girls, but I would like to see from you something different, more powerful, more tragic, which would go as far as a crime." Oikonomidis implies that Xenopoulos was producing reruns of the same storyline. Xenopoulos' response to Oikonomidis is also found in this dedication: "Your comment made me think of the old legend of Tereza Varma-Dacosta. Here is the 'different' girl you asked for!"[19] The message the dedication holds is that Xenopoulos clearly wanted to appear responsive to his readers' requests, particularly to influential persons such as Oikonomidis. This supply-and-demand approach only managed to provide fodder to

16 Greece in general, under Ottoman rule, did not have a hereditary aristocracy. However, the Ionian Islands under Venetian rule were an exception. After union with Greece in 1864 their titles were not officially recognized and they evidently lost their economic status too.

17 Later I will be comparing *Tereza* with such themes in August Strindberg's famous play *Miss Julie*.

18 Xenopoulos, *Τερέζα Βάρμα-Δακόστα*, 463.

19 Presumably this legend is a story that was orally circulated.

those critics who believed that Xenopoulos was a hack writer generating works only to please his readers—writing superficially and without substance.[20]

However, the following letter written by Xenopoulos to his friend Petros Charis (also editor-in-chief, from 1933 to 1987, of the literary periodical *Nea Estia*) presents a different side to Xenopoulos' purpose for this novel.

My dear Charis,

With this letter you will receive a copy of *Tereza Varma-Dacosta*. I thought that I had sent it to you, that you knew of it, but better late than never. It is a work which I wrote with exceptional attention. Just imagine that, after I had worked on it from the beginning, I copied out the first draft by hand, I made new corrections on that, and after it was published in *Ethnos*, I looked at it again so as to publish it in book form! It is written in a deliberate manner, word for word. Never have I cared more about language and style. Because the theme could take it, I wanted to show with this *psychological-social* romance novel how far our modern language could go, or more precisely, I wanted to see how far I could go with this new language. For me the result was satisfying: I managed to say all that I wanted to say, including the detailed aspects—*psychological, scientific, philosophical*—maintaining almost complete language uniformity; and so much so that the dialogue would not appear, as with other works, to be written in another language, but instead it is in keeping with the text and, I would say, it is stylized in the same way. I wanted to draw your attention to these things. As for the essence of the work, it is not necessary for me to say anything. You will see for yourself how *original, different, and deeply artistic* this romance is. I think that I did not do it justice by not making the main title "The Middle Ages today"—the subtitle; and also from an inexcusable oversight I did not send it when it came out to *a few individuals who understand about these things, and nearly all of them appear not to know of it.* Only Voutieridis happened to purchase a copy from Kollaros and he wrote an enthusiastic but short review in *Anagennisis.*"

Yours,

G. Xenopoulos, Athens, July 3, 1929[21]

---

20 Such critics prompted him to write the following essay: Xenopoulos, "Η διασκεδαστική τέχνη," 3–20. Briefly, the essay shows his firm beliefs: that any form of literary art need not be boring to its readers; that it can entertain and inform at the same time; that just because it appears entertaining, it does not follow that it is not a "work of art." He believed that art is essentially entertainment.

21 Xenopoulos, *Τερέζα Βάρμα-Δακόστα*, 11 (my italics). The letter appears as a prologue in posthumous editions of *Tereza Varma-Dacosta*; it is in the 1968, 1971, and 1972 editions by publishers Biris, Athens, but not in the 1984 edition by Vlassis Brothers, Athens.

My aim here is not to guess from this letter what Xenopoulos' intentions were in the novel. Nonetheless, the letter's analysis will reinforce my argument that there was certainly more to this work than literary scholarship has claimed. This enigmatic letter, which does everything but tell us what the "essence" of the work is, attempts to lure its reader, Petros Charis, into reading the novel. Xenopoulos clearly touches on certain aspects of the novel, including the language he uses, its reception and originality, as well as his classification of it.[22] Examination of this letter provides an understanding of how Xenopoulos attempted to attract the average reader to this novel, as well as a small circle of scholars and writers who he felt were in the know regarding such a novel.[23]

His intention was also to appeal to this elite group who knew the "fine details" regarding the "psychological, scientific, philosophical" matters of the day. He views his novel as "deeply artistic," intending to prompt the attention of his literary friend Charis.[24] Even his remark about Voutieridis' review being enthusiastic though short is a clear indication that Xenopoulos believed his novel had not been sufficiently examined and was possibly underestimated.[25]

He goes to painstaking lengths with the language he uses; he is pleased with his ability to successfully utilize the modern Greek language (the demotic) in matters psychological, philosophical, and also scientific, and so to maintain uniformity in the language and style. Relevant to the earlier

22 Through his letters, Xenopoulos frequently promoted his work and fiercely defended it when the press or literary critics gave negative reviews. He was often criticized for his lack of originality and constant reuse of themes. For one such view see Mousmoutis, "Δυο επιστολές," 516–517. In Greece, Xenopoulos is considered perhaps as one of its most prolific writer of letters; this includes "Letters" in the children's magazine *Children's Guidance*, letters to other periodicals and newspapers, and also personal letters about his work. Curiously, one can draw a parallel between Xenopoulos and Charles Darwin, who also spent a great part of his life "answering letters, justifying and explaining his views to friends, relations, and 'bitter opponents.'" See Burkhardt et al., *Correspondence of Charles Darwin, 1858–1859*, 7:xv.

23 See Xenopoulos, "Η διασκεδαστική τέχνη," 8. He believed that a successful literary product should be able to reach the average person (he gives the examples of a doorman or a dressmaker), as well as the highly educated, such as the Greek poet Kostis Palamas.

24 This is perhaps in response to those critics who considered his work superficial. One such harsh critic, Timos Malanos, believed Xenopoulos wrote outside the literary movements at the time and was not well-informed about any contemporary issues. This only manages to illuminate Malanos' lack of knowledge; it appears he had little if any contact with the European literary trends that portrayed Woman and the New Woman. See Malanos, *Δειγματολόγιο*, 92–102. (Note that my intention in this study is not to assess whether Xenopoulos' work is "superficial" or not; this was an issue which seemed to preoccupy the critics of Xenopoulos' time. It is merely my goal to reveal how he used Darwinian and post-Darwinian ideas at the time.)

25 See my analysis of Voutieridis' study in the next section of this chapter.

part of the century, the use of the demotic had been disputed in formal or academic writing. Pertinent to this study, it is his use of the post-Darwinian language in *Tereza* which would have created a problem for him. He had to reach those scholars who would have known Darwinism well, as well as the general readership. Interestingly, Darwin had a similar problem when writing the *OS*; his problem was coining words for new concepts.

Although Xenopoulos did not disclose any themes, he classified the novel as a "psychological-social romance." He also stated that in his novel he "managed to say everything, even the finer details." He alerts us with his comment that *Tereza* is "original and different." The implication here might be that this work differs markedly from the fifty or so short stories and novels he had written prior to this, and also differs from any other prose written in modern Greek at that time.

Xenopoulos expressed regret at not giving the novel the title of "The Middle Ages today"—the actual subtitle—that he says may have done it justice. This comment may also reflect his bitterness at his society's inability to appreciate the work; in other words, that Greek society was "medieval" in its inability to accept new ideas. There is also a hint in the opening lines that Charis too may have ignored this work. Xenopoulos indicated that those "few" who he felt "understood about such things" were unaware of, or ignored its existence because he did not send them copies; there may also be a hint that they should have noticed it anyway. Contrary to Xenopoulos' own enthusiasm regarding his novel, the letter reflects its poor reception in the Greek literary world at that time.

Literary Criticism on *Tereza Varma-Dacosta*

Although Xenopoulos acknowledged that *Tereza* was not as popular as some of his other novels, such as *Margarita Stefa* (*Μαργαρίτα Στέφα*) (1906), he considered it a personal favorite. He also included it among six of his many serialized works that he considered superior in literary skill.[26] In a 1925 letter to the Alexandrian critic Timos Malanos, Xenopoulos defends his writing stating that among his romance novels he considers *Tereza Varma-Dacosta* as his highest achievement. He goes on to defend *Tereza* together with three

26 Xenopoulos, *Η ζωή μου*, 344.

of his plays, as follows:[27] "Nevertheless, believe me that even these works are so significant for our country and for this place in time—yes, even these small works; that it is with the greatest disrespect that they talk about me the way they spoke about me recently, all together, like a chant, those Alexandrian critics."[28]

Literary scholarship on modern Greek prose was at its early stages at the time Xenopoulos wrote this novel, and it remained undeveloped in the twentieth century regarding much of Xenopoulos' work, *Tereza Varma-Dacosta* being no exception. Few though they may be, there is a need to discuss those critics who have studied *Tereza Varma-Dacosta*. The bulk of the commentary is directed to one or a combination of the following: general comments including an evaluation of the novel and a summary of the story line; its social themes in relation to the classes and socialism; the folkloric perspective dealing with the Zakynthian setting of the novel.[29] Less prevalent are the comments on the thematic use of the sexual instinct, the psychology of women, the features of degeneration such as atavism, and Tereza's change due to her environment.

Although the literary historian Linos Politis and the critic Petros Charis considered *Tereza Varma-Dacosta* one of Xenopoulos' best works, many scholars did not mention it at all in their accounts of Greek prose fiction.[30] Those commentators who considered the inequality of the classes a key theme in this work include Aristos Kambanis, Dimitris Kostelenos, and Konstantinos Vassis.[31] Kambanis, for instance, acknowledged *Tereza* to be a major work and claimed Xenopoulos was a "glorifier of the sexual instinct;

27 Malanos, *Δειγματολόγιο*, 92–102. The letter was presumably written in 1925, as Xenopoulos mentions in it that at that time *Tereza* was being published in *Ethnos*. The three plays are *Το μυστικό της Κοντέσσας Βαλέραινας* (The secret of Contessa Valerena) (1928), *Το ανθρώπινο* (What is human) (1927), and *Δεν είμ' εγώ, ή λογική* (It is not me, or is it the logic) (1928).

28 Ibid., 99.

29 I have not included works which only manage to summarize the obvious story line, or cite text without comment.

30 Politis, *History of Modern Greek Literature*, 176. Charis, *Έλληνες πεζογράφοι*, 7:93. Among those who exclude *Tereza* are: Beaton, *Introduction to Modern Greek Literature* (he mentions only *Rich and poor*, published 1919); Vitti, *Ιστορία της νεοελληνικής λογοτεχνίας* (who also only mentions *Rich and poor*); Dimaras, *Ιστορία της νεοελληνικής λογοτεχνίας* (he does not mention any of Xenopoulos' books by name).

31 Kambanis, *Ιστορία της νέας ελληνικής λογοτεχνίας*; Kostelenos, *Σύγχρονη ιστορία νεοελληνικής λογοτεχνίας*; Vassis, "Γρηγόριος Ξενόπουλος," 37. See also Karantonis, *Νεοελληνική λογοτεχνία: φυσιογνωμίες*. He dealt generally with the representations of women, the sexual instinct, and the classes in Xenopoulos' works but did not mention any by name. Petros Charis also dealt with Xenopoulos' use of the sexual instinct in a general manner. See Charis, *Έλληνες πεζογράφοι*, 1:130–131.

the instinct that is a leveler of the social classes." Indeed he believed that Xenopoulos hated the aristocracy and that this was reflected in the novel through the narrator Stephanos, who turned to socialism.[32]

Kostelenos also highlighted that Xenopoulos was familiar with "the particular social conditions of Zakynthos, where the remnants of feudalism remained strong even up to the beginning of our century."[33] Kostelenos argued that the message Xenopoulos wanted to impress on his readers was "the marriage of the classes in society" and hence the "natural abolition of the classes with the power of Love."[34]

Even the more recent study of Konstantinos Vassis sees *Tereza* as a story about the class differences in Zakynthos, adding that it is also "a love setting more daring than Xenopoulos' usual."[35] In the same review he likens it to the works of Edgar Allan Poe.[36] Unconvincing is Petros Charis' implication that Xenopoulos' intention in writing *Tereza* was to return to a folkloric portrayal of society, in particular of a past Zakynthian society at its peak.[37]

In relation to the female representations in *Tereza*, Apostolos Sahinis indicated that it stands apart when considered among Xenopoulos' fiction on "erotic passion," emphasizing Xenopoulos' risqué style.[38] Although Sofia Mavroeidi-Papadaki likened Tereza's changed psyche to that of the "the psychological make-up [of] Lucrezia Borgia," she did not really comprehend Tereza, describing her as "paradoxical."[39] In fact, she viewed Tereza through rose-colored glasses: "uplifted and justified in the end, ... beautiful both mentally and physically."[40] Furthermore, in a presentation of an anthology of Xenopoulos' works, Georgia Farinou-Malamatari makes a distinction between biological and moral degeneration, acknowledging only

32 Kambanis, *Ιστορία της νέας ελληνικής λογοτεχνίας*, 297.

33 Kostelenos, *Σύγχρονη ιστορία νεοελληνικής λογοτεχνίας*, 176.

34 Ibid.

35 Vassis, "Γρηγόριος Ξενόπουλος," 37.

36 Vassis' allusion is relevant, as a number of Poe's works have been examined in the context of the *femme fatale* of earlier times and not associated with the era of the New Woman.

37 Charis, "Πρώτοι απολογισμοί, Α'," 128.

38 Sahinis, *Το νεοελληνικό μυθιστόρημα*, 264.

39 Mavroeidi-Papadaki, "Η γυναίκα στο έργο του Ξενόπουλου," 81. Lucrezia Borgia (1480–1519) was a member of a powerful Renaissance family. She has been represented in literature and art as a person of wantonness, vices, crimes, and cast as a *femme fatale*. See Praz, *Romantic Agony*, 227. Praz clearly places Lucrezia Borgia as a *femme fatale*, corresponding to the "La Belle Dame sans Merci," describing her kind as "always the same type of unrestrained, imperious, cruel beauty." However, these representations are not associated with the New Woman or aspects of her.

40 Mavroeidi-Papadaki, "Η γυναίκα στο έργο του Ξενόπουλου," 81.

the presence of the latter in *Tereza Varma-Dacosta,* as well as in *The plebeian (Ποπολάρος)* (1924)—"the degeneration of the Heptanesian aristocracy appears inherent in the biological *Leeward nights (Απάνεμα Βράδια)* or the 'moral' degeneration of the girls of that class (*The plebeian, Tereza Varma-Dacosta*)."[41] She overlooks the seeming biological degeneration in *Tereza* (in line with Tereza's evolutionary reversion to a beast) that is associated with the moral degeneration she mentions.

Feminist literary commentators dealing with representations of women in Xenopoulos' fiction have excluded *Tereza Varma-Dacosta* from critical analysis, focusing on other novels, such as *The Three-Sided Woman* and *Stella Violanti (Στέλλα Βιολάντη)* (1914).[42]

Angelika Psarra has indicated that with respect to modern Greek literature, antifeminist depictions of women "remain unresearched" with few exceptions.[43] So that relevant to this study, no critics appear to have been able to associate Tereza with the representations of the New Woman as a degenerate.

In his *Short history of modern Greek literature,* Ilias Voutieridis manages somehow to touch on these issues when he indicates that the story has "something of the medieval, at the beginning of the twentieth century," and also that "it is a deep analysis of the psyche of the female character."[44] It is also worth dwelling on a separate study by Voutieridis.[45] In contrast to Xenopoulos' aforementioned letter to Petros Charis, Voutieridis does not particularly discern anything "original" or "different" about the novel, viewing the theme of *Tereza* as "nothing but the age-old, veiled, and quiet war between the aristocracy and the populace."[46] He goes on to say that this social war theme has been done before. Voutieridis revealed that with this

41 Farinou-Malamatari, "Γρηγόριος Ξενόπουλος," 308. *Απάνεμα βράδια* (Leeward nights) was published in serialized form in 1938 and in book form in 1984 by Vlassis Brothers.

42 See Varika, *Η εξέγερση των κυριών,* 113–115, 118–121; Thrylos, "Ο πεζογράφος Ξενόπουλος," 103–109; Thrylos, "*Η τρίμορφη γυναίκα* κινηματόδραμα," 2. Thrylos pointed out in her newspaper review that the play *The Three-Sided Woman* is very closely based on the novel with the same name. Dealing only with his urban novels, Doulaveras did not include *Tereza* in her "Τα δυο φύλα." Note that Doulaveras indicated in a footnote (91) that in *Tereza* and in other works (*Rich and poor* and *Honorable and dishonorable*) "the writer openly supports the theories of social Darwinism."

43 Psarra, "Το μυθιστόρημα της χειραφέτησης," 474–475.

44 Voutieridis, *Σύντομη ιστορία της νεοελληνικής λογοτεχνίας,* 348.

45 Voutieridis, "Μελέτες," 376–378.

46 Ibid., 377. This was a topic in the works of the Heptanesian writers, and it is associated with various social rebellions of the Ionian Islands that had been under Venetian rule.

novel "Xenopoulos ... steers us to the dark abyss of the female psyche."[47] Voutieridis' observation of Tereza typifies his views on women: a gynophobic sentiment carried by many male intellectuals at the fin de siècle, part of a chronic antifeminist mentality.

Yet little has been discussed regarding the novel's depiction of female sexuality. In fact, Xenopoulos mentioned that the sexual instinct, a most important feature of the human species, was a theme which he used in his earlier novels involving the psyche.[48] To date, this novel has not been placed in the context of others on the same theme.

Essentially, the commentary on *Tereza* shows a broad spectrum of views, most of which have not dealt at all with Tereza as a representation of the New Woman. Only Voutieridis touched on what he saw in Tereza as "the dark abyss of woman's psyche."

Consistent with Xenopoulos' letter to Petros Charis, there is a "prototype" theme in Tereza which needs to be addressed, and it is not the eternal dichotomy of the classes, but the representation of Woman, her psyche and sexuality, within an evolutionary discourse. The next section aims to examine the mutable nature of Tereza in reference to the way the woman, in particular the New Woman, was often represented in literature and art in the late nineteenth and early twentieth centuries.

Transforming Tereza

Perceptions and images

The textual examination of *Tereza* reveals numerous post-Darwinian concepts displayed at thematic, stylistic, and language levels. There is no doubt that, as Xenopoulos indicated in his letter to Petros Charis, he wanted in his novel "to say it all, even the finer details—psychological, scientific, [and] philosophical." This comment shows that Xenopoulos wanted to cover issues which he thought were important at the time. As a well-read scholar he was greatly influenced by many international trends, so that Darwinism would not be the only topic he would cover. In the novel he showed his

47 Ibid., 378.

48 Xenopoulos, *Η ζωή μου*, 345.

knowledge of Nietzsche through the dialogue between Tereza and Stephanos on the Superman and on the differing moralities of masters and of slaves.[49] Philosophies like those of Nietzsche and other contemporary intellectual trends were often interwoven into fiction. It is important to note that when Xenopoulos published the serialized form (1925) and the book form (1926) of *Tereza*, surprisingly, he did not know Freud's theories.[50]

Like many of his European peers, Xenopoulos narrated what the European visual artists of the fin de siècle created in relation to the post-Darwinian woman. A furious interchange of ideas and imagery between artists and writers was taking place. New technology in photography made the images of painters' work accessible to all, through popular magazines, books, and even postcards, so that writers like Xenopoulos could transcribe the images. He used postcards of supposed "ladies of leisure" to create his images of women.

By the end of the nineteenth century and at least till 1940, in the medical sciences, anthropology, and literature, the discourse regarding woman's close proximity to animals further down the evolutionary scale was well and truly "Darwinian based." At the same time it was highlighted that woman's sexual instinct made her a degenerate predator whose intention was to destroy the man Darwin's theories were taken on by Lombroso, Nordau, and Lankester, who all initiated some aspect of these perceptions. Even by 1927, popular discourse claimed that women were still considered to be more prone to insanity and that hysteria and emotional instability were "perculiar [*sic*] to women" due to their closer proximity to the primitive in nature.[51]

In this Zakynthian tale, through the eyes of Stephanos, Tereza's life is shaped by Xenopoulos' response to adaptational evolutionary theory, specifically Darwin's theory of evolution of the species involving transformation, regression, and extinction.[52] Tereza Varma-Dacosta, the granddaugh-

---

49 Xenopoulos, *Τερέζα Βάρμα-Δακόστα*, 56-58. For a summary of these aspects of Nietzschean philosophy see Marías, *History of Philosophy*, 362–364.

50 Xenopoulos, *Η ζωή μου*, 345.

51 Fielding, *Woman: The Eternal Primitive*, 21, 45. Note also that the very word "hysteria" comes from the ancient Greek word meaning "womb."

52 One must be reminded that there is an overlap with other theories which were pre-Darwinian, such as Lamarckism. As Pick notes: "Darwin himself and many of his followers were Lamarckian at least in certain of their assumptions, notwithstanding the crucially non-Lamarckian key assumption of 'natural selection.'... [That] French theories of heredity often produced confusingly contradictory connotations in relation to the environment ... to trace out differences in the later-nineteenth century in terms of a radical bi-polarity of Darwinism and Lamarckianism raises as many questions as it resolves." Pick, *Faces of Degeneration*, 100–101.

ter of the aristocrat Frederikos Varmas, whose family fortune had declined, undergoes a transformation; this occurs when she and her mother are banished from her wealthy father's home after he has taken on the maid as his mistress—and so mother and daughter take a drop in social standing by moving to grandfather Varma's old mansion. Despite her fears of not assimilating to the decaying, eerie, and damp conditions, gradually Tereza does adapt. The consequences are her transformation, where according to Stephanos, she slowly metamorphoses into a sexual predator, a half-woman, half-beast Lamia, with near fatal consequences for her lovers. Well-known in classical mythology, the Lamia, among other things, was a seducer of young men whom she later killed; and was depicted with a serpent's body and beautiful head. It appears that only a change of environment will save Tereza, allowing her to revert to her original state. She finally marries the consumptive aristocrat Tzortzakis Sirmas.

This transformation perceived as regressive has all the hallmarks of the degeneration popularized in the Western literary world at the end of the nineteenth century up until at least the 1930s. Tereza's regression leads to the looming probabilities of extinction.

Viewed through the eyes of the narrator Stephanos, it is his commentary with its associated themes, vivid imagery and language, and its repetitiveness that provides the basis of this analysis. His commentary is colored by unmistakably gynophobic and misogynistic undertones, despite the fact that he is physically obsessed with Tereza.[53]

Transformation and atavism (regression)

This post-Darwinian story line of reversion from human to beast, and in particular reversion of a woman to beast, was pervasive in the arts and literature and, relevant to this study, was occurring in Western fiction right up to the 1940s; it still remains a popular technique in fiction. In terms of feminist issues, this method was used in fiction to depict the New Woman, the *femme fatale*, and the hysterical woman.

53 He is referred to as an unreliable narrator. See Farinou-Malamatari, "Γρηγόριος Ξενόπουλος," 310. First person narration is usually unreliable.

Xenopoulos artfully and painstakingly develops Tereza's transformation, taking close to fifty pages of this 150-page novel. The transformation, seen through Stephanos' eyes, is on two levels: there is, first, a "visual" metamorphosis, where she is perceived to have turned into various animal or monstrous forms, in particular the Lamia; and there is the transformation of her personality into a sexual and hysterical degenerate. Both levels are developing in parallel and yet both are interrelated. Prior to the transformation, through Stephanos' narration we are presented with a person who is beautiful and intelligent; however, she already has the "predisposition" to regress due to the decaying aristocratic lineage on her mother's side and due to her views on morality between the classes. Chapter 8 of *Tereza Varma-Dacosta* begins with a comment by the narrator: "And yet! I was forgetting that one *adapts* to everything and I was overlooking *the large, the vast assimilating power of the environment. Gradually* even Tereza *adapts* to that 'medieval' house; and at the same time, simultaneously, she was experiencing its influence, she was becoming drenched *involuntarily* in its potion and she was *metamorphosing*."[54]

Darwin used this type of transformation to explain evolution by natural selection from a bear to a creature to "as monstrous as a whale."[55] The message in the passage above effectively alerts the reader to the commencement of Tereza's transformation. Stephanos is surprised at how well Tereza has adapted to her surrounds in the old mansion. Yet his comment does not stop there, as he ponders and marvels at "the large, the vast assimilating power of the environment." Slowly she adapts, simultaneously experiencing its influence, coming under its spell despite herself; and being metamorphosed. Xenopoulos was a self-confessed naturalist, so the milieu plays a great role in his writing. However, the aforementioned passage is not just about environment but about *change* within that environment. The environment brings the change into effect; and it does this without Tereza consciously willing it—it happens "involuntarily." Stephanos' perception would seem to reflect Darwin's adaptive evolutionary theory; compared to Lamarckism, where "the *instrument*[s] of change" are "intention or will"—

54 Xenopoulos, *Τερέζα Βάρμα-Δακόστα*, 97 (my italics).
55 Darwin, *OS*, 215.

that is, change is voluntary.[56] Furthermore, it is more likely to be Darwinian because the mechanism of change is unwilled—that is, involuntarily. Beer aligns unwilled selection with natural selection.[57] Most importantly by the turn of the century, discourse associating woman's reversion with her close links to nature and lower life would be considered thematically under the umbrella of Darwinism.[58] Also, placing this transformation into the broader picture of the actual devolution[59] of woman in terms of the struggle between the sexes, the theme in literature has tended to be interpreted by means of Darwinian-based modes of thought.

In the language of evolutionary discourse, the great propensity for adaptation as the only driving force of evolution due to change in the environment (from Tereza's well-to-do family home to the decaying old mansion), together with Tereza's lack of free will, all lead to her metamorphosis or transformation. This we view through the life cycle of the individual Tereza. It was common practice for creative writers to respond to evolutionary theory utilizing the life cycle of an individual rather than dealing with a species.[60]

In the ongoing analysis of Tereza's transformation it will be revealed that her changes were due to what were then considered inherent and degenerate traits. These were her degenerate aristocratic background and her role as an inherent variant of womanhood, the New Woman. No doubt, the aforementioned passage on Tereza's transformation and assimilation is reminiscent of the style used by Karkavitsas in his novel *The fair maid* with the transformation and assimilation of his protagonist Anthi (discussed in chapter 3). The difference is that in her transformation Anthi becomes the positive regenerate New Woman, whereas Tereza becomes the degenerate version, Anthi's alter ego. Concerning the passage on Tereza, it is worth noting that in the original 1926 publication the word for "surrounds" or "environment" ("περίγυρος") is written "Περίγυρος," thus given the importance of upper

56 Beer, *Darwin's Plots*, 19–20 (italics in original). As Beer indicates, conscious endeavor is an agent of evolutionary change, and intention is the key to Lamarck's concepts.

57 Ibid., 8, 196.

58 For one such perspective see Dijkstra, *Evil Sisters*, 74–75.

59 The term "devolution" is discussed in detail in chapter 1.

60 Beer, *Darwin's Plots*, 98. See chapter 3 of my study on the application of the scientific concepts of ontogeny and phylogeny to the life cycle of an individual in creative writing.

case (the status often given to words such as "Nature"). It is clear that the change to the lower case in the later editions reflects a lack of appreciation of its original importance.

Physiognomy reflecting the metamorphosis

Xenopoulos used naturalistically, in a stylistic manner, a form of physiognomy throughout *Tereza*. This needs to be clarified at this stage, in terms of pre- and post-Darwinian ideas. The following passage is one example of Xenopoulos' use of physiognomy in terms of the relationship between the house and Tereza's nature: "In *that medieval house* Tereza was becoming *a medieval girl*. Something dark, hard, savage, untamed was waking in her soul. The influence of that environment, into which one incident of bad luck had caused her to fall suddenly, precipitated the display of some kind of atavism, which might have come later or maybe never."[61]

Although dating back to Aristotle, physiognomy was popularized in literature by Johann Caspar Lavater in his *Essays on physiognomy* (*Physiognomische Fragmente*) (1775), and in the earlier nineteenth century (pre-Darwin) it was reinforced by Lamarckism. It is widely accepted that one aspect of Lavaterian physiognomy "is the treatment of the novel's setting as an expression of the character of an individual or a group [hence] the idea of the relationship between human nature and environment" and that it "was to become central to nineteenth-century scientific thought and thematic in the naturalist novel."[62] Further to this, Lavater indicated that a person's habitat, including their house and possessions, such as furniture and clothes, reflected a person's character traits. Hippolyte Taine showed traces of this style in his *History of English Literature* (*Histoire de la litterature anglaise*) (1863).

As mentioned in chapter 4, Darwin's *EE* paved the way for uses of physiognomy in the sciences, medicine, and criminology. Physiognomy was now validated due to a Darwinian etiology with concepts such as a person's facial traits mirroring their personality; this line of thought was absorbed into literature.[63] Max Nordau and Cesare Lombroso would discuss at length the

61 Xenopoulos, *Τερέζα Βάρμα-Δακόστα*, 105 (my italics).

62 Tytler, *Physiognomy in the European Novel*, 252. The theory was frequently used by Balzac and Dickens.

63 Greenslade, *Degeneration*, 90–91. Lombroso-Ferrero, *Criminal Man*. In the new introduction to the latter volume, Leonard D. Savitz maintained that the sources that "seemed to coalesce and culminate in Lombroso's theory of the born criminal" were Darwin's *OS* and *DM* on evolutionary thought; based on

relationship between physiognomy and craniotomy with what they called the "stigmata" or the signs of physical, mental, and moral degeneration.[64] Also, social physiognomy, where the appearance of a person or his environment reflected the person's social background, was frequently used in nineteenth-century fiction, especially in novels regarding "social stratification" and "class consciousness."[65]

Writers such as Balzac and Dickens, particularly pre-Darwinism, used physiognomy extensively in a thematic manner in nineteenth-century literature. However, as *Tereza* was first published in 1925, and due to its broader key Darwinian themes, it can be argued that the use of physiognomy, specifically relating the characters to the environment, is more of a stylistic technique meant to reinforce the major post-Darwinian themes.

Hence in the previous passage, Tereza's reversion or atavism is reinforced stylistically by Xenopoulos' use of physiognomy. Tereza's character is becoming "dark, hardened, savage" and "untamed," which is also a reflection of the old medieval house. In this study, references to such use of physiognomy will be made when they are relevant to the Darwinian theme. Tereza's atavism presents as a motif of a bleak Middle Ages, and triggers the makings of a *femme fatale*.

In the spirit of Lavaterian physiognomy, Tereza continues to take on the character of the old castle. Everything one could associate with a medieval decaying house is taken on by her character. The fear that she had for her surrounds in the castle disappears, and she starts searching it only to find medieval paraphernalia, including an old dagger, with which she pretends to lunge at Stephanos: "And she laughed so differently, which made me think that in that medieval house Tereza had also started to be … a medieval girl. On the other hand, I felt at that point so vividly of the allure of Tereza, playing with a Toledan stiletto, that I said to myself: 'I wish she could make me her lover just once and she could then slaughter me! …'"[66]

the *DM* Savitz argued that "some men are closer to their primitive ancestors than others: a belief which is a central feature of Lombroso's subsequent theorizations." Also according to Savitz, the second source was B. A. Morel's *Treatise on Degeneracy* (1857), where he described "the pathologic phenomenon of the primitive degenerative type, who represented a peculiar interplay of heredity and environment" (vii).

64 Nordau, *Degeneration*; Lombroso and Ferrero, *Female Offender*.

65 Tytler, *Physiognomy in the European Novel*, 246.

66 Xenopoulos, *Τερέζα Βάρμα-Δακόστα*, 98 (suspension points in original).

She begins to use these medieval items, initially out of necessity then out of choice; she dresses like a medieval lady in robes that accentuate her body; changes her hairstyle to suit; reads the old books; uses the old crockery. Her lethargy is due to her degenerate state. The change is not just superficial:

> It was not only her appearance and her beauty that were changing—even her soul was becoming altogether more in harmony with all these things. No, she was not the "intellectual girl" that I first met in the brand new, bright Dacosta house, but neither was she the "wild girl" that I used to meet in the brilliant-white neat villa at Saint Kyriaki. Her always alert and obliging soul seemed now to have become heavy due to some unconquerable sleepiness, and at times it appeared to me that she was asleep. On the other hand, it was as if within her there had awakened some other qualities which *in the past were dormant.*[67]

She would become very religious and adopt superstitious beliefs commonly associated with the Dark Ages, changing from the enquiring open mind to the closed fanatic. Stephanos was quite aware of her atavistic traits when he mentioned that they were "a mark also of her degeneration," and commented similarly regarding her ease in talking about taking lovers.[68]

The serpent woman

In 1893, Xenopoulos published the short story "The bracelet" ("Το βραχιόλι"), introducing the reader to a notion which was found later in his works and is relevant here to *Tereza*—that is, the snake as a metaphor for Woman.[69] The short story is entirely about a bracelet in the shape of a snake and about its connection with the woman who wears it. The misogynistic first person narrator buys a bracelet for his fiancée Maria to replace her own snake-like bracelet, which she always wore, and which annoys him initially because it looks cheap. He often refers to each bracelet as the snake: "And at the first opportunity … I gave her a little bracelet, elegant, beautiful, from coral and gold, snake-like as well because—I don't want to hide it—after the story of Eve, I can't find a more *suitable symbol for woman.*"[70]

67 Ibid., 99 (my italics).
68 Ibid., 135. Nordau also refers to the "stigmata" of degeneration as brandmarks. See his *Degeneration,* 17.
69 Xenopoulos, "Το βραχιόλι," 326–332.
70 Ibid., 326 (my italics).

Maria refuses to take off her own bracelet and eventually only agrees to wear her fiancé's if she can keep her own on as well. The fiancé becomes obsessively jealous, because Maria will not take the first bracelet off and he believes that it is associated with a previous lover. Apart from the association between the bracelets and the snake-like qualities of Maria, the story is full of various forms of the snake symbolism associated with the bracelets and with jealousy, love, and being possessed. So that similarly, when another phase of Tereza's transformation takes place, Xenopoulos depicts her as a seductive woman with serpentine qualities. He is also preparing her for the form that she will soon take—that is, the snake-like Lamia. The discussion in detail on Tereza as the Lamia will be dealt with later in this chapter. Tereza's physical appearance has matured: "You would say that she understood, that she sensed her body, that *gigantic and elastic body* which in the past, though it was so noticeable to others, for her had been nonexistent. It was as if her nerves had been overstretched, had made her *sensitive*, in fact, *hypersensitive* to things which in the past she would not have noticed; and she was now touchy, always strange and difficult."[71]

Her python-like appearance exhibited traits of hypersensitivity, suggestive of those of a snake when touched. Another perspective regarding Tereza's neurotic and hypersensitive traits reflected the "nervous and mental maladies," hysteria and neurasthenia which Max Nordau's book *Degeneration* indicated as part of the fin-de-siècle degenerate condition. Xenopoulos had of course read Nordau's book, as he mentions it in his novel *The night of degeneration*, which will be discussed in chapter 6.

With the "Woman Question," this neurosis became predominantly a female condition, and surfaced in the writings of Sarah Grand, Thomas Hardy, and George Gissing.[72] Medically this condition was considered to be caused by the woman's inability to cope with the stresses of life; it was also associated with the New Woman's threat to man. It was suggested that if women used their brains too much and also tried to assert their independence, then they would succumb to this condition.[73] Hysteria and neurasthenia would be the dialogue of woman's excessive or inadequate sexual activities. Even in Greece, in 1912 coeducational schooling was considered

71 Xenopoulos, *Τερέζα Βάρμα-Δακόστα*, 99 (my italics).

72 For commentary on these see Greenslade, *Degeneration*, 136.

73 Ibid.

to be detrimental to the female brain and body, because the alleged male-oriented schooling was too difficult for females to handle: "They are taught the same lessons as the boys, whereas we know what a difference there is between them in their perception and their judgment and their endurance and their interest, indeed at a period where for both genders the manifestation of puberty is so different age-wise and physiologically. So they rote learn, become overfatigued, destroy brain and body; and with such conditions neither housewives nor women nor man can be educated."[74]

Tereza becomes increasingly aware of her own body, sexually. Through Stephanos' eyes, Xenopoulos provides his reader with risqué, voyeuristic scenes of Tereza's autoeroticism; all common features of late nineteenth-century decadent literature and art. After one such scene, Stephanos comments: "These things she did not do only so as to drive me mad. They were the markings of a wild, somewhat belated frenzy. Tereza, who now understood and sensed her body, was agitated and tormented by it."[75] On one such alluring occasion, when Stephanos is observing her in the garden of her grandfather's house one day, her commonplace autoerotic scenes take on these snake-like qualities:

> And she lay straight down and started to roll about relentlessly. Her clothes were becoming messed up; *branches were entwining her, girdling her, entangling her,* and although she was *shaking her feet every now and then to free herself and so push forward twisting and turning,* the sticks would often catch her skirts and raise them to her waist. What utter madness! Never had she shown me her body so shamelessly as with this *frenzied writhing* in the greenery and dead wood. Later she stopped near a tree, she lay down straight, on her back, and she closed her eyes as if she was dizzy. I approached her but she did not move at all. I spoke to her but she did not answer. She made out she was asleep ... Well, this time the invitation was not a trick. As she relaxed, with her eyes closed, her chest heaved. But the rest of her body lay so motionless on the ground, so dissolute; you would say that every bit of strength had left it. She was surrendering herself, it was clear.[76]

74 Education Association [Εκπαιδευτικός Όμιλος], "Ανάγκη ... ν'ανοιχτούνε στα κορίτσια κι' άλλοι δρόμοι," 594.

75 Xenopoulos, *Τερέζα Βάρμα-Δακόστα*, 107.

76 Ibid., 108 (my italics; suspension points in original).

On the whole, the italicized words suggest—and more so to the predisposed reader who is aware of a Lamia's features—the image of an animal that has snake-like qualities. Through Stephanos' eyes, we view Tereza lying down and wallowing or writhing incessantly; then rolling about to become entangled in the tree branches. The shaking every now and again of her legs to disentangle conjures the spasmodic movement of a beast, wriggling to free itself. There is the image of a beast, potentially snake-like, with a mode of movement—that is, while still lying down, of "push[ing] forward twisting and turning."[77] Finally, Tereza is exhausted and appears to be sleepy, leaving Stephanos clearly aroused by the frenetic scene. In this passage Xenopoulos' style is reminiscent of those used by European writers in that period. In addition, by placing her here in the garden-temptation context (as in Eden), Xenopoulos portrays woman's oneness with nature, her closeness to it, a style popular in the literary and visual art worlds in the late nineteenth and early twentieth centuries.

Voutieridis was on the mark when, in his study of *Tereza,* he commented that Xenopoulos' style was similar to Flaubert's. In Flaubert's *Salammbô* (1862), Salammbô performs an erotic dance where she sheds her clothes one after the other, allowing a python to wrap itself around her whole body. Moreover, in Genesis Eve is tempted by the snake, and in art and literature Eve becomes one with the snake and ultimately becomes the snake herself. Representations of "Woman, the eternal Eve," the eternal feminine, associated with snakes or with snake-like attributes, include Salammbô and Lamia. In Rider Haggard's *She* (1887), Ayesha's hypnotic snake-like qualities attracted the men around her; she would also regress into a monkey. In Bram Stoker's *The Lair of the White Worm* (1911), Lady Arabella March transforms into a snake form whose metamorphic power seems darkly intrinsic to womanhood itself. Notably, passive women, or those not shown to be emancipated, were not as a rule shown to "mutate," unless of course they significantly changed in their thinking. The Russian writer Maxim Gorky published *The Artamonov Business* in 1925, the year when Xenopoulos wrote the serialized version of *Tereza*. Gorky endowed his protagonist,

77 Xenopoulos uses the words "στριφογυρίζοντας," which here I have translated as "twisting and turning," but it could also mean "in a rotating movement." Also he uses the word "κύλισμα," which I have translated as "writhing" but could also be "rolling."

Paula Menotti, with movements likening her to "the primal serpent"[78]: "The woman's movements became faster, more furious: she writhed as if trying to leap down from the piano, but unable to; her suppressed cries became more and more loathsome and evil: what was particularly disgusting to see was the way her legs writhed, together with the jerkings of her head which sent her hair flying first over her breasts, then lashing over her back like the tail of a snake."[79]

Nikos Kazantzakis utilized the metaphor of the serpent in his decadent *Serpent and Lily* (1906) to describe the smothering erotic passion for a woman, which overwhelms man and allegedly leads to the eventual "fall" of man.[80] In various disciplines, including literature, the age-old theme of progress is popularized post-Darwinism. This occurs in Kazantzakis' play *The master builder* (1910), noting Henrik Ibsen's play of the same name (1892).[81] Man could not strive for a higher ideal, whatever that might be, because Woman was in the way. In his writing Kazantzakis often used the theme of "woman as the temptation" impeding some sort of progress. This would reflect his personal struggle in life towards a higher spiritualism.

A misogynistic approach was common in the post-Darwinian era of the decadent school of writing at the turn of the twentieth century. While there is clear evidence that Kazantzakis was strongly attracted to women, he revealed much later in the epilogue of his autobiographical *Report to Greco*: "We did not allow women, even the dearest, to lead us astray. We did not follow their flower-strewn road, we took them with us. No, we did not take them, these dauntless companions followed our ascents of their own free will."[82]

The post-Darwinian interpretation of wild, animalistic Woman only managed to perpetuate and cement in society the diachronic view of the Woman Eve, the temptress—no less than within the Church. The religious ascetic male's struggle to perfection with its reward in heaven, or the afterlife, is strongly reliant on the male's ability to disassociate or even isolate himself from women, mentally and physically. Counter-feelings of resentment, gynophobia, or even misogyny arise when the presence of a woman "obscures" this noble path to perfection. In the case of Kazantzakis, his path

78 Dijkstra, *Idols of Perversity*, 306.
79 Gorky, *Artamonov Business*, 208.
80 Kazantzakis, *Serpent and Lily: A Novella with a Manifesto.*
81 For Kazantzakis' play see Psiloritis [Nikos Kazantzakis], *Ο πρωτομάστορας.*
82 Kazantzakis, *Report to Greco*, 494.

was no longer religious in the traditional sense but a philosophical or spiritual quest for perfection and cosmological enlightenment. As he suggests in the previous passage, it is different for women; the temptation is not an issue, and Woman's purpose for existence is dependent on and dominated by that of Man's.

Femme fatale

In the novel, Tereza abuses Stephanos for responding to her with kisses. He is offended by this and believes that he now no longer loves her, even though he lusts after her: "I couldn't do without Tereza anymore! For me even the torment which I still suffered near her was pleasure. ... I ceased entirely loving Tereza herself. ... I desired, oh how I desired her body!"[83] Stephanos toys with Tereza, unwittingly planting the idea in her mind to secretly take as a lover the young and handsome barber's assistant Kyriakos, and then either slay him or throw him down a trapdoor. Tereza is a potential *femme fatale*, the fatally seductive woman.

> "Tereza, tell me the truth now. Would you want to be in the old days, when there were no laws for you lords and ... you could take in Kyriakos; you could take him to your bed or here on this divan so that you could enjoy him all night, and at dawn with the crowing of the rooster you could slaughter him with your stiletto or you could drop him through a trap door? ..."
> "I would have done it!" She yelled. "I swear that I would have done it!"
> "Why then don't you do it? ..."
> "Your laws! ... Now you see there are your laws! ..."[84]

Having read Nordau, Xenopoulos presumably knew of Caesar Lombroso's work, such as *The female offender* (1895), since Nordau mentioned him in his book. Had he read Lombroso's book, it would have been a validation for Tereza's lawlessness and highly erotic nature: "The atavistic diminution of secondary sexual characters ... shows itself once again in the psychology of the female criminal, who is excessively erotic, weak in maternal feeling, inclined to dissipation, astute and audacious, and dominates weaker beings some-

83 Xenopoulos, *Τερέζα Βάρμα-Δακόστα*, 109.
84 Ibid., 111 (suspension points in original).

times by suggestion, at others by muscular force. ... Added to these virile characteristics are often the worst qualities of woman: namely, an excessive desire for revenge, cunning, cruelty, love of dress, and untruthfulness."[85]

Despite her rejection of Stephanos as a sexual partner (because of his lower-class origins), Tereza's animalistic instinct of sexual selection attracts her to the Adonis beauty of Kyriakos (in Darwinian style, like a peahen to a peacock). She cannot resist Kyriakos' beauty, so she does carry out Stephanos' suggestion, throwing him down a trapdoor in her grandfather's house. When the youth disappears and Stephanos suspects this, she regrets her actions, and they both rescue the virtually unscathed Kyriakos, who thinks his fall was accidental.

The Lamia and vampire: revamping of the vital essence theory

As the evolutionary transformation of Tereza continues, the reader learns that this seductive woman with serpentine qualities has developed other features and now has a name, Lamia. The Lamia—a mythical beast, classically Greek or Roman, having the face and breasts of a woman and the body of a serpent—has appeared in various forms, ranging from classical antiquity to modern times. A short overview will help situate Xenopoulos' representation of her in his novel.

According to Lemprière's *Bibliotheca Classica* (1788), this monster could not speak, and so it lured men and children with its pleasing hissing, and then devoured them.[86] Also, according to Angelopoulos' *New dictionary of Greek mythology* (2000), among other forms Lamia could take the form of a vampire, drinking people's blood and devouring their hearts.[87] The same source indicates that Lamia was a beautiful girl from Corinth, who beguiled young men and then tore them up.

The famous poem *Lamia* (1820) by John Keats examines in a pre-Darwinian manner the classical Lamia, who is snake-like in relation to eroticism and beauty. Images of metamorphosis occur in the poem, though of an Ovidian nature.[88] Numerous poems associated with age-old representations

85 Lombroso and Ferrero, *Female Offender*, 187.
86 Lemprière, *Bibliotheca Classica*, 646.
87 Angelopoulos, *Νέον λεξικόν της ελληνικής μυθολογίας*, 401.
88 Murray, *Poems of John Keats*, 233–253.

Figure 5.1. Herbert James Draper (1864-1920), *Lamia*, ca.1909 (oil on canvas). Private Collection / Photo © Whitford & Hughes, London, UK / Bridgeman Images.

of dangerous, blood-sucking female creatures appear in modern Greek folklore. An example of the Lamia in Greek folklore is the folk ballad "The shepherd and the Lamia."[89] Salomé and the Lamia are also coupled in the poem of the early 1900s *Salomé* (*Σαλώμη*) by Emilia S. Dafni,[90] while the poem of Kostis Palamas "The Lamia" epitomizes the mythological character Lamia with the hair of snakes, as a deceiver and destroyer of young men.[91]

89 For the ballad "Ο βοσκός και η Λάμια" see Ioannou, *Το δημοτικό τραγούδι*, 52–53. For further representations of the Lamia in modern Greek folklore see Stewart, *Demons and the Devil*. Stewart makes frequent references to citings in studies by N. Politis, such as in *Νεοελληνική μυθολογία* and *Παραδόσεις*.

90 Rekas, *Echoes of Old Athens*, 177–182.

91 See Palamas, "Η Λάμια," 394–395.

More specifically, at the turn of the twentieth century in the literary and art worlds the Lamia was associated with the feminists or the New Woman. Apart from maybe the work *Salomé* by Dafni, the Lamia of none of the aforementioned is associated with the late nineteenth-century to early twentieth-century representations of woman, that is—the New Woman or the Eternal Feminine. Angelika Psarra notes that the poem The (evil) woman (Η γυναίκα [η κακή]) (1903) by Kleanthi Vassardaki misogynistically depicts Woman as a Lamia, a sphinx, a she-wolf, a deadly flame, and a leech. The poem is found in Kornelia Preveziotou's antifeminist woman's magazine *Vosporis* (*Βοσπορίς*).[92] According to Dijkstra, "the Lamia of Myth was thought to have been a bisexual, masculinized, cradle-robbing creature, and therefore to the men of the turn of the century perfectly representative of the New Woman who, in their eyes, was seeking to arrogate to herself male privileges, refused the duties of motherhood, and was intent upon destroying the heavenly harmony of feminine subordination in the family."[93]

The Lamia, who had always been associated with the fatal women or *femmes fatales* throughout history, was in this period a clearly adverse representation of the turn-of-the-century New Woman and sexual woman.[94] In addition, Dijkstra draws attention to Eva Nagel Wolf's image of woman in 1919 as the "Lamia, the serpent goddess, with all the sinuous grace and the tantalizing haunting memory of a beautiful woman, [with] a serpent's head."[95] Similarly, English artist Herbert James Draper's representation of the Lamia in 1909 (fig. 5.1) shows her association with a serpent on her forearm. Although her lower body is human, there is the allusion of her serpentine origin due to the shed snake skin flowing around her waist.

The following passage from *Tereza* is the reader's first encounter where the beast is referred to by name as Lamia. Stephanos' reaction to Kyriakos' appearance now is significant. Lamia appears in Stephanos' eyes to have the ability to "suck the beauty" out of Kyriakos, leaving him physically drained of energy and of his good looks. There is already a vampiric quality surfacing

92 Preveziotou, *Βοσπορίς*, 357–358.

93 Dijkstra, *Idols of Perversity*, 309.

94 See Praz, *Romantic Agony*, 250. His important study examining the *femmes fatales* found throughout history, specifically in nineteenth-century literature, includes the Lamia. Further on the *femme fatale*, her origins, depictions, and literary responses see Milful, "Songs of the Siren." Earlier in this chapter I mentioned that Mavroeidi-Papadaki likens Tereza to Lucrezia Borgia; she is not far wrong when one considers that both Borgia and Lamia were *femmes fatales*.

95 Wolf, "Elenore Abbott, Illustrator," xxvii.

in the passage: "I could see Kyriakos in the *reddish light* shed by a low, late-rising half-moon. He appeared to me to be *ugly, livid, thin, with weary eyes*; everything but an Adonis or Ganymede. And I thought: See! The Lamia who did not take his life, did however take his beauty..."[96] The vampire imagery is generally linked with the moon, which is considered an accomplice to the vampire's blood-sucking episodes. Prior to his attack by Tereza, Kyriakos is seen to be energetic, and the moon is "fully bright" and "full of mirth" (123). In contrast to this, after the event the moon is reflecting a "reddish," "bleak, half-light," and Kyriakos has lost his energy and appears drained (122–123). In Western literature and art, the depiction of snake-like qualities of women was succeeded by the resurgence of the vampire story. The old Gothic vampire was now being reworked to suit the sociohistorical context highlighting woman's sexuality and the New Woman. Concurrently, the "scientifically" upgraded "vital essence" medical theory would provide creative writers with the evidence they needed to fuel the vampire depiction of the New Woman.[97]

It was said that humans have a limited amount of vital essence or fluids, a form of energy which needs to be utilized sparingly. In the male, seminal fluid was considered the most concentrated form of this vital essence. Late nineteenth-century medicine elaborated on this theory, which dated back to classic times, indicating that this essence circulated through the body via the human's blood.[98] The fluid was needed to maintain a healthy body and mind. Numerous theories emerged under the guise of science, theorizing this elixir's physiology and the symptoms it caused when it had been wasted. It was widely accepted that frequent wasting of this vital fluid, such as through sex (especially in males via semen loss), caused the male to show signs of anemia, such as paleness; also signs of muscular weakness, mental fatigue, reduction in mental capacity, flabby or shriveled skin, and even effeminacy. These symptoms became a metaphor for the degenerative effects of female sexuality.

Therefore, Stephanos' comment "she drained him of his beauty," which will become a repeated and panic-stricken comment in the narrative, is sim-

96 Xenopoulos, *Τερέζα Βάρμα-Δακόστα*, 122 (my italics).

97 Analysis of the section on the vital essence theory is based on Dijkstra, *Evil Sisters*, 49–81.

98 One could reasonably speculate that, although metaphysical, Bergson's *élan vital* (his creative and primordial impetus of life) could have been influenced by one or other such theories.

ply the result of the sexual act, and refers to the draining of the vital essence from Kyriakos' body. As the story progresses, Stephanos makes further and more frequent references to the Lamia with repeated images of the de-energized Kyriakos. This is a reflection of Stephanos' escalating fear of the Lamia. The fear arises from the two contradicting effects he knows that Tereza has on men—that is, the intense attraction that he feels for her and the repulsion in case he loses his vitality.

Stephanos will again highlight Kyriakos' symptoms, comparing him before and after his affair with Tereza. Stephanos clearly makes a statement here by repeating his view about the Lamia and by placing these comments at the end of the chapter: "But how different was he from the previous time that I saw him! Then, in the joyous light of the full moon, he flew like a bird, like an angel. *Now black and shriveled* in the reddish, bleak half-light, *he dragged his feet like a sickly old man.* The Lamia had drained him of his beauty."[99] Further to this, Stephanos sees Kyriakos' state as "ugly, thin, shriveled, pale, with listless eyes, and without his adolescent freshness which previously had made him marvelous" (128). He initially reasons it is due to his separation from Tereza and his traumas associated with her. In the following passage, by depicting Tereza as a vampire bat biting into Kyriakos' vein and sucking his blood, Xenopoulos continues to adhere to the vital essence theory: "I couldn't get out of my head the idea that it was from the draining of a lustful and insatiable love. In my imagination I could see Tereza attaching her mouth to his body, opening the vein with a bite, and sucking, drinking his blood, similar to the apparition—that is, of the huge blood-sucking vampire bat. That is why he now looked pale, flagging, and shriveled. And the Lamia, which had *sucked his beauty* out of him, again threatened his life."[100]

Development of this vital essence theory alerted intellectual men of the earlier decades of the twentieth century to be "economical" in terms of their vital essence. Thanks to Darwin's theories in the *DM*, it was considered that man, the creative and intellectual gender, was more likely than woman to reach superior mental achievements, and hence needed this vital fluid primarily to "create" intellectually. Woman was considered intellectually infe-

99 Xenopoulos, *Τερέζα Βαρμα-Δακόστα*, 123 (my italics).
100 Ibid., 128 (my italics).

rior to man (Darwin's theory), and so her function was only to reproduce, leaving the thinking to the male; loss of this vital fluid through the female's menses and reproduction rendered her even more brainless.

A male who gave in frequently to a sexual woman was depleted of his vital essence and was also considered a weakling himself since, as part of the superior male species, he was supposed to show strength of character and use his vital essence for his personal progress and that of society. Men who showed such weaknesses were themselves on a par with the sexually degenerate New Woman. Their characteristics were feminized or effeminate, and so they were considered as degenerates alongside this brazen masculinized New Woman. Man's fear of being in this position further perpetuated the gynophobia (as seen by Stephanos' tone in the aforementioned passages) and misogyny felt by the intellectual hegemony. The misogyny is explicit when Stephanos imparts the following comment directly to the reader: "And 'between us men,' a woman is always—a woman" (127).

Late nineteenth-century medical science rekindled the views of the "medieval church fathers" who declared "woman's vicious hunger for men's precious seminal fluids," sparking "the 'scientific' origins of the early twentieth-century male's fascination with the concept of woman as vampire."[101] Further evidence of this burning attraction and fear is seen in the following passage, when Stephanos and Tereza become lovers. His fear is due to Tereza's obsession to kill Kyriakos, as she sees him as a potential blackmailer: "Even though I again loathed my lover's soul because of these things, her insatiable body, though, continued to bewitch me. In the end she made me fearful also ... oh, the poor thing! ... Was she trying a new medieval crime against his life [Kyriakos'], this frenetic girl who had become the Middle Ages today?"[102]

Again, literary and art representations of the sexual woman and New Woman as vampire were very common in the several decades commencing the twentieth century. Porter Emerson Browne's *A Fool There Was* (1909) dealt with John Schuyler, a successful businessman and husband who is lured by a woman with vampire-like qualities. She destroys his business, family, and social life, and leaves him for her next "victim." The message was

101 Dijkstra, *Evil Sisters*, 62, 66.

102 Xenopoulos, *Τερέζα Βαρμα-Δακόστα*, 126–127.

a warning not to give in to the seduction of the flesh, or society would degenerate.[103] Others included F. Scott Fitzgerald's *The Beautiful and the Damned* (1922); Bram Stoker's masterpiece *Dracula,* which reflects aspects of the New Woman,[104] and Stoker's *The Lady of the Shroud* (1909), which portrays its female protagonist as a vision with Lamia qualities.[105]

The spider and praying mantis

Although Stephanos will continue to refer to Tereza as the Lamia through to the end of the novel, he also observes her in other beast-like or half-beast-like forms, such as a voracious spider or praying mantis. Stephanos finally tells Kyriakos that Tereza's intentions were to kill him, that she might still do so, and that he should leave the town. Stephanos dramatizes Kyriakos' fear with the following imagery on Everywoman: "I had not seen such fear before! It was a new thing to the soul of one inexperienced, that is the fear of Woman, *eternal Woman, the wild and savage Beast, the wanton Animal with the dark and primitive instincts, inherited, you would say, from that fearsome Phalanx, the Religious one*; which has always a prayer-like stance, but which tears to pieces its male and eats it after it has used it for its pleasure and for fertilization."[106] The "prayer-like stance" of the "Religious one" indicates that Xenopoulos is referring to the praying mantis. However, he appears to have conflated this (intentionally or not) with the Phalanx spider.[107] Xenopoulos may be confusing the spider with the praying mantis; however, the implications are the same: the message is that woman is again placed closer to nature than man and is situated further down the evolutionary scale. As far as Stephanos is concerned, Kyriakos' fate is set—that is, once she has had her pleasure with him, she will set out to kill him.

103 Dijkstra, *Evil Sisters*, 20. *A Fool There Was* was also presented as a play in 1909 and as a movie in 1915.

104 Pykett, *Reading Fin de Siècle Fictions*, 14–15.

105 Auerbach, "Magi and Maidens," 29.

106 Xenopoulos, *Τερέζα Βαρμα-Δακόστα*, 132 (my italics).

107 Phalanx (Φαλάγγι or σφαλάγγι) is a name given to various spiders of the genus Pholcus. They are thought to be venomous to their specific prey. See Stamatakos, *Λεξικόν της νέας ελληνικής γλώσσης*, 2664. Female spiders of the genus Pholcus are known to often eat their males after copulation. It is worth noting that in the early 1900s woman was also likened to the praying mantis (*Mantis religiosa*). When it is about to pounce on its prey, it holds its forelegs in a stance resembling hands folded in prayer. Interestingly, it is this position that the spider takes in Xenopoulos' text also, and possibly the reason why he refers to the spider as "the religious one."

A second generation of post-Darwinian biologists such as Rémy de Gourmont provided the data for many novelists to identify woman as a pre-evolutionary, voracious creature of an entomological nature, often utilizing the image of the spider or praying mantis. The following passage is from Gourmont's book *The Natural Philosophy of Love* (*Physique de l'amour—essai sur l'instinct sexuel*) (1903), which deals with the praying mantis and is analogous to Xenopoulos' aforementioned passage: "The mantis who eats her husband is an excellent egg-layer who prepares, passionately, the future of her progeny. … The male is bashful; at the moment of desire he limits himself to posing, to making sheep's eyes, which the female seems to consider with indifference or disdain. Tired of parade, he finally decides, and with spread wings, leaps trembling upon the back of the ogress. The mating lasts five or six hours; when the knot is loosed, the suitor is, regularly, eaten. The terrible female is polyandrous."[108]

The New Woman became Everywoman, who innately had the features of a spider or praying mantis (or a vampire) and was directed by her sexuality and her female function of reproduction. In the same fashion that the female, as the vampire, "craved" the vital fluid of the male because of its regenerative qualities, according to Gourmont the female, as the spider, ate her mate after copulation so as to receive his vital fluids. Hence the spermatophage characteristics of the spider, which entomologists were showing in terms of anthropomorphic traits, were to become a direct reflection of the way women behaved in society.

The study of Woman in terms of her pre-evolutionary entomological traits was pervasive by the 1920s. This is seen with the publication in 1927 of the comprehensive *Woman and Love*, a study by the German doctor Bernard Bauer. His study of the female psyche in terms of the predatory and cannibalistic nature of spiders and praying mantis (and vampires) continued to reflect the misogynistic sentiments of many intellectuals of his time. They believed that a woman who was unsatisfied sexually, like her female entomological counterparts of the lower animal kingdom, was ill-tempered and dangerous; and in the extreme case she became the "ultimate vampire."[109] It is of no surprise then that the decade this large study was being explored

108 Gourmont, *The Natural Philosophy of Love*, 98–99. For further representations of Woman as a spider or praying mantis around 1900, see Dijkstra, *Evil Sisters*, 68–69.

109 Dijkstra, *Evil Sisters*, 215.

Figure 5.2. Xavier Mellery, *Fall of the Last Autumn Leaves*, ca. 1890 (watercolor, ink, charcoal and black chalk on paper mounted on cardboard, silver-gold background). © [Royal Museums of Fine Arts of Belgium, Brussels / Photo d'art Speltdoorn & Fils, Bruxelles].

was also the time when Xenopoulos was writing *Tereza*, thus highlighting further that these ideas were popularized and flowing freely between science and the literary world.

The literary world was absorbing the "scientific data," creating numerous literary representations of the human female as one of these creatures; for example, the African woman in Joseph Conrad's *Heart of Darkness* (serialized in 1899; published as a book in 1902) is depicted as a praying mantis.[110]

110 Conrad, *Heart of Darkness*, 127–128.

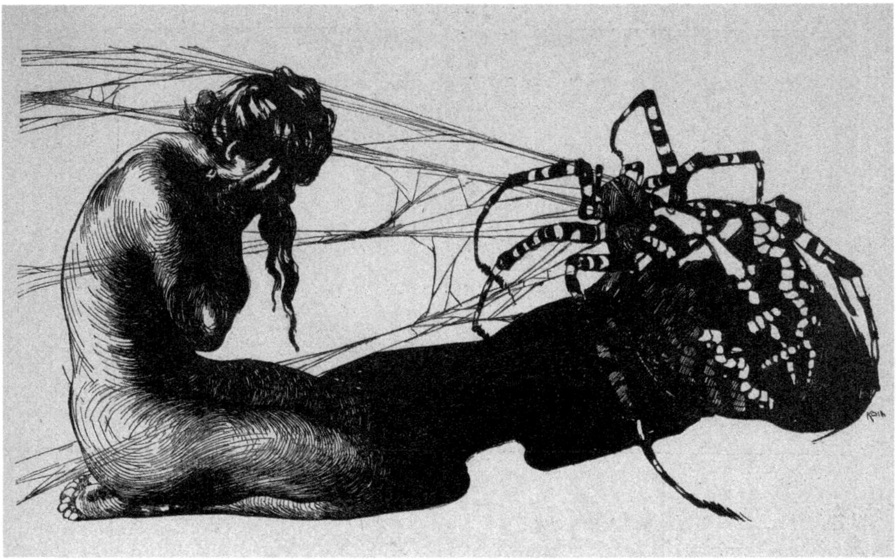

Figure 5.3. Alois Kolb, *Sex and Character*, Reproduced with permission from Heidelberg University Library, *Jugend: Münchner illustrierte Wochenschrift für Kunst und Leben*, no. 51, 1903, 937.

Again, art and literature interacted, as seen in Xavier Mellery's ca. 1890 *Fall of the Last Autumn Leaves* (see fig. 5.2), illustrating three women as spiders in a web, ready to devour some unsuspecting creature, most likely male. Creative writers would commonly use this imagery as a metaphor for the "predatory nature of women."[111] The spider spinning her web of death for "the unfortunate victim of love" is elaborately described in Kazantzakis' novella *Serpent and Lily*.[112] See also figure 5.3 for Alois Kolb's 1903 *Sex and Character*.

The Medusa

For fear of being found out by Stephanos about her lovers, Tereza attempts to stab him with her medieval dagger. Stephanos will escape but is overcome by the same fear which previously enveloped Kyriakos:

111 Bade, *Femme Fatale*, 16, 22.

112 Kazantzakis, *Serpent and Lily*, 59.

> Only then did I realize all the fear, the great fear, the mystic fear of that needy man. The same [fear] seized me also! With eyes closed and eyes open, in the light and in the dark, I couldn't see anything but a savage Tereza, with her naked breasts, with the flare of hate in her blue-green eyes, and with the flashing blade in her hand, ready to plunge it into my heart. Her beauty, unchanged, made it even all the more horrifying and terrifying. She reminded me of the *Medusa* and it appeared to me that *instead of hair, she also had live gold snakes*. I wouldn't look at the Varma's house for fear that I might see her at some window and freeze.[113]

Stephanos also sees a different side to her in the form of the Medusa. Generally speaking, in Greek mythology Medusa was one of the Gorgons, who had a head of snakes; whoever looked upon her gazing eyes was turned to stone. Praz provides representations of the Medusa as the fatal woman in nineteenth-century literature.[114] In the sexual woman and the New Woman at the turn of the century, Dijkstra notes that "with her bouffant of snakes, paralyzing eyes, and bestial proclivities, [she] was the very personification of all that was evil in the gynander."[115] Stoker's Lucy in *Dracula,* in a bestial form at one point is depicted as a Medusa. Unlike the classic Medusa, who acted only through her gaze, the turn of the century Medusa was active, as in Carlos Schwabe's 1895 artwork *Medusa,* who was considered an important example of "woman as a predatory sexual being."[116]

Ghostly presence in the mansion

Tereza's transformation is complete, and so is the transformation of the old mansion. The grandfather's old residence has become a physiognomical expression of its resident Tereza. Stephanos likens the medieval mansion to a multi-eyed monster which embodies the "nesting" Tereza:

> And the same medieval palace appeared to me now to be hideous! Its black façade was grimacing in my eyes like a bad, brazen-faced Giant, multi-eyed, multi-mouthed, and fearsome. And I couldn't even think or imagine any-

113 Xenopoulos, *Τερέζα Βαρμα-Δακόστα,* 139 (my italics).

114 Praz, *Romantic Agony,* 23–45.

115 Dijkstra, *Idols of Perversity,* 309. The gynander is the masculinized female, which was a depiction of the New Woman.

116 Ibid.

> more without horror its insides, those dark sanctums, the endless halls. … I was running, and the entire palace with the church and the enclosed garden, an enormous ghost, doubly eerie was running behind me leaning, lopsided, crumbling so as to reach me and flatten me. Tereza, who was nesting inside it, was now its soul. She brought it to life; she haunted it; she steered it towards me …[117]

Fearing that Tereza would kill him, he leaves his home town to live in Athens. Stephanos later hears that Tereza and her mother had returned to her father's home after Dacosta's mistress had run away with a younger man. Tereza then married the consumptive Count Tzortzakis Sirmas, and according to Stephanos it was "for his wealth and for his name" (135). Stephanos believed one of the reasons Tzortzakis married her was because he was tricked by her "female cunning" (143), another sign of Stephanos' misogyny.

The transformation—a "fait accompli"?

Fourteen years after his traumatic episode with Tereza, Stephanos is given the opportunity to see her. He once again fears that he will see the Lamia he vividly remembers, so he hesitates and prefers to watch her from a distance, without being seen. As Dijkstra notes, "man's need to maintain his distance from the animal-woman" is reflected in the art and literary world.[118] Stephanos is shocked to see "the unimaginable vision," describing a complete change in Tereza: "The dark circles under her eyes—those marks of sexual rage from the previous years—had disappeared completely. Her nostrils were no longer so red, shiny, and swollen, or so spasmodically flaring as they were then, betraying her wantonness. The expression of her physiognomy was entirely that of a child; and a sweetness, a purity, and a pleasantness flowed throughout her personality."[119]

Questioning this unexpected "metamorphosis" to her original state, Stephanos speculates that it is due to the change of environment:

117 Xenopoulos, *Τερέζα Βάρμα-Δακόστα*, 139–140.
118 Dijkstra, *Idols of Perversity*, 321.
119 Xenopoulos, *Τερέζα Βάρμα-Δακόστα*, 146.

> What had happened to the frenetic Lamia of the medieval palace? How had Tereza Varma-Dacosta changed and metamorphosed like that, soul and body? ... It would have been the change of the environment first of all. The Dacosta girl would have reverted to the way she had been as soon as she left the terrible Varma's residence, returning to live in her father's house. There were no old things there or darkness or suggestions or mysteries. Everything was new, bright, and well-lit; none of those things which terrified her, froze her, heated her, or drove her to madness in that dark ruin of her grandfather. There were no traps, chests with stilettos, or books with secret recipes; no Ganymedes metamorphosed into a young barber. The instincts, the sexual and the other instincts which awakened in her suddenly in the Varma's residence, had *gradually* become dormant in the Dacosta house.[120]

He almost makes his presence known to her, but a sudden flashback of the dagger makes him wonder if what he sees now is just deception: "In a little while I lost from my sight this metamorphosed Tereza, never to see her again. Hence I never found out if her metamorphosis was deep and real as it had seemed to me at the beginning, or superficial and fake as I had feared in the end."[121]

Dual personalities

Xenopoulos presented two extreme sides of Tereza's personality and appearance, which were manifested before and after the transformation, reflecting a bipolar nature. Prior to the transformation, Tereza was seen as a beautiful, intelligent, self-assured, and unperverted woman. The transformation revealed a dark side of Tereza, where she presented as a frenetically uncontrolled, sexually perverse, and deadly monster. This atavistic regression occurred supposedly because of the change in the environment, when she moved out of her well-to-do home and into her grandfather's decaying old mansion; the latter symbolically epitomizes the perceived degeneration of the human race which obsessed society then. She appeared to revert back to her original self when she moved back home and then married into money.

120 Ibid., 146–147 (my italics).
121 Ibid., 147.

Particularly among mutation fantasies, this duality of personalities was a common tool in the later nineteenth century and well into the twentieth. It served to demonstrate the dual nature of humanity, which can be exhibited using various dichotomies. These would include the good and evil, the regressive and progressive, the masculine and feminine, and numerous others. Apart from those novels that have already been mentioned in this study, perhaps the most important exemplum is R. L. Stevenson's *The Strange Case of Dr Jekyll and Mr Hyde* (1886).

Clearly, regarding Tereza the dichotomy of good and evil is juxtaposed alongside the dichotomy of the non-feminized woman and the degenerate representation of the New Woman. Angelika Psarra refers to woman in the literature around 1900 as being represented by the face of Janus—that is, in the form of "male-female terror from the one side and the feminine ideal from the other."[122] The fact that Tereza displayed chameleon-like qualities by transforming back and forth in this dichotomy perhaps also reflects the perceived highly mutable nature of woman. In creative writing, the character of the New Woman, particularly the decadent version, would lend itself to metamorphic possibilities, whereas passive or unemancipated woman did not have the capacity to transform.[123]

Eternal battle of the sexes

Intellectuals around 1900 and well into the century linked the theory of evolution with intellectual modes of thought and philosophies, such as those of Nietzsche, Nordau, and Lombroso, to show that the male was heading towards "superman" status and needed to do so in order for society to take, in evolutionary terms, a progressive road.[124] Unfortunately, the female as predator drained the male's vital essence for reproduction, resulting in the loss of his creative and also his physical capacities. She was considered to be in a constant struggle with the male and was driving the male and society down the path of regression and extinction. This will follow with a discussion by Xenopoulos on the struggle for existence between the sexes, in relation to Strindberg's play *Miss Julie*.

122 Psarra, "Το μυθιστόρημα της χειραφέτησης," 461.

123 Auerbach, "Magi and Maidens," 23. See Auerbach's study also for further literary representations of transformation.

124 Dijkstra, *Evil Sisters*, 73. He includes Freud, Wagner, Alfred Jarry, and also Adolf Hitler as those influenced by this mode of thought.

Extinction

Creative writers of the late nineteenth century and up to at least the 1930s commonly used the term "extinction" to denote the decline of a family with an aristocratic bloodline. It was of course used by Darwin and other scientists to denote the death or destruction of an animal or plant species. It was also used in literature to denote the extinction of a "species," which is relevant to this discussion, since this "species" may have been the degenerate representations of the New Woman. In the post-Darwinian era, extinction had been a component of early degeneration theories such as that of Benedict Morel (1809–1873); eventually, after the 1880s extinction was viewed as predominantly Darwinian. This did not follow for Lamarck's theory. D. R. Oldroyd states clearly that "Lamarck, unlike Darwin, was not primarily concerned with the problem of the origin of discrete species. And although change occurred constantly, organisms did *not* become extinct in Lamarck's system."[125] Furthermore, it should also be noted that Beer highlights in a literary sense: "Darwin's theory required extinction," unlike "Ovid's assertion in *Metamorphoses*," where "'Omnia mutantur, nihil interit.' Everything changes, nothing dies."[126]

Extinction, an important part of the evolutionary process, was a concept that was on Xenopoulos' mind when he gave his lecture at the 1908 Athenian premiere of August Strindberg's play *Miss Julie*.[127] Here Xenopoulos used, to a large extent, Strindberg's own preface to the play.[128] Strindberg was clearly a misogynist, which reflected in his play with his negative representation of the New Woman. It will be argued here that this play had a significant influence on the novel *Tereza*, due to its themes of extinction, primarily of the New Woman and also to a lesser extent of the aristocracy. Certainly critics will argue that the play influenced Xenopoulos' novel *The Three-Sided Woman*, and this will be discussed in chapter 6.

125 Oldroyd, *Darwin's Impacts*, 33 (italics in original).

126 Beer, *Darwin's Plots*, 104.

127 Xenopoulos, "Ο Αύγουστος Στρίντμπεργ," 166–173. The lecture was originally published in *Panathinaia*, no.17 (1908–1909): 140–146.

128 Strindberg, *Miss Julie*, xxv-xxxix. The story line is similar to that of *Tereza* in that it is about an aristocratic young woman who falls morally by having sexual relations with her father's valet. However, unlike Tereza Julie finally feels obliged to end her own life, hence her extinction. Whereas Tereza as *femme fatale* attempts to kill the men with whom she has had affairs.

Tereza's persona and Strindberg's Miss Julie

In the novel Stephanos ponders Tereza's genetic make-up in relation to her ravishing beauty, indicating that she was neither like her father nor her mother. So he already saw her as some type of "variant." "Defective by nature" due to her aristocratic background, she was doubly marginalized because she was a woman. More specifically, she was a woman who was assertive, intelligent, and outspoken—a potential contender for a New Woman of the late 1800s. However, her perceived "medieval ideas" of a social Darwinian nature regarding the aristocracy's biologically determined superiority also rendered her a potential degenerate.

In his 1888 *Miss Julie,* August Strindberg sees the female "species" as a weak, short-lived deviation which will, by the Darwinian laws of the struggle of existence and natural selection, inevitably become extinct. He claimed that this will be achieved either by some form of self-destruction (in the case of Miss Julie, suicide) or by marrying a male who himself is some form of degenerate; for instance, weak-mindedness in males was considered a form of degeneration.[129] Similarly, because of her deviant nature Tereza's own aristocratic ancestral line was doomed. If she married the "weak-minded" Stephanos, then according to the thinking of the time she would have produced these "asexual" offspring, which would not have produced further; hence the line would have become extinct.

Returning to her father's house and marrying Tzortzakis for his noble line, Tereza reverted to her pre-beast phase. However, the aristocratic line on Tzortzakis' side was also doomed because of his tuberculosis, considered an "inherited trait" and a mark of degeneration which would lead to extinction.[130] Consequently, he would not have been considered an ideal mate to continue a bloodline. Note, though, that at the end of the story it is implied that he may have been cured. He would still have been marked with degeneration due to its "inheritability."

In an evolutionary sense, Stephanos associated Count Frederikos (Tereza's grandfather), the final patriarch of the Varma bloodline, with his ances-

---

129 Extinction of the New Woman as a literary theme is found in Kate Chopin's *The Awakening* (1899). Edna also takes her own life for similar moral reasons.

130 Tuberculosis and syphilis were considered hereditary diseases up to at least the 1940s. See Greenslade, *Degeneration,* 294.

tors, primitive men, and simultaneously with the degeneration and decay of the old mansion: "There is a 'living medieval man'! I would say to myself. He came straight from those thick darknesses, from the murky Night of Humanity which on departing left its last veils to drag over those ruins."[131]

So as to perpetuate the family bloodline, Count Frederikos wanted Tereza to marry Stephanos even though he was of a lower class. The Count knew he was of good stock, so he offered to make him his "son," giving him and his heirs the Varma's name along with the title of count (46). The Varmas bloodline had degenerated to such an extent (probably due to economic ruin) that the only thing that could save it from extinction was "artificial selection"—that is, the "infusion of outside blood," which comes in the form of Stephanos.[132] Relevant to the class strata, as Beer indicates, extinction becomes "a class parable," as only the aristocracy would have been concerned about the ending of noble family lines.[133] There was fluidity of the class system in Greece, and Stephanos confirmed this with his comments that at the time there was a "mixing" of the classes. Nevertheless, Tereza had no intention of marrying outside the aristocracy, whose members tended to marry among themselves. Such unions would ultimately lead to problems in biological traits of the offspring. Degenerate inherent traits along with a fall in finance would lead to a devolution or regression of the bloodline and result in extinction.

Social physiognomy relating the continuing devolution of the Varma family with the decay of the Varma house is commented on by Stephanos, as narrator. He describes in great detail the decaying nature of the old Varma mansion (19–24, 82–91). The family bloodline is on the road to extinction. Through Stephanos' eyes, the degeneration of the house will appear to continue along with that of the aristocratic family: "We climbed the old staircase which appeared to be even more decayed" (82–83).

131 Xenopoulos, *Τερέζα Βάρμα-Δακόστα*, 42–43.

132 Artificial selection, a Darwinian term, was frequently used in Victorian literature for this purpose. Beer refers to it extensively in *Darwin's Plots*.

133 Beer, *Darwin's Plots*, 120. She also indicates that Thomas Hardy used extinction in this way in his novels. Generally speaking, in Victorian society, with social levelling the aristocracy was under threat. Due to education the middle classes were on the rise, in contrast to the economic ruin of the aristocracy. Only those Greek islands such as Zakynthos, with a strong nobility, could be considered to come close to this socioeconomic trend. The rest of Greece, parts of which were still under Ottoman rule during the setting of this story, could not be included and would have been under a different structure. For further general reading on classes in Greece see Hadziiossif, "Class Structure and Class Antagonism," 3–95; for further reading on the English aspect see Alden, *Social Mobility in the English Bildungsroman*.

Xenopoulos' lecture on Strindberg's *Miss Julie* may be close to what Strindberg said himself in the preface of his play. Xenopoulos' message on Strindberg's modes of thought and comments indicate that he was in complete accordance with his ideas. He also attempted to justify Strindberg's "antifeminist and misogynistic" attitude, which he appears to validate with biological determinism and evolutionary theory.[134] Xenopoulos' talk becomes also a valuable analytic tool, because it transparently shows Xenopoulos' own interpretation and understanding of the preface and the play. His analysis is accurate, apart from one omission. Strindberg mentions his use of Darwinian thought by name and elaborates on it, whereas Xenopoulos discusses the evolutionary issue without giving it a name.[135]

Referring to Miss Julie in terms of her variant and deviant nature, Xenopoulos reasons: "According to the playwright, the beings of this species who have diverged from their gender and are by nature half-women, are destined to be defeated in the struggle for life and become extinct. Hence Julie, too, is defeated as you will see, in first close combat with her man, a strong, complete one, true to his gender and to nature; and the sick woman dies due to the defeat and becomes extinct, and at the same time her generation becomes extinct—an age-old generation of royal origin, which is fatefully extinguished as soon as a half-woman entered into her evolution—that is, Julie's mother."[136] Xenopoulos indicates that Miss Julie's mother had "rebellious instincts" (168), which were considered by both Xenopoulos and Strindberg to be the cause of the extinction.

In his lecture Xenopoulos suggests that the catastrophe is mainly due to evolutionary reasons, though he does say that Strindberg also examines the physiological, psychological, and social reasons. Further to this Xenopoulos surmises: "John (Jean) originates from the populace, he does not have an ancestry but he is, he says, capable of becoming an ancestor. Julie originates from the aristocracy. In her blood flows royal blood, but already the degeneration of her generation has taken place and it [the generation] will be extinguished. In the struggle between those two creatures, which are so dissimilar, we see the battle of the classes. The one rises continu-

134 Xenopoulos, "Ο Αύγουστος Στρίντμπεργ," 168.

135 By contrast, when discussing the morality associated with the aristocrat and the servant in the play, Xenopoulos does mention that it is of a Nietzschean nature. Ibid., 170–171.

136 Ibid., 168.

ously with all the momentum of its young and healthy blood. The other falls."[137]

In relation to the gender issues and extinction in the play, Xenopoulos declares: "He presents an aspect of the great, the eternal struggle between male and female, between man and woman. Strindberg is of the opinion that the two genders are two opposing army camps. ... On the other hand, in *Miss Julie* the man devours the woman because here John is the strong one."[138] Xenopoulos likens Miss Julie and her generation to a rotting tree which must die (173). Prior to Darwin, unilinear rather than branching schemes of evolution were recognized; and although pre-Darwinian family-tree diagrams had the branching tree format, they were not aligned to human evolution. The post-Darwinian image of descent has been likened to branching trees; and this in turn has become an image of kinship relationship and taxonomic classification.[139] The family tree metaphor has been used extensively in literature, and is associated with Darwin's web of affinities.[140] Janet Beizer indicates that the family tree doubles as a tree of degeneration.[141] This doubling of the family tree as a tree of degeneration is described in detail in *Tereza* (44–46).

New Woman writer versus decadent artist

Linda Dowling refers to the New Woman writer and the decadent writer as "twin apostles of social apocalypse," because their representations of woman are diametrically opposed to each other.[142] Ann Heilmann argues that "feminist writers, who saw themselves as agents of moral renovation, were the antithesis of fin-de-siècle decadence."[143] In other words, the dichotomy in evolutionary terms was regeneration versus degeneration or eugenics versus retrogression. Further to this, feminist writers of the New Woman genre were against the decadent representations of the aesthetes of the nineteenth century. According to Heilmann, the marked difference between decadent writers and New Woman writers was that the first emphasized

137 Ibid., 170–171.
138 Ibid., 170.
139 Alter, *Darwinism and the Linguistic Image*, 5–6.
140 Beer, *Darwin's Plots*, 156–168
141 Beizer, *Ventriloquized Bodies*, 246.
142 Dowling, "The Decadent and the New Woman in the 1890s," 57–58.
143 Heilmann, *New Woman Fiction*, 48.

"form and brilliancy of style" and the latter were concerned with "content and woman's social reality."[144] With this she differentiates the decadents' motto, "art for art's sake," from that of the feminists, who she argues had a social purpose.[145] Citing Elaine Showalter, Heilmann argues that feminist writers "challenged the sexual stereotyping of women reinforced by the aesthetic fetishization of the *femme fatale*."[146]

Based on this line of thought, one would suggest that *Tereza* does contain motifs of decadence, despite the fact that Xenopoulos strongly denied he was a decadent writer in his autobiographical *My life as a novel*.[147] In fact, in his letter to Petros Charis discussed earlier in this chapter, he mentions that he spent a great amount of time on the language and style of the novel. He would also be criticized for the antifeminist treatment of his protagonist in *The Three-Sided Woman*. His response to this was: "And first I want to be good with Art and then with Feminism."[148]

Xenopoulos was certainly well aware of the New Woman novels. He had read the feminist writer Kalliroi Parren's post-Darwinian trilogy and even reviewed it in the literary periodical *Panathinaia*.[149] Xenopoulos maintained that the style in Parren's novels was "simple-minded," "sloppy," and "entirely journalistic." Perhaps this reflects Heilmann's earlier comments that such writers were more concerned with the content than the style.

In addition, the young female protagonists Maria and Anna in Parren's trilogy, who were depicted as feminist representations of the New Woman, could easily pass as the antithesis of the decadent representations of Tereza. One example comes from *The witch*: "Although not beautiful, Maria Myrtou was good-looking, fresh-complexioned, and charming. Her life's struggle since early childhood ... had made her character so active and efficient that her assertiveness, that sustaining power of her life, was reflected in her glance and countenance, in her posture and in her words. She knew how to will. She was a person of great patience, modesty, and reserve. ... She was always extremely simple but at the same time always elegant and graceful."[150]

144 Ibid.

145 Ibid.

146 Ibid.

147 Xenopoulos, *Η ζωή μου*, 345.

148 Xenopoulos, "Η *Τρίμορφη γυναίκα*, ο φεμινισμός και ο ανθρωπισμός," 3. For further discussion surrounding this article see Doulaveras, "Τα δυο φύλα," 299–309.

149 Xenopoulos, "Γράμματα, τέχνη, επιστήμη: φιλολογία," 364.

150 Passage cited by and translated in Anastasopoulou, "Feminist Discourse," 13.

As Heilmann aptly conceptualizes regeneration,[151] it follows that Parren's New Woman is associated with (re)generation, whereas Xenopoulos' Tereza is associated with degeneration or devolution. One can speculate whether Xenopoulos' representation of Tereza is a reaction to Parren's representations of the New Women in her trilogy.

The rereading of the novels

Although Xenopoulos may have received his inspiration from Strindberg's *Miss Julie*, which focuses on the New Woman and class struggle, the forty-year span between the two must be noted even though the time setting of the two was the same, and the themes similar. Xenopoulos did have the advantage of this extra period which brought about new modes of thought based on new intellectual trends. By the 1920s, literature based on the biologization of the classes was less common than it had been at the time Strindberg wrote his play. For instance, the traditional nineteenth-century Lavaterian style of physiognomy, in particular social physiognomy, was not utilized as much in the 1920s as a primary theme. So that with the enormous scientific and pseudoscientific data on women accumulating around the 1900s and well into the twentieth century in association with the "Woman Question," creative writers became prolific in producing work related thematically to women.

In terms of critical analysis of literature, Dijkstra has mentioned that the themes of "gender-ideological-fervor-of-the-turn-of-the-century evolutionary theory" have tended to be brushed aside.[152] Other critics have agreed with this, such as Lyn Pykett, who acknowledges the "rereading of discourse on degeneration,"[153] such as Bram Stoker's *Dracula*; this novel, she points out, has been the basis for "discussions of literary representations of and responses to the Woman, and also for discussions of the fin-de-siècle representation of sexuality."[154] In Greece however, commentators have tended not to recognize gender themes from an evolutionary perspective, which signifies a gap in literary scholarship, and hence requires investigation. The

151 Heilmann, *New Woman Fiction*, 188.

152 Dijkstra, *Evil Sisters*, 154–155.

153 Pykett, *Reading Fin de Siècle Fictions*, 14–15. See also includes essays in this book by Nina Auerbach, Sandra Gilbert, and Elaine Showalter.

154 Ibid., 15.

next chapter of this book further examines gender issues from a post-Darwinian perspective.

Returning to *Tereza,* there is no doubt that other themes, such as that of class, remain structural features of the narrative, but it appears that Xenopoulos made it quite clear that Stephanos' final rejection and hatred for the aristocracy was secondary to his miserable surrender to Tereza—that is, the New Woman as the primitive, sexual predator.

CHAPTER VI

NEW WOMAN BIOLOGY, DEGENERATION/REGENERATION, AND THE DESCENT OF MAN

As you say—the plain fact that man bears the evidence of a former hermaphrodite type are [*sic*] as indisputable—as they are [*sic*] carefully ignored.
CHARLES KINGSLEY[1]

It is to the women of the country we must look in this great eugenic movement.
MRS. ALEC-TWEEDIE[2]

The New Woman was represented in many ways. On the one hand, she was considered both the cause and result of degeneration and, on the other, she was important in the discourse on the regeneration of a nation. In the late nineteenth to early twentieth centuries, the discourse of evolutionism became the scientific grounding for the social discourse on the inequalities of gender, race, and class. So as to debate these social issues, the language of biological evolution became the language of social evolution. Moreover, social evolutionism had "emerged ... into a Darwinian milieu."[3] Biology reflected society, and the corresponding terminology intermingled evolution with progress (though not always synonymous), "retrogression" and "atavism" with "degeneration" and "decline," and "natural selection" and "artificial selection" with "eugenics" and "reform." As indicated by Bert Bender in *The Descent of Love*, the terminology utilized in society and

1 From a letter to Charles Darwin, dated November 8, 1867. See Burkhardt et al., *Correspondence of Charles Darwin, 1867*, 15:422.
2 Alec-Tweedie, "Eugenics," 857.
3 Stocking, *Race, Culture, and Evolution*, 120.

in literary discourse included words such as: "instinct," "chance," "genius," "higher and lower races," "savage," "barbaric," "degenerate," "primitive," "ancestor," and also "attraction," "unconscious."[4]

The dominant conservative view of society was that all forms of degeneration, whether biological or moral, had to be removed for fear of humanity's decline. This included the New Woman and her variants (except in her regenerative form), and biological or "inherited" abnormalities in humans.

Darwin's theory of natural selection and his support for artificial selection in the *OS* certainly played a key role in the above ideas. It was however the *DM* which focused and expanded his theories on humanity. By viewing heritable characteristics within the human species as analogous to those of other animals, he examined the issues pertaining to gender, including sexual differences, sexuality, race, and class. These ideas crossed over to the literary world and influenced writers like Xenopoulos. From her perspective, Gillian Beer remarks: "It is likely that the *Descent,* even more than the *Origin,* was the seedbed for later Victorian writers, such as George Gissing, Grant Allen, H. G. Wells, and for New Woman novelists, like Sarah Grand, Mona Caird, and George Egerton."[5]

Besides Darwinism, there were social, historical, and cultural reasons for the birth of both positive and negative versions of the New Woman novel. The perceived fin-de-siècle degeneration of the human species, the Women's Movement, and later World War I had impact on the literary representations of the New Woman. The New Woman degenerate reflected the male's fear of the woman crossing into his domain, thus threatening his long-lived privilege and sense of self-worth in a changing patriarchal society. She was unlike her alter ego, the regenerate New Woman, who wanted to save humanity from degeneration and extinction.

Bearing these observations in mind, the aim of this chapter is to focus mainly on two novels that Xenopoulos wrote in the 1920s which deal with the post-Darwinian themes associated with degeneration and regeneration, and also with issues of sexuality, inheritance, and eugenics.

Following on from chapter 5, the first of the two sections of this chapter examines the biologizing of the New Woman, and the discussion centers

4 Bender, *Descent of Love*, xi. The list I have given in the text is only a portion of Bender's list. By "unconscious" I mean from a biological perspective, and not in Freudian terms.

5 Beer, *Darwin's Plots*, xxiv.

predominantly on Xenopoulos' novel *The Three-Sided Woman* (*Η τρίμορφη γυναίκα*) (1922). The second section of this chapter investigates the treatment of a period of social decline in Xenopoulos' novel *The night of degeneration* (*Η νύχτα του εκφυλισμού*) (1926). Both novels appear to have a didactic element that is associated with the theme of regeneration in society.

My approach in this chapter concentrates on the biological sciences, in particular on the evolutionary biology of Darwinism, rather than on feminist and other social theories. The employment of evolutionary biology is not an uncommon method, even in our times, for probing cultural issues and even gender issues.[6] I will be referring to aspects of Darwin's *OS* and *DM* in my examination of Xenopoulos' work. The chapter reflects the influence of this intellectual trend on Xenopoulos' work through various themes, including the New Woman, women and sexuality, society's decline through degeneration, and its reform through regeneration and eugenics. With reference to eugenics, literary commentators have overlooked the eugenicist themes and motifs in Xenopoulos' narratives, as they have similarly done in other works of modern Greek and non-Greek literature. Primarily, this is due to eugenics' later association with the horrors of Nazism and the Third Reich. However, its earlier association in the first decades of the twentieth century with the New Woman and with health reform had little to do with Hitler's genocidal implementation of a crude eugenics in the 1930s and 1940s.

THE NEW WOMAN, DEGENERATION/REGENERATION: THE CASE FOR *THE THREE-SIDED WOMAN*

Xenopoulos' novel *The Three-Sided Woman*[7] exemplifies the New Woman and her variants derived from a Darwinian model. The story is set around 1900. Nitsa Gazeles, a young Athenian girl, has just returned from Switzerland, where she had been sent by her father to study. She goes about leading

6 For instance Grosz takes a Darwinian approach in her book *The Nick of Time*.

7 *The Three-Sided Woman* was first published in serial form as *Τα τρελοκόριτσα* (The wild girls) from April 9, 1917, to July 3, 1917, in the Athenian newspaper *Nea Imera*. It was published as *The Three-Sided Woman* in the newspaper *Ethnos* between April 28, 1922, and August 8, 1922, in book form by Kollaros in 1924 with Xenopoulos' introduction, and then again in 1930. An English translation by Barbara Kent is also available, published by Vlassis Brothers in 1992. Kent's translation will be used in this study. I also make reference to pages in the following Greek edition: Xenopoulos, "Η τρίμορφη γυναίκα," in *Άπαντα*, vol. 8 (Athens: Biris, 1972), 213–436.

the same freer life which she experienced in Switzerland. Her family warns her that unlike other places in Europe, Athens is not ready to give its daughters such freedom. As a consequence of her freedom, Nitsa finds herself in some moral dilemmas. In spite of this, the story ends on a happy note when Nitsa marries Kleanthis, the man she actually loves, which saves her from moral degeneration and destruction.

In the last few decades there has been significant work on the feminist and antifeminist aspects of certain of Xenopoulos' novels.[8] The history of the New Woman and New Woman novels in Greece has been assessed with reference predominantly to feminist writer Kalliroi Parren, but also with some references to Xenopoulos.[9]

Earlier commentators, such as Alkis Thrylos (1896–1971), have interpreted Nitsa as a distorted form of the emancipated woman of around 1900. They have criticized Xenopoulos for this representation, as potentially damaging the cause of the feminist movement in Greece.[10] I argue that Xenopoulos represents Nitsa as a masculinized degenerate who he attempts to portray as a New Woman. Several commentators have also acknowledged Nitsa's masculine traits, and those of other female protagonists in some of Xenopoulos' work.[11] Farinou-Malamatari indicates that all of Xenopoulos' female protagonists who rebel and who either end up dying or marry after changing their rebellious ways have a "masculine temperament."[12] However, she does not connect this to the New Woman genre. To the best of my knowledge, there has not been an investigation of the Darwinian influence in the representation of the New Woman within modern Greek literature. Some biological connotations have been made, but no commentator has taken the studies further.[13] In this work I will use Xenopoulos' *Three-Sided Woman* as a case study.

---

8 See for instance Doulaveras, "Τα δυο φύλα"; Varika, *Η εξέργηση των κυριών*, 112–121.

9 See Psarra, "Το μυθιστόρημα της χειραφέτησης," 409–486; Anastasopoulou, *Η συνετή απόστολος*; Anastasopoulou, "Feminist Discourse and Literary Representation," 10–20.

10 See further discussion on this rivalry between the feminist Alkis Thrylos and Xenopoulos in Doulaveras, "Τα δυο φύλα," 299–399.

11 Ibid., 237, 308.

12 Farinou-Malamatari, "Γρηγόριος Ξενόπουλος," 313.

13 Anastasopoulou maintains that Otto Weininger's *Sex and Character* may have influenced Xenopoulos when writing novels such as *Stella Violanti* and *The Three-Sided Woman*. She attributes to Xenopoulos the view that by nature women are unable to protect themselves morally. This is due to what she claims is Weininger's premise: that women never mature and that they are prey to their sexual hormones. See Anastasopoulou, *Η συνετή απόστολος*, 344.

Xenopoulos' preoccupation with women protagonists in many of his novels was a trend typical of many Western literary writers in the early twentieth century. Virginia Woolf in her essay *A Room of One's Own* reflects on the vast proliferation of women-oriented novels written in the first decades of the twentieth century: "Have you any notion how many books are written about women in the course of one year? Have you any notion how many are written by men? Are you aware that you are, perhaps, the most discussed animal in the universe? ... Why are women ... so much more interesting to men than men are to women?"[14]

Although many of the writers were also women, Woolf brings to light the overall literary world's preoccupation with women, their sexuality, femininity, morality, psyche, emancipation—themes captured in almost all of Xenopoulos' fiction. It is possible to speculate that many of Xenopoulos' novels which explore the different types and different degrees of "immorality" are not just frivolous stories of love. Instead, one could suggest that they use "the transgressive woman as both a trigger and a focus for a range of narratives of uncertainty about gender, class, marriage, and the family."[15] I will be exploring this idea with reference to Xenopoulos' novels.

A post-Darwinian model of the New Woman: sexuality and sexual selection

Xenopoulos often wrote about the sexual instinct and sexuality in his novels, and these themes were often about women. This can be seen even in his novel *My son and my daughter*, where he examines the lives of a brother and sister, focusing primarily on the daughter. In 1939, in his autobiographical novel he wrote on the sex instinct and on how after the dominant period of literary naturalism, he came about writing his psychological novels: "Later, before I had read Freud—it is just ten years since I got to know him—when I myself arrived at a comparable theory and I formed the belief that the sexual life is the primary life of man; it is this with which I was mainly preoccupied in my psychological novels. Indeed, the awakening of the sexual instinct, sometime very early, was one of my favorite themes. And depending on the

14 Woolf, *Room of One's Own*, 40–42.

15 This approach by writers was not uncommon. Lyn Pykett makes this observation in the writings of Mary Elizabeth Braddon. See Pykett, *The "Improper" Feminine*, 102.

theme, my novel became either more sedate or more open. ... With regards to the open novel, I mean serious, true, *scientific*, if I can say it like this."[16]

One can presumably say that up until 1929 Xenopoulos was not influenced by Freud's theories in his novels. While he does not acknowledge it in the aforementioned passage, it is very likely that he was influenced by his Darwinian readings on the sexual instinct in the *DM*. Two thirds of the approximately 700-page *DM* are entirely on the subject of sexual selection and sexual differences in animals and humanity. On sexual selection among humans, Darwin deemed it important enough to add the following to his chapter 20 in the second edition: "No excuse is needed for treating this subject in some detail: for, as the German philosopher Schopenhauer remarks, 'the final aim of all love intrigues, be they comic or tragic, is really of more importance than all other ends in human life. What it all turns upon is nothing less than the composition of the next generation. ... It is not the weal or woe of any one individual, but that of the human race to come, which is here at stake.'"[17]

Lawrence Birken describes Darwin's work: "Darwin redefined biology itself. ... Darwin discovered desire as the fundamental ground of both sexes. By making this discovery, he became the real founder of sexology, the forerunner of the sexologists. ... With Darwin, a new domain of knowledge appeared, one inhabited by desiring individuals."[18] Birken continues by saying that Freud was one of Darwin's "intellectual progenies," extending Darwin's work by defining women's sexuality.[19]

Darwin was the first to embark on scientific analysis of the differences in sexual selection, and of sexual secondary characteristics between males and females. Although he defined sexual selection in the *OS*, it is his *DM* which mainly develops his analysis. Darwin's discourse on the way males and females court and on the differences in their sexual nature initiated further scientific analysis and captured the imagination of the literary world. Themes such as desirability, sexual selection, and the "awakening" of the sexual instinct were taken up by many literary writers whose work was

16 Xenopoulos, *Η ζωή μου*, 345 (my italics).

17 Darwin, *DM*, 2nd ed., 2:653. Darwin is referring to the following article: Asher, "Schopenhauer and Darwinism," 323.

18 Birken, *Consuming Desire*, 7. This idea is also noted in Bender, *Descent of Love*, 16; see also Bender, "Frank Norris on the Evolution and Repression of the Sexual Instinct," 76.

19 Birken, *Consuming Desire*, 7.

seen as Darwinian. As indicated by Bender, prior to Darwin's *DM*, "unwelcome [were] any public discussions of sexual biology ... to the Victorian sensibility."[20] In fiction, the negatively-viewed New Woman was often strongly correlated with the sexually immoral woman, as can be seen in many of Xenopoulos' works.[21] This deviant New Woman showed all the hallmarks of a degenerate, physically and morally, unlike her alter ego, the feminists' version of the New Woman, who was attempting to attain moral and economic freedom.

Darwin's *DM* provided abundant ammunition for the validation of Victorian antifeminist ideology.[22] The following passages from Darwin's *DM* will show this phenomenon:

> Sexual selection depends on the success of certain individuals over others of the same sex in relation to propagation of the species; whilst natural selection depends on the success of both sexes, at all ages, in relation to the general conditions of life. The sexual struggle is of two kinds; in the one it is between the individuals of the same sex, generally the male sex, in order to drive away or kill their rivals, *the female remains passive*; whilst in the other, the struggle is likewise between the individuals of the same sex, in order to excite or charm those of the opposite sex, generally the females, which no longer remain passive, but select the more agreeable partners.[23]

The male's role is active, and, while the female may select, her power of choice could also be taken by the male: "Man is more powerful in body and mind than woman, and in the savage state he keeps her in a far more abject state of bondage than does the male of any other animal; therefore it is not surprising that he should have gained the power of selection."[24]

When Darwin refers to the "mind" as he does in the above passage, or when he refers to "mental differences" throughout the *DM*, he is referring to the intellectual, emotional, and moral aspects. Clearly from the following passage, Darwin sees the female as taking a truly passive role, emphasizing that she is not presumed to demonstrate her sexuality to the same degree as

20 Bender, *Descent of Love*, 8. Note also that Xenopoulos in his autobiography takes on a defensive tone regarding his more "explicit" novels.

21 For instance, this can be seen with Xenopoulos' protagonist Roza in his novel *The Downfall*.

22 E. Richards, "Darwin and the Descent of Woman," 61.

23 Darwin, *DM*, 2:398 (my italics).

24 Ibid., 371.

the male: "When the sexes differ from each other in external appearance, it is the male which, with rare exceptions, has been chiefly modified; for the female still remains more like the young of her own species, and more like the other members of the same group. The cause of this seems to lie in the males of almost all animals *having stronger passions than the females.* ... The female, on the other hand, with the rarest exceptions, is less eager than the male. ... She generally *'requires to be courted'; she is coy, and may often be seen as endeavouring for a long time to escape from the male.*"[25] To add to this, as Richards notes, Darwin saw the male as a "natural polygamist."[26] Darwin's views of sexual selection present as androcentric, like the antifeminist approach adapted by some literary writers.

In *The Three-Sided Woman* the "awakening" of Nitsa's sexual instinct occurs swiftly when a stranger steals a kiss from her in Geneva, whereas years later it is the detestable Dardouvas who also catches her off guard with a kiss, which has a more lasting effect: "It was that kiss of Dardouvas! The memory of it was at once repulsive and pleasurable. It both delighted and disgusted her. For she often thought of it and tried to recapture the sensation she had experienced at the time and to relive it: his prickly moustache, the ardor of his breath, and something wet, cold, and slippery which his warm lips left on her cheek, with a peculiar sound which, in her confusion, deafened her ..."[27] Nitsa goes on to contemplate that a kiss from someone she actually could love would not be as thrilling as one received from any of the detestable men she had already been kissed by; perhaps a rather distorted behavior that only affirms her "degenerate nature."

In the late nineteenth and early twentieth centuries, women were viewed in society and in literature through the influence of a "sex science" and psychoanalysis, which were rooted in the Darwinian biological and medical sciences.[28] Utilizing an evolutionary framework, the New Woman and the homosexual were represented as "freakish sports of nature—regressive or degenerative forms—or conversely, as highly evolved types."[29] Otto

25 Darwin, *DM*, 1:271–273 (my italics).

26 "Natural polygamist" is Richards' terminology. See E. Richards, "Darwin and the Descent of Woman," 70. The actual section on male polygamy in the animal kingdom is Darwin, *DM*, 2:265–271.

27 Xenopoulos, *The Three-Sided Woman*, 112 (suspension points in original). (Xenopoulos, *Η τρίμορφη γυναίκα*, 295.)

28 Pykett, *Engendering Fictions*, 20.

29 Ibid., 25.

Weininger in his book *Sex and Character* (1903) referred to emancipated women as sexually intermediate forms who were masculinized in some way.[30] In a similar manner, these intermediate forms that were male, were feminized. Homosexuality in a woman was considered to be an outcome of her masculinity, and so presupposed a higher degree of development.[31] Theories of social evolution derived from Darwinism represented women as "either symptoms of cultural degeneration and decadence, or as forms of resistance to cultural crisis and as points of cultural renewal and regeneration."[32]

According to Trotter and Pykett, the discourse from which this phenomenon originated was the nineteenth-century biologist Edwin Lankester's *Degeneration: A Chapter in Darwinism* (1880).[33] Lankester's book, based on Darwinian ideas, was drawn upon in creative literature well into the 1930s.[34] However, Darwinian evolution was not the same as progress, and could just as easily regress as progress: "It is clearly enough possible for a set of forces such as we sum up under the head 'natural selection' to so act on the structure of an organism as to produce one of three results, namely these: to keep it in statu quo; to increase the complexity of its structure; or lastly, to diminish the complexity of its structure. We have as possibilities either Balance, or Elaboration, or Degeneration."[35]

Darwin's theories on the difference between males and females could be interpreted as reaffirming society's patriarchal hierarchy. Darwin's *DM* declared woman's inferiority because of her "arrested" development, which claimed to have left her at a lower point on the evolutionary scale than man, and hence more primitive. Darwin states on the female: "in the formation of her skull, [she] is said to be intermediate between the child and the man"; and her "faculties are characteristic of the lower races, and therefore of a past and lower state of civilization."[36] According to the *DM*, the male's superiority was also reflected in his more developed mental powers.[37]

30 Weininger, *Sex and Character*, 64. Weininger mentions Darwin's influence in his book on a number of occasions, including to prove the "maleness of genius" (71).

31 Ibid., 66.

32 Pykett, *Engendering Fictions*, 25.

33 See Trotter, *English Novel in History*, 111–112; also Pykett, *Engendering Fictions*, 25.

34 Pykett, *Engendering Fictions*, 25.

35 Lankester, *Degeneration*, 28–29.

36 Darwin, *DM*, 2:317, 327.

37 See the discussion in chapter 3 on Darwin's "Difference in the Mental Powers of the Two Sexes."

In terms of sexual differences and instinct, sexual selection, degeneration, and hierarchy, Darwin's analysis could be used to justify and perpetuate antifeminist sentiments at the end of the nineteenth century. This continued well into the first decades of the twentieth century. This biological trend, along with the Women's Movement, generated a vast literary response by way of the New Woman novels. The language of evolutionism and, in particular, of degeneration and progress was entrenched in the writings of the New Woman fiction of the late nineteenth and early twentieth centuries. The "scientific" discourse of degeneration in terms of the genders and sex was vast and included extremely popular works, such as Richard von Krafft-Ebing's *Psychopathia Sexualis* (1886), Max Nordau's *Degeneration*, Havelock Ellis' *Man and Woman* (1894), his *Sexual Inversion* (1897), and Edward Carpenter's paper *The Intermediate Sex* (1896).[38]

Variation and reversion as degeneration

As a biological concept, variation can be traced to Darwin's *OS*, his *DM*, and to his 900-page *The Variation of Animals and Plants under Domestication* (1868). One of the cornerstones of Darwin's theory of natural selection was the postulate of variation. Darwin understood that variation was primarily a factor associated with heredity, even though he did not understand the mechanism of inheritance. Mendelian heredity came later.[39] In his 1988 book *Consuming Desire*, Birken emphasizes how, for a time, many cases of variation were interpreted as "degeneration":

> Darwin's great contribution to and also his break with bourgeois culture lay in his recognition for variation among, and thus possession of idiosyncratic characteristics by organisms. In this sense, Darwinism was the last and greatest ideology of individualism. But by putting variation at the center of evolutionary change, Darwinism threatened to relativize variation; any idiosyncratic organic characteristic *might* be selected for. As Darwinian thought was absorbed into the productivist complex of values bequeathed by the Enlightenment, the imposition of developmentalism led to the stigmatization of many variations as "degenerations." Conversely, as the productivist

38 For a more comprehensive list on degeneration in general see Pick, *Faces of Degeneration*, 20.
39 For biological variation see further Mayr, *Growth of Biological Thought*, 681–682.

> ideology itself has receded, really only in this century, the more radical side of Darwin's thought has come to the fore and what was earlier seen as degeneration has come by degrees to be viewed as variation.[40]

Woman's masculine behavior and appearance were viewed as a deviation from femininity and its corollary, womanhood. In a similar manner, any physical or mental signs of weakness in a man were seen as effeminate and hence degenerate. These perceptions were a consequence of the blurring of gender spheres and often of the fear that women and men would not be able to fulfil their biological roles of perpetuating humanity.

Darwin's themes of reversion[41] and his concept of woman's natural inferiority provided the premise for antifeminists; any deviation from the natural roles prescribed by Darwin for women was a sign of degeneracy. The crossing of the boundaries that kept the male and female spheres separate was also a sign of degeneration. This degeneration, whatever form it took in the individual, was seen in some way to contribute to the fall of humanity—either because civilization would not work effectively to populate the world or because deviant traits, considered hereditary, would be passed on to offspring and thus lead to humanity's degeneration and possible extinction. "For the woman to take on ... masculine qualities was actually a sign of reversion, sinking back into the hermaphroditism of that indeterminate primal state—just as it was a clear sign of analogous degeneracy in the male to show himself to be effeminate."[42]

The theme of the "wild women," which translated from society to literature, derives from the Darwinian idea of a savage state.[43] One can only speculate as to whether Xenopoulos had in mind a savage or primitive state of woman when he chose the original title *Τα τρελοκόριτσα* (The wild girls), or whether he wanted to refer to the variant forms of woman, in a Darwinian sense, when he chose the title *Η τρίμορφη γυναίκα* (The three-sided woman). The Darwinian roles of "active" male and "passive" female were the bench-

40 Birken, *Consuming Desire*, 65 (italics in original).

41 Darwin believed that with variations of characters (or characteristics) within a species some characters could "revert to some of the characters of an early progenitor." For this see Darwin, *OS*, 195. See also his *DM*, 1:122. Also in his *DM* he refers to "injurious characters which tend to reappear through reversion, such as the blackness of sheep" (1:173).

42 Dijkstra, *Idols of Perversity*, 212–213.

43 Ibid.

mark for accepted societal behavior. Deviations from these roles resulted in perceptions such as "the predatory woman, the autoerotic or lesbian woman ... who ... chose to conjoin herself only with other women in an orgy of degenerative, self-extinguishing regression into the absolute of femininity, a perverse journey back into the primordial earth."[44]

Nitsa's appearance and behavior

In the late nineteenth and early twentieth centuries, based on evolutionary ideas many intellectuals in the sciences and the literary world saw feminism as a form of masculinizing degeneracy. This view spread through society and depictions of these women were seen in journals and magazines. See figures 6.1 and 6.3, which illustrate the masculinized New Woman alongside the feminized male. Otto Weininger in his book *Sex and Character* proclaimed: "A woman's demand for emancipation and her qualification for it are in direct proportion to the amount of maleness in her."[45] In *The Three-Sided Woman,* Kleanthis describes Nitsa's external qualities: "her almost masculine energy"[46]; her "flinging her masculine hat"[47]; her "husky, boyish voice"[48]; her emancipation[49]; and what people perceived as her "tomboyish good looks."[50] The narrator equates Nitsa's new confidence in the street with masculinity, commenting on her behavior: "During this period—the two or three months following the new phase in her 'life'—Gazeles' daughter had taken on an even more boyish air, with her cheeky, *impertinent look.* No one seeing her in the street could fail to be struck by it or not turn round to give her a second look. Some people, indeed, not knowing who she was, were emboldened by her air and took to following her and saying things to her."[51]

Xenopoulos' attempt to portray the naïve Nitsa as a New Woman is demonstrated when Nitsa joins a literary circle which, unbeknown to her, doubles as a house of ill-repute. This naïvety causes her problems when "clients"

44 Ibid., 273–274.
45 Weininger, *Sex and Character,* 64.
46 Xenopoulos, *Three-Sided Woman,* 60. (Xenopoulos, *Η τρίμορφη γυναίκα,* 256.)
47 Ibid., 67. (Xenopoulos, *Η τρίμορφη γυναίκα,* 261.)
48 Ibid., 75. (Xenopoulos, *Η τρίμορφη γυναίκα,* 267.)
49 Ibid., 100. (Xenopoulos, *Η τρίμορφη γυναίκα,* 287.)
50 Ibid., 160. (Xenopoulos, *Η τρίμορφη γυναίκα,* 331.)
51 Ibid., 197 (my italics). (Xenopoulos, *Η τρίμορφη γυναίκα,* 359.)

of the house, such as Dardouvas, mistake her for a prostitute. Meanwhile, Nitsa is reminded of Miss Julie's fate, and on one occasion this temporarily inhibits her from having an affair with the street boy Tilemachos: "At that moment at least, her idea was that she must stop right there, so that she would not come to grief like Countess Julie."[52] Despite this, she does change her mind and has a fleeting affair with him. Her mother shows great concern for this behavior and sees that society finds it unacceptable: "You see, you see what you have done now, with your crazy ideas? With your dressing like a boy, going around the streets with boys and wandering around on your own, no one wants to know us anymore."[53]

In Xenopoulos' representation of this masculinized New Woman, Nitsa did not flaunt her femininity. According to the narrator, "Nitsa had never particularly been a clothes-conscious girl. She dressed all anyhow, without much thought, wore no jewelry, and did not make the most of her natural charms."[54] Also, in terms of biological degeneracy Nitsa appears to have inherited "frailties" from her father (he is having an affair): "It was not at all unlikely that some such dangerous seed was hidden in his daughter's character, waiting to spring up later—although it might be argued it had already begun to emerge ... This was for sure. Both brother and sister showed signs of moral weakness, in their different ways, and this they owed to their father."[55] I prefer to translate the last sentence from the original Greek passage as: "The certainty was that the brother and sister showed some kind of *degeneration*, with different tendencies and signs, and this they owed to their father."[56]

In a similar fashion Xenopoulos had described the hereditary degeneracy of Miss Julie in his lecture in 1908.[57] According to Xenopoulos, in Miss Julie's case, her degenerate nature was inherited from her mother: "Julie had already inherited from her mother the rebellious instincts and a hatred for men";[58] and "Julie comes from the aristocracy; from her veins flows royal

52 Ibid., 207. (Xenopoulos, *Η τρίμορφη γυναίκα*, 366.)

53 Ibid., 209. (Xenopoulos, *Η τρίμορφη γυναίκα*, 368.)

54 Ibid., 232. (Xenopoulos, *Η τρίμορφη γυναίκα*, 386.)

55 Ibid., 144 (suspension points in original).

56 In Greek: "Το βέβαιο είναι πως και τα δύο αδέλφια παρουσίαζαν κάποιον *εκφυλισμό* με διαφορετικές τάσεις κι εκδηλώσεις, κι αυτόν τον χρωστούσαν στον πατέρα τους." Xenopoulos, *Η τρίμορφη γυναίκα*, 319 (my italics).

57 Xenopoulos, "Ο Αύγουστος Στρίντμπεργ," 168–170.

58 Ibid., 168.

blood, but the degeneration of her generation has already taken place, which will make her extinct."[59]

The Three-Sided Woman is Xenopoulos' response to the New Woman, in the same way that Strindberg wrote his play about Miss Julie. Strindberg presented the New Woman as deviant. Miss Julie committed suicide because she could not live with the thought of having dishonored her gender and class. Jean, her valet and lover, comments on her sexual indiscretion in disgust, claiming that she made the first move in the seduction: "Where do you think you are—farmyard? Whorehouse? [I've heard of you emancipated women, free-thinking, free-loving, free everything, picking up schoolboys, potboys.]"[60]

As indicated by Ann Heilmann, antifeminist novelists "featured heroines temporarily infected by New Woman ideas, but ultimately rescued by good Old Men. Unregenerate New Women were severely punished for their transgressive behavior."[61] Usually they were killed off by the writer, often by some form of suicide. Nitsa in *The Three-Sided Woman* asks for forgiveness from her father and admits her mistakes to Kleanthis. In this manner, Nitsa's repentant ways facilitate her marriage. Had she attempted to continue her previous lifestyle, it is likely she would have been expelled in some way from the story. Xenopoulos does this with so many of his rebellious women. This occurred with Xenopoulos' protagonist Stella in *Stella Violanti,* who refused to abide by her father's order to marry someone of his choice and died imprisoned in her bedroom.[62] Similarly, the morally degenerate Roza in *The Downfall,* despite trying to start a new life, is deserted at the altar.[63] The thought of having to return to prostitution for a living determines her to take her own life.

Xenopoulos' novel *The Three-Sided Woman* is his response to the New Woman of Athens at the turn of the twentieth century and is a parody of the feminist version. As previously mentioned, he depicts her as mannish in her demeanor, her voice, her appearance, including her dress; and he pres-

59 Ibid., 170.

60 Strindberg, *Miss Julie,* 27. The sentence in the square brackets, which is included in this edition, is only found in Strindberg's original drafts. For more on Xenopoulos' lecture of 1908 on Strindberg's play *Miss Julie* see chapter 5.

61 Heilmann, *New Woman Fiction,* 24. "Old Men" refers to the traditional Victorian masculine stereotypes.

62 Xenopoulos, *Στέλλα Βιολάντη: έρως εσταυρωμένος,* 501–545.

63 Xenopoulos, *Ο κατήφορος,* 9–354.

ents her as a frivolous, brazen, degenerate female. He undermines the representation of a feminist New Woman by highlighting her pathetic attempt to make a living for herself, painting wrapped sugared almonds. He further mocks her when she is only too quick to give up her alleged financial independence so as to avoid "the hard struggle to earn her living," as the narrator ironically puts it.[64]

One begins to realize that Xenopoulos' intentions in this novel are to parody the serious, inspiring, and regenerative New Woman described by feminist writers, such as Kalliroi Parren's depiction in her trilogy *The books of dawn*. Note that Parren's work also has Darwinian and post-Darwinian content, and although it is not known if she had read Darwin herself, he is mentioned in the *Ladies' Newspaper*, where she was editor-in-chief.[65]

A change of view is noted in Xenopoulos' introduction to *The Three-Sided Woman*, that he wrote in 1923, five years after the original in serial form, *The wild girls*. He attempts to be more sympathetic to the character Nitsa and contradicts his novel's narrator by saying that "she was not, by nature, depraved or degenerate."[66] In 1921 in the *Children's Guidance*, Xenopoulos wrote an "Athenian Letter" entitled "Feminism," about how he believed in the equality of sexes. But he warns: "The thing which I could never put up with is the shamelessness of some girls in the name of Feminism—the *impertinent* look, the very eccentric and unconventional dress, the smoking, the cuzo, and the rest. This is the comical form of the emancipated woman, of the feminist, of the manly woman which we see at times at the theatre and makes us laugh so much. I urge our girls to turn away from this form. No one should be a laughingstock, for any reason and for any idea!"[67] He uses similar language in the aforementioned passage as he does to describe Nitsa.

In his prose Xenopoulos addressed a number of varieties of the New Woman, typifying the post-Darwinian literary response to women at the turn of the twentieth century and in its first decades. *Tereza Varma-Dacosta* presents one version, as described in chapter 5. Other examples are *Stella Violanti: From the kitchen to the harem* and *Litsa's wedding* (1929). However, Nitsa in *The Three-Sided Woman* stands out among his female protagonists as the mascu-

64 Xenopoulos, *Three-Sided Woman*, 109. (Xenopoulos, *Η τρίμορφη γυναίκα*, 293.)

65 Parren, ed., "Το κάλλος και η εργασία," 1; also, in the following she mentions Herbert Spencer: Parren, ed., "Αι αδικούσαι και αι αδικούμεναι," 3. The articles are anonymous.

66 Xenopoulos, *Three-Sided Woman*, 8. (Xenopoulos, *Η τρίμορφη γυναίκα*, 215.)

67 Xenopoulos, "Φεμινισμός," 140 (my italics).

linized version of the New Woman.[68] Xenopoulos' original title for the serial version further reflects his perception of the New Woman.

The little known novel *Her lover* (*Η ερωμένη της*), published in 1929 by Dora Rozetti (pseudonym, real name is Nelly Kaloglopoulou-Boyiatzoglou) (1909–1989), received a review by Xenopoulos in the same year in the periodical *Nea Estia* (*Νέα Εστία*) on its lesbian theme.[69] In 2005, the novel was reprinted with Xenopoulos' review, where he spoke highly on the theme of the novel, which is about a lesbian relationship between the narrator Dora and her lover. He describes Dora as "ugly, eccentric, and a bit of a tomboy."[70] The novel is presumably one of the first modern Greek literary representations of a lesbian. The fact that Xenopoulos was the only person to comment on the novel at the time reflects his active preoccupation with the various representations of women in the early twentieth century, in particular the variant types of New Woman.[71]

It is important to note that at the same time as Xenopoulos was publishing his *Three-Sided Woman*, Victor Margueritte published his scandalous and bestselling *La garçonne* (1922), a New Woman novel, and that this was acknowledged by critics in Greece. "La garçonne" is a play on "le garçon" (the boy) and denotes a "boyish girl" or "mannish young woman."[72] The novel *La garçonne* tells the story of a young modern woman who leaves her middle-class family to lead a promiscuous life, has a lesbian affair, and finally marries, giving up her independent life. In 1924, in an article in the Athenian newspaper *Dimokratia* (*Δημοκρατία*) Kostis Bastias wrote on the play version of *The Three-Sided Woman*: "It does not concern us if the new work, by a coincidence not unusual in our literary chronicles, is similar to the 'Garçonne' of Paul [*sic*] Margueritte. Besides, the heroine of the new work and the Garçonne, in my opinion, are quite different in their psychology."[73]

68 Similarly, he describes Roza in *The Downfall* as having a preference for dressing "like a boy," but also dressing in a provocative manner. See Xenopoulos, *Ο κατήφορος*, 299. This occurs when Roza is presented as becoming liberated, freeing herself from the moral restrictions of her patriarchal society.

69 Xenopoulos, "Ντόρα Ρωζέττη: *Η ερωμένη της*," 1035–1037. Note that Rozetti was a medical doctor from Alexandria, who later lived in Athens. In her late teens and early twenties she actively contributed to Xenopoulos' *Children's Guidance* under the pseudonym "Πανταχού Παρών" (Omnipresent).

70 Xenopoulos, "Ντόρα Ρωζέττη: *Η ερωμένη της*," 1036.

71 It is interesting to note as well that it was Xenopoulos who in 1903 introduced the poetry of the Alexandrian Greek Constantine Cavafy to Greece. It is well documented by Cavafy scholars that a significant number of his poems have homosexual undertones.

72 Petit et al., *Dictionnaire français-anglais*, 597. (It can also mean an unmarried girl.)

73 Bastias, "Η τρίμορφη γυναίκα," 2.

Bastias fails to mention the similarities between the two protagonists, who are both masculinized. Xenopoulos does see this and also emphasizes Nitsa's tame behavior compared to that of the French novel's protagonist. On Nitsa, Xenopoulos will maintain: "She does not have ... the taste for vice, as the Garçonne had or acquired in Paris. She did not do any of the horrible things which are described in Margueritte's book. At the awakening of her instinct she simply liked to sample a most natural, legal pleasure; and as she was lively and daring by nature, she wanted to sample it as often as possible."[74]

It is no coincidence that a novel on a mannish woman in France should be published at the same time as a similar one in Greece. These novels were a very common event in European literature.[75] The degenerate variants of the New Woman of the early twentieth century in modern Greek literature have been noted by critic Angelika Psarra: "Because in that period there would appear texts in the Greek language which also used the negative stereotypical depictions of the New Woman in order to directly and explicitly discredit emancipation, attributing to her [the New Woman] the role of a social plague. Adopting Parren's method but pursuing the opposite results, some writers would also resort to the language of didactic literature so as to defame the Women's Movement, to ridicule and to foil its activities."[76]

In Kalliroi Parren's trilogy *The books of dawn*, which I touched on in chapter 5, in the period around 1900 Anna is portrayed as the wholesome New Woman of Athens. In one of the novels, *The witch*, Anna engages in a dialogue with Kostas and the narrator:

> "Today there is fencing and that event interests me greatly. At school I was the best fencer and if they accepted women I would certainly take part."
> ... Kostas stopped.
> "So, you are a fencer?" He asked her surprised.
> "Yes, I love the sword as much as I love all exercises. I don't run of course. But in an endurance run I could compete with anyone."
> Kostas thought: so all that they had written in the newspapers on the American female duelists was true!

74 Xenopoulos, "Η *Τρίμορφη γυναίκα*, ο φεμινισμός και ο ανθρωπισμός," 3.

75 See the literary studies Roberts, *Civilization without Sexes*; and Roberts, *Disruptive Acts: The New Woman*.

76 Psarra, "Το μυθιστόρημα της χειραφέτησης," 471.

Figure 6.1. The masculinized New Woman. Untitled illustration, *Phil May's Illustrated Winter Annual*, 1894, 7. Courtesy of the State Library of NSW.

> And meanwhile that girl looked anything but a manly woman. She was young, charming, lovely, dressed very elegantly, and so calm!—It appeared she was made to get on with everyone.[77]

Parren's representation of Anna, the New Woman, is in stark contrast to that of Xenopoulos' Nitsa of the same period. Considering her books as propaganda, Xenopoulos viewed Parren's protagonists, such as Anna, as "imaginary characters"[78] in contrast to his character Nitsa, who he believed was a real representation of the New Woman of Athens in 1900. It appears that Xenopoulos' degenerate representation of Nitsa is the antithesis of the

77 Parren, *Τα βιβλία της αυγής: η μάγισσα*, 139.
78 Xenopoulos, "Γράμματα, τέχνη, επιστήμη: φιλολογία," 364.

Figure 6.2. "Passionate Female Literary Types," *Punch*, June 2, 1894, 255. Courtesy of the State Library of NSW.

regenerate protagonists in Parren's novels. For instance, Parren depicted Anna's sexual awakening in a pure and positive way, which is unlike Xenopoulos' portrayal of Nitsa's similar episode. Parren's New Women had the ability to create their own identity, were not masculinized in temperament and appearance, and were able to determine their own future. Xenopoulos' New Women were presented as degenerate, masculinized, unintelligent, dependent, and weak.

In 1924 in the newspaper *Dimokratia*, Alkis Thrylos maintained that Xenopoulos' *Three-Sided Woman* (the stage production) was a misrepresentation of what feminism was about: "*The Three-Sided Woman*, by unconsciously intensifying the confusion which unfortunately exists so abundantly in Greece, could easily have significantly harmed Feminism's Struggle. The Greek public, which is uneducated, could easily have accepted

Figure 6.3. "What It Will Soon Come To," *Punch*, February 24, 1894, 90. Courtesy of the State Library of NSW.

as feminism whatever Mr. Xenopoulos presented and then it [the Greek public] would be justified in hating it [Feminism's Struggle]."[79]

In general, Xenopoulos had a chauvinistic attitude towards women writers such as Parren and Rozetti, and argued that they were all sloppy with their writing. Xenopoulos would view the woman writer as "careless, disorderly, rough ... in her writings."[80]

79 Thrylos, "*Η τρίμορφη γυναίκα* κινηματοδράμα," 2.
80 Xenopoulos, "Ντόρα Ρωζέττη: *Η ερωμένη της,*" 1036.

Saved by marriage to avoid degeneration: eugenics and the New Woman

During the interwar period, the decline of society and race due to the perceived degeneration of women through their pursuits into the male sphere had become a chronically controversial issue. Antifeminist literary representations of woman as the degenerate New Woman sought to make her "species" extinct if she was unrepentant of her immoral ways. Killing her achieved what was deemed necessary to rid society of a threat to moral, spiritual, or physical progress. An alternative fate for this New Woman was for her to marry and thus be "saved" morally and continue to contribute in the "home sphere."

In society the emancipated woman was seen as drawn away from her role as the wife and mother, especially if her female role was threatened in any way. Eugenicists considered women to be the saviors of humanity. Women were thought to be improving the health of the nation by keeping the family unit healthy, controlling birthrates in the lower classes, and producing more children in the middle and upper classes. Philosophically, humanity was still seen through a Lamarckian perspective with society placing an emphasis on social conditions.

In the first decades of the twentieth century it was a commonly held belief that the ideas in a "novel, like a disease," could spread from myth to real life.[81] There were real fears of the literary "garçonne" transversing over to society. Writers were able to appease "both cultural and gender anxieties" by redeeming their mannish free women, leading them back into the home sphere,[82] as in the case of Monique in *La garçonne* and, of course, Nitsa in *The Three-Sided Women*.

This woman's return to the home sphere was to contribute to the eugenicist's move to save the nation and the race. Francis Galton's quest to improve humanity and save it from physical, moral, sexual, racial, and literary degeneration was his vision through negative and positive eugenics.[83] The onus was often left to the women, and the ideals of this reformist action were often

81 Roberts, *Civilization without Sexes*, 46.

82 Ibid., 47.

83 Also see the introduction and chapters 2 and 4 for eugenics' applications to literature.

very close to the antifeminist ideals that they had been fighting against. Eugenicists and natalists, particularly after influenza pandemics and other disease outbreaks, urged women to remain in the home sphere, with those who were not in the lower classes urged to have children. "Eugenicist undertones" are frequently found in the many variations of most New Woman novels.[84]

Eugenics did have an impact on Greece, as evidenced by the presence of various publications promoting womanhood, motherhood, how a woman should behave morally, and how she should keep herself physically—all with the prime message that the regenerate New Woman, who had returned to her biological role, would save her family, nation, and humanity from decline, resulting in a utopian future. For instance, Greek eugenicist Irinaios Asopios' well-known literary periodical *Attikon Imerologion* (*Αττικόν Ημερολόγιον*) often included articles which praised motherhood and womanhood as the virtues of the nation and humanity. In 1896, in one such article on maternal love Zaharoula M. Lamni wrote: "Maternal love is an enduring feeling, because it is virtuous. It partakes of the infinite which gives wings to our souls. In this way it generates the family, the nation, and the human race."[85] At this point it is worth recognizing Parren's trilogy *The books of dawn* on the regenerate New Woman as a narrative of eugenic discourse, in terms of promoting womanhood and motherhood for the betterment of society and the nation.[86]

Thus, Xenopoulos was well aware of the implementation of eugenics at various levels to save humanity from decline (moral and biological); for he was forever warning women to avoid being "masculinized" in case they lost their womanhood. Xenopoulos would reflect the common view that certain levels of feminism were acceptable, but outside the traditional careers feminism was seen to create a masculinizing degeneracy in women.

A prime example of this was that he could never come to terms with women writers. As late as 1940, Xenopoulos revealed his sentiments that "the women that write creatively are ... manly women."[87] (See figure 6.2, where

84 Heilmann, *New Woman Fiction*, 85.

85 Lamni, "Περί της μητρικής στοργής," 473.

86 In her studies on Parren and her trilogy, Anastasopoulou does mention a declining society with Darwinian elements in the novels and she does make references to the novels' mention of the New Woman's role in race regeneration. However, Anastasopoulou does not acknowledge this role of the New Woman as being within the eugenic trend of the period. See Anastasopoulou, *Η συνετή απόστολος*.

87 Mylogiannis [G. Peristatikos, pseud.], *Πορτραίτα*, 227.

the older woman on the right is an example of how literary women were depicted in 1894 in the magazine *Punch*.) Implied in Xenopoulos' comment is that he probably felt women were a personal threat to his own writing. One could propose that, as a writer of woman themes, he felt that women writers on the same topic would have a better understanding of the female psyche. In many of his novels Xenopoulos had an "informing-the-readers approach" on many issues, including on women. In such a situation he would have felt threatened by woman writers.[88] Xenopoulos' comment would not of course have been unusual among male writers. This sexual discrimination aroused a perception of unstable gender identities for women, disadvantaging their work. This caused women to write under male pseudonyms, to include writers such as George Eliot, George Egerton, and in Greece the literary commentator Alkis Thrylos (pseudonym of Eleni Ourani).

As mentioned earlier, in the *Three-Sided Woman* Nitsa is "saved" from moral degeneracy by getting married. In his introduction to the novel Xenopoulos adds: "And so the bitter struggle, the great tragedy of Nitsa Gazeles ends with her salvation."[89] In the novel Xenopoulos compares her to Miss Julie, the degenerate New Woman who Strindberg sends to her death by suicide. In addition, Xenopoulos' Stella in *Stella Violanti*, a symbol of the New Woman to a degree and also of the Persecuted Virgin,[90] dies due to her rebellious nature in refusing to marry someone she did not love.

The salvation or destruction of the New Woman figure in literature is revealed here in various novels. In *La garçonne* Monique Lerbier is absolved of her immoral ways by seeking forgiveness from her partner and marrying him; this coincides with the symbolic return to her feminine ways by growing out her short haircut. Kate Chopin's Edna in *The Awakening* also dies,

88 One example is Xenopoulos' novel *Ο αρραβωνιαστικός μου* (My fiancé) (1923), where Xenopoulos, as a female narrator, has problems describing men from the female perspective.

89 In the main text I have given my English translation of Xenopoulos' Greek version: "Κι η δεινή πάλη, το μεγάλο δράμα της Νίτσας Γαζέλης τελειώνει με τη σωτηρία της." See Xenopoulos, *Η τρίμορφη γυναίκα*, 216. Barbara Kent's translation reads: "So the bitter struggle and tragedy of Nitsa Gazeles came to a very happy ending." See Xenopoulos, *Three-Sided Woman*, 9. I have not used Kent's translation, "a very happy ending," since the phrase does not convey the message that Nitsa is considered "saved." It is important to use the word "salvation," which alerts the reader to the depiction of the literary New Woman and the New Woman in society.

90 The "Persecuted Virgin" in literary terms was one who, although she had no sexual contact with anyone, did not conform morally to her social expectations, was tormented for it, and often died. See Dijkstra, *Idols of Perversity*, 25–63.

drowning. The old woman Frankojannou in Papadiamantis' *The murderess* can also be seen as a New Woman, representing womanhood/nature who takes control of the female births on the island and subsequently drowns.[91] Eugenic feminists felt it their responsibility to reduce births among the poor who were overpopulating society and sending it into "decline." The villagers in Kazantzakis' *Zorba the Greek* (first edition, 1946) view the widow as a *femme fatale* who allegedly lures a young man, resulting in his suicide; as a consequence of this, they stone her to death. As discussed in chapter 5, the *femme fatale*/New Woman in Xenopoulos' *Tereza* is a changed woman after she has married. However, her marriage to the "degenerate" Count can only result in her family's extinction. Many of Xenopoulos' "rebellious" and hence so-called degenerate female protagonists follow either a fate of destruction or one of salvation through marriage.[92]

It is likely that *My son and my daughter* (*Ο γιος μου κι η κόρη μου*) (1921)[93] was Xenopoulos' attempt to appease the feminists after *The Three-Sided Woman*.[94] He appears to take an entirely anti-Darwinian stance in this novel, by indicating that woman is the same as man biologically and, in particular, sexually. The narrator in the first person is Tzortzis Doxaras, father of Nanis and Kiara, his son and daughter respectively. From birth to marriage he observes systematically, in a "scientific" manner, the differences between his two children in terms of sexual desires, and comes to the conclusion that: "My daughter had the same right as my son—that is, to enjoy her life, her youth, or indeed to be able to stray, 'as a human, arrayed in flesh and dwelling in the world,' due to her warm erotic temperament."[95] The father will uphold that society had robbed the female of her natural right: "They don't have the most natural right of humanity—that is, to get bored. And when they do, they pay for it with some defamation which sometimes brings with it real misery."[96]

91 On *The murderess* see further chapter 2.

92 See also Xenopoulos' novel *Η τιμή του αδελφού* (The brother's honor) (1915). Argyro's fate due to her immoral deeds is death at the hands of her brother.

93 The novel was first published in serial form in the newspaper *Ethnos* from March 27, 1921, to October 10, 1921. In a revised form with a new prologue it was published again in serial form in the periodical *Nea Estia* (1928–1930). In my study I use the 1984 edition in book form by the Athenian publishers Vlassis Brothers.

94 Doulaveras, "Τα δυο φύλα," 305.

95 Xenopoulos, *Ο γιος μου κι η κόρη μου*, 180.

96 Ibid., 204.

Xenopoulos wavers in his views, if this comment is compared with his "Athenian Letter" on "Boys and girls" discussed in chapter 3. This "Letter" has a Darwinian approach, exploring some similarities but emphasizing the differences of brain and behavior between the two sexes. Pykett maintains that D. H. Lawrence, as a male New Woman writer, "deployed an inherently contradictory discourse on and of gender throughout his writings, and he used those discourses in contradictory ways."[97] The same could be said about Xenopoulos. One could say that gender writers were constantly being bombarded with scientific and other intellectual ideas because the topic of gender was so controversial. These ideas were fluid and often at odds with each other, and so were the opinions of the novels' readership. One could presume that this played a role in the contradictory nature of such writers.

Generally speaking, literary critics have not considered Xenopoulos' theme of sexuality in his novels on women, either on a biological level or as part of the wider literary response to the New Woman from the late 1800s to the first decades of the twentieth century. The protagonist Nitsa in *The Three-Sided Woman* has all the hallmarks of the degenerate masculine version of the New Woman as perceived by Xenopoulos at that time; it is his antifeminist representation of the New Woman, and so should be placed in the context of other Greek and, more generally, European New Woman literature.

The night of degeneration

Xenopoulos' novel *The night of degeneration*[98] has elicited little significant commentary associated with the themes here, perhaps because it was only in 1994 that it was published in book form and also because of its harsh hereditarian approach related to mental health. Set after 1911, the story takes place in Athens in the milieu of Max Nordau's concept of degeneration as depicted in his book *Degeneration*.[99] The main characters Phoibos Vramis and Merope Kara-

97 Pykett, *Engendering Fictions*, 127. Pykett does not really have an explanation for this phenomenon that she sees in D. H. Lawrence's writing.

98 *The night of degeneration* was first published in serial form in the Athens newspaper *Ethnos* from April 18, 1926, to September 6, 1926. In 1994 it was issued by the publisher Vlassis Brothers, which is the version used here.

99 Nordau published the book in German, entitled *Entartung* (1892). The narrator in *The night of degeneration* mentions that Phoibos had read Angelos Vlachos' Greek translation *Εκφυλισμός* (1911), a shortened translation of Nordau's *Degeneration* (181). Phoibos had also read the unabridged French edition translated in 1894. It was translated into English in 1895.

manlis display Nordau's stigmata of degeneracy, in particular the stigma of neurasthenia, with Phoibos' condition more severe than Merope's. Despite a doctor's warning, they marry and have a severely retarded child, both physically and intellectually. Phoibos' condition is degenerative and eventually he becomes insane and is institutionalized.[100]

This novel reveals that Xenopoulos was very much attuned to the "social diseases" and the eugenic discourse associated with their management in Greece. Relevant to Greece, though not exclusive to it, the interwar period gave rise to eugenic-based reforms on how to deal with the hygiene and health of the nation. Concerned with the inheritance of psychiatric illness and other alleged degenerate conditions, the Greek medical community endeavored to establish reforms to eradicate or minimize them. In 1925 for instance, there was an attempt to introduce a law allowing doctors and psychiatrists to carry out "psychiatric examinations in order to verify the mental and physical health of couples prior to marriage."[101]

I argue that the novel is Xenopoulos' representation of the perceived decline in society at the turn of the twentieth century. Xenopoulos uses the discourse of Nordau's degeneration and equates hereditary disease with degeneration.[102] By exploring these diseases using a medical and Darwinian discourse of the period, he attempts to send a message of regeneration to society. The message is eugenic-based since it promotes rational reproduction in society, with individuals choosing their partner to have children not according to love or sexual selection, but instead based on the health of the partner. Eugenic narratives include Sarah Grand's *The Heavenly Twins* (1893), George Egerton's *The Wheel of God* (1898), and Mona Caird's *The Stones of Sacrifice* (1915) and *The Great Wave* (1931).[103] Also, in George Meredith's *The Egoist* (1879) there is hope for race improvement utilizing Darwin's law of sexual selection.[104]

100 There is an ironical allusion to the Greek mythic characters with the same names. Phoibos was another name for the god Apollo and Merope was one of the seven Pleiades nymphs, who was shamed because she married a mortal.

101 Trubeta, *Physical Anthropology*, 260.

102 Nordau defines degeneracy as "a morbid deviation from an original type." See Nordau, *Degeneration*, 16.

103 Richardson, *Love and Eugenics*. Richardson examines the application of eugenics to society and fiction through the promotion of rational reproduction and the New Woman.

104 See also the marked effect of eugenics in American literature in Cuddy and Roche, *Evolution and Eugenics*.

The application of eugenics to society and fiction has been viewed through the promotion of rational reproduction and the New Woman. Some critics viewed Grand's work as just booklets on scientific and sex issues, considering them as no more than pedagogical discourse in disguise.[105] The Darwinian message in relation to the human species' indifference to issues of eugenics in marriage, as well as the need to direct humanity, inspired writers to encourage selective breeding or rational reproduction. It is in this vein that I examine Xenopoulos' novel *The night of degeneration*. It was promoted that the only way to curb the pressure of hereditary illness was by rational reproduction, which was the selection of a partner based on healthy hereditary traits and not on unconscious selection.

Neo-Darwinism

With Weismann's germ plasm theory, Galton's studies on the inheritance of gifted traits, and later Mendelian theory, theories I have already discussed, came the prominence of heredity and the dominance of nature over nurture. This eliminated the Lamarckian view that whatever one gained in life could be passed on to one's offspring. Which posed the question: can the individual's experience of life, such as the endeavor to better one's self, change his or her fundamental nature? The dominance of nature over nurture provided the climate where humanity was predestined to a fate which could not be modified, leading to fears of moral and physical decline.

Prior to this fear of decline in society, Darwin foreshadowed breeding in all living things, and in the *OS* implied an importance for artificial selection: "Slow though the process of selection may be, if feeble man can do much by his powers of artificial selection, I can see no limit to the amount of change, to the beauty and infinite complexity of the co-adaptations between all organic beings."[106]

In the final pages of his *DM* Darwin stresses the importance of careful choice of one's partner and this is all in "the advancement of the welfare of mankind."[107] Inspired by Darwin, Galton, who had coined the term eugenics (which he derived from the Greek "*eugenes*, namely good in stock, hered-

105 Richardson, *Love and Eugenics*, 124–125.
106 Darwin, *OS*, 152.
107 Darwin, *DM*, 2:403.

itarily endowed with noble qualities"),[108] defined eugenics in 1883 as "the science of improving stock, which is by no means confined to questions of judicious mating, but which, especially in the case of man, takes cognizance of all influences that tend in however remote a degree to give to the more suitable races or strains of blood a better chance of prevailing speedily over the less suitable than they otherwise would have had."[109]

Genius: heredity versus environment

The main character in *The night of degeneration*, Phoibos Vramis, a law student, is described as a degenerate (8, 42–43); his effeminate appearance is consistent with Nordau's diagnosis of degeneracy. Initially, as the son of the great poet Costas Vramis, he showed great promise in his life: "His mind worked powerfully. He had a great memory; he could follow the most complex reasoning, and he had the belief that, whichever Science he chose, he would become great in it. There were times that he imagined that he could become a great poet. *So long as he wanted it!* All those around him had the same belief. ... Phoibos Vramis was something exceptional! He was great like his father. Indeed, why not even greater? *Evolution!* But before he even finished high school, this 'exceptional' child had already started to lose his memory and his intellectual ability."[110]

Was Xenopoulos here, through his narrator, parodying Phoibos' aspirations to become great by means of Lamarckian will? Or was Xenopoulos, again through his narrator, mocking evolution as a process to perfection? The narrator does not appear to believe in a one-way evolution, suggesting that evolution can appear to regress. With his father's disappointment there was a sense of hope that Phoibos might be at least "normal": "He would say, does he also have to be a genius? It's enough that he is a good boy and with enough brains to live. So he should live, the poor thing! ..." (42–43). Xenopoulos will contemplate the subtleties of the concept of genius, a recurring idea in his novels, one which occupied intellectual minds, including those of

108 Galton, *Inquiries into Human Faculty*, 17. He rekindled the link between the two terms "nature" and "nurture," known in the Ancient Greek period as "φύσις" and "νόμος." See also Galton, *English Men of Science*.

109 Ibid.

110 Xenopoulos, *Η νύχτα του εκφυλισμού*, 42–43 (my italics).

scientists, anthropologists, and literary writers, not only outside but also inside Greece.[111] As Tom Gibbons succinctly describes: “The Age of Evolutionism was fascinated by ‘genius’ and sought anxiously to discover how it arose, how it was transmitted, how its transmission could be controlled in the interests of evolutionary progress, and whether it was not in fact a treacherous aberration which would lead directly to the nightmare of evolutionary decline.”[112]

Galton had maintained that genius and other gifted traits, including those of poets, were inherited and not an acquired trait through nurture. So, according to this, as seen in the passage before last, the Lamarckian factor of “conscious effort” did not make a genius of Phoibos. Instead, his hereditary fate was that of mental decline. Xenopoulos depicts Phoibos’ downturn, employing Nordau’s diagnosis of artists, writers, and geniuses as degenerate.

Sexual selection: by choice?

Phoibos often wondered how he became attracted to someone like Merope, whom he met by chance. He realized his attraction to Merope was due to “the power of selection,” and it is noted in the following passage which alludes to Darwin’s sexual selection:

> Just by looking into her eyes, he realized that she was for him. There was no mistake about this. It was the instinct which spoke within him, the *power of selection*, that mystery which pushes two individuals towards union when they are suited. Merope had also heard the same mysterious language from within herself. [...] But what was it which harmonized and connected those two individuals so perfectly?—Giannis and Foula would have simply answered: “They are both mad.” Maybe they weren’t all that wrong. Only this thing cannot be defined with one such comment. It is much more complex, more sophisticated, more wondrous ...[113]

The external appearance, to which Phoibos is attracted, allows sexual selection to act in its purest form—that is, the factor of desire through

111 As discussed in chapter 3, Xenopoulos considered the concept in relation to his writing. I mentioned in chapter 4 that the concept of genius in association with poets was also hotly debated in Greece.

112 Gibbons, *Rooms in the Darwin Hotel*, 51.

113 Xenopoulos, *Η νύχτα του εκφυλισμού*, 167 (my italics; suspension points in original).

beauty. In Darwin's *DM* beauty in humans was a major force in the unconscious selection of a partner.[114] So that when Phoibos first saw Merope, he knew that he desired her. Xenopoulos' use of a pause, as seen at the end of aforementioned cited passage, is often associated with a subtle form of highlighting that there is a further underlying issue or idea associated with the phenomena observed by the narrator. Here it could be that the narrator leaves it open for the reader to follow Phoibos' thoughts in the direction of sexual selection, highlighting also the complexity of the process which was foretold by Darwin; and, in particular, foreshadowing the danger which Phoibos would fail to see.

Darwin's preoccupation with the numerous factors associated with sexual selection is exhibited in the much longer second part of the *DM*, which is devoted to this topic. He often refers to sexual selection as follows: "In itself an extremely complex affair, depending as it does, on ardor in love, courage, and the powers of perception, taste, and will of the female. Sexual selection will also be dominated by natural selection for the general welfare of the species. Hence the manner in which the individuals of either sex or of both sexes are affected through sexual selection cannot fail to be complex in the highest degree."[115] Furthermore, Darwin warns that "the courtship of animals is by no means so simple and short an affair as might be thought."[116]

When Merope realizes that her son is severely retarded, many of her beliefs are shaken, and she turns to science for answers: "She started to believe more in Science and less in God; because she did not believe anymore in the existence of an immaterial, free, immortal soul."[117] Xenopoulos had great interest in the conflicting ideas of science and God, and this is described in chapter 3 in relation to his "Athenian Letters." Similarly, in Xenopoulos' final paragraph of his *Rich and poor*, his narrator enigmatically comments on what followed Popos' death: "He had passed in a state of non-existence, transient like the Life he had lived—transient also like the Life he would live again ..."[118] Merope rationalizes Phoibos' mental illness thus:

114 Darwin, *DM*, 2:355, 371.
115 Darwin, *DM*, 1:296.
116 Ibid., 262.
117 Xenopoulos, *Η νύχτα του εκφυλισμού*, 231.
118 Xenopoulos, *Πλούσιοι και φτωχοί*, 425 (suspension points in original).

> Because what else was this "degeneration" but a disease of the nervous system and of the brain? And how could she ever again accept an immaterial, free, immortal soul when the body made of it [the soul] what it wanted? Whereas if it [the soul] was like that, then it would have had the power to direct the body. She knew that many believed this—that is, that the soul molds the body the way it wants. But with Phoibos she saw the opposite. And as much as she tried to remember, so as to compare, she could not find a soul which molded its body. She could not say this of any of the people she had known.[119]

She is referring to the dominance of the physical body rather than the psyche, emphasizing a hereditarian view in line with nature over nurture. Again, what can be seen in the above passage (through Merope) is Xenopoulos' questioning of the relationship between mind and body, a theme he takes up also in *Rich and poor*. There Popos maintained that "the soul makes its garment."[120] The relationship between mind and body is of course a timeless issue, which was further popularized with Darwin's theories that were considered materialist.

Degeneration, inheritance, and atavism

Merope contemplates the duality of the soul and body. She draws the reader away from God and introduces chance as the cause for the child's existence and also for his eventual demise. The narrator comments: "Her child, again because of Phoibos, had been born in this state. But what characteristics had the father passed on to him? Physical or mental? Physical of course, because the doctor examined the body, the cranium, the skeleton, the eyes, and found it to be 'degenerate.' And that body again would have molded the soul—a soul of an animal, of a beast-like human. How was it possible to imagine her husband anymore as a creature of God, with a free and immortal Soul, when the child she bore him was an imbecile, a monster!"[121] She questions here the traditional religious belief that humans are all created perfect in the image of God. And further, her rejection of the child and her wish for its death is confronting: "'Nothing exists!' she would say in her despair. 'Neither God nor Soul. And who do I beg now to take this wretched

119 Xenopoulos, *Η νύχτα του εκφυλισμού*, 232.
120 Xenopoulos, *Πλούσιοι και φτωχοί*, 42.
121 Xenopoulos, *Η νύχτα του εκφυλισμού*, 232

creature which will be the sorrow and curse of our life. Don't worry about it! It too will live for as long as the *blind forces of nature* want it to. *A chance occurrence brought it into the world, a chance occurrence can take it! ...*'"[122]

Merope questions her biological choice of husband and attempts to educate the reader on one's existence as being the chance product of natural selection. She alludes to the doctor's eugenic advice: "She still pondered over her insistence at the time on marrying Phoibos, to become united and to have children with him. *She knew of natural selection*; and she wondered: Did Nature want the birth of an Apollonius? And is that why as soon as Phoibos and Merope saw each other, they immediately felt that mutual inclination which in no time made them unite? His doctor had told him that, if he wanted to have healthy children, he had to marry a country girl."[123] Darwin indicated that "sexual selection will be dominated by natural selection for the welfare of the species."[124] Hence they are interconnected. So that when the narrator speaks of natural selection, the sense of attraction implies sexual selection. Phoibos had told Merope of the doctor's advice to choose a "eugenic" partner rather than to have children with Phoibos. All she knew was that the desire or attraction, which was narrated in terms of sexual selection, was the basis for their union. Further to this, the doctor diagnoses Phoibos with hereditary neurasthenia and the possibility of insanity later. He reveals the bleak outlook if Phoibos tries to have children: "Either he would be sterile—that was also probable—or he would have entirely degenerate children; because naturally he would marry a girl with some 'predisposition,' like nearly all the girls of the large cities and of the upper classes are in these times. If however he married a healthy girl of the common people, or a village girl, the danger would be less; and even less if he found a woman from another uncivilized race, a Fellahin peasant, e.g., a Georgian, a Sudanese, a Senegalese, or an Annamite. Only in this way would he have healthy, viable children and he would continue his lineage."[125]

Women living in cities of urban decay were not considered to have the robust nature of their country counterparts, and had a predisposition to "bodily degeneration,"[126] whereas higher-class women were seen as having

122 Ibid. (my italics; suspension points in original)

123 Ibid., 232–233 (my italics).

124 Darwin, *DM*, 1:296.

125 Xenopoulos, *Η νύχτα του εκφυλισμού*, 95.

126 Pick, *Faces of Degeneration*, 191. Also, cities—as "centers of decay"—were the causes of national decadence (189–203).

inbred abnormalities. Merope's well-to-do upbringing would have placed her in the latter category. Why does he marry her? "They were compatible. Because she was also degenerate to a degree. She recognized this now, now that she saw that it was not within her powers to restore the health of the seed which she had accepted. So Nature did not want healthy children to come from the Vramis family. Nature had condemned the family to death."[127] The harsh reality of the child is exemplified in the following racist description: "And he was dark-skinned, very deeply dark-skinned, although Phoibos and Merope were fair. He had inherited the coloring of his grandfather but even much darker, and changed. His blackness had something earthy, of the soil, like that of some Africans. But also his head reminded me of an African, and his lips were thick and almost hanging. He looked like a child from another race and *he looked like some harsh game played by Nature; an oddity of a most peculiar atavism*."[128]

In the novel, "the spirit of the period" is depicted in terms of society's decline (the symptoms being the degeneracy of the art world): "How could anyone react if it was Destiny? ... All one could do was to cross one's arms and wait for death—the Night of Degeneration."[129] Reaction to "Destiny" also refers to biological determinism, the belief that nothing one can do could overcome nature. Weismann eliminated the role of environment from evolution. Hereditarianism becomes a common theme in Xenopoulos' novels.[130] Giannis, Phoibos' cousin, has a message regarding "the night of degeneration":

> The doctor had said to Phoibos that if he wanted to have healthy and fertile children, he had to take on a wife from another race, half savage, uncivilized—one of our country girls, a Fellahin, a gypsy, or an Arab. And he [Giannis] said something fantastic to Foula, something, which if that critic had heard it, he might have been embarrassed at rushing into despair:
> "Phoibos did not do that," he said to her. However, civilized Humanity will do it and will be saved. Indeed, it seems to me that it has started doing it.

127 Xenopoulos, *Η νύχτα του εκφυλισμού*, 233.
128 Ibid., 235 (my italics).
129 Ibid., 274–275.
130 For instance see his "social trilogy": *Rich and poor, The honorable and dishonorable*, and *The fortunate and the unfortunate*.

> Later it will do it more and without its wanting to. It will be forced to do so. This way a new Humanity will result. And with this the Night of Degeneration will be distant ... [131]

The aforementioned paragraph is the last in the novel. Xenopoulos again used his last paragraph for an ideological message which he ends with a pensive pause. Throughout the novel the repeated allusions to the doctor's advice culminate in a decisive message. The novel serves as a didactic warning to society that it should beware of its reliance on sexual selection when choosing a partner for the procreation of children. The only way to curb the pressure of hereditary illness was by rational reproduction which was based on the selection of a partner with healthy hereditary traits, and not on unconscious sexual selection.

These premises were part of the eugenics movement which hoped to save society from "racial suicide" and hence extinction.[132] Giannis strongly alludes to eugenicist intervention, firstly, on an individual level by rational reproduction, by way of individuals artificially selecting partners of a "wild strain"; and then collectively, humanity would enforce this in some way—possibly a eugenics trend, which he indicated was already occurring. This, he claimed, would save humanity from the "night of degeneration"—that is, from biological decline—and regenerate it so that it could continue to progress (in Darwinian terms).

As discussed in this book, Greece was contributing to the intellectual discourse on eugenics. Further to this, in 1934 Moysidos will write on eugenic sterilization, outlining its beginnings, methods, and practice. He will elaborate on the views of various countries including the United States, Canada, Switzerland, Germany, Britain, and Greece. His study will show that unlike some of the other countries, Greece was very hesitant about sterilizing, and the Church would not accept it either. Moysidos did refer to its use in other countries for those with psychoses, Huntington's disease, epilepsy, alcoholism, syphilis, and for "degenerates."[133]

131 Xenopoulos, *Η νύχτα του εκφυλισμού*, 275 (suspension points in original).
132 Racial suicide was a term used commonly well into the second decade of the twentieth century. See Nies, *Eugenic Fantasies*, 70.
133 Moysidos, "Ευγονική αποστείρωσις," 1–68.

Eugenics, beauty, and sexual selection

Darwin maintained that sexual selection was the likely cause of racial difference. He demonstrated that different races have different tastes in beauty, so that selection of partners according to the different tastes would create the physical diversity of various races.[134] He claimed that male choice in selective breeding resulted in upper-class Europeans attaining the best physical beauty: "There is, however, reason to believe that sexual selection has effected something in certain civilised and semi-civilised nations. Many persons are convinced, as it appears to me with justice, that the members of our aristocracy, including under this term all wealthy families in which primogeniture has long prevailed, from having chosen during many generations from all classes the more beautiful women as their wives, have become handsomer, according to the European standard of beauty, than the middle classes; yet the middle classes are placed under equally favourable conditions of life for the perfect development of the body."[135] Here Darwin was in agreement with Galton, who viewed the English upper class as the most handsome because the male could freely choose from the most beautiful women over many generations.[136] So that beauty in terms of eugenics becomes a class issue. This concept was easily grasped by the literary imagination, including by Xenopoulos.

Characteristic of Xenopoulos' writing is his use of a eugenic subtext, which is exhibited either as a theme or as a motif. In his novel *Leeward nights*, hereditary degeneracy has produced sickly but good-looking children in the aristocratic Kantis family. Of these, Stephanos marries the robust lower-class Nina to save the next generation:[137]

> The children of the Kantis family, with blue blood flowing in their veins, unadulterated through many generations—almost three centuries—displayed the inevitable degeneration of the old Venetian aristocracy. They were delicate, fair, almost pale, and they resembled flowers in a garden

134 Darwin, *DM*, 2:355, 370–371.

135 Ibid., 356.

136 Jann, "Darwin and the Anthropologists," 286–306; Galton, "Hereditary Talent and Character," 165.

137 See Xenopoulos, *Απάνεμα βράδια*. This edition published in 1984 is the first time the novel has appeared in book form. The novel first appeared in serial form in the newspaper *Athinaika Nea* (*Αθηναϊκά Νέα*) from June 13, 1938, to September 26, 1938.

> bed. But precisely those meagre faces, those faded colors which reminded one of the evening horizon of the west; those large flaming eyes, like those of someone who has a fever, with the dark circles underneath; those slim snow-white hands, where one could clearly discern the small blue veins, gave them a distinct charm.
> The most beautiful of the three was the eldest. "Genteelness became a person—and he is Stefanos Kantis."... Tall, slender, willowy, with light blue eyes, with the sweetest of smiles, and with something sickly about him—many were afraid that he was consumptive, but he was very healthy. ... He only loved one woman in his life, Nina Darletos. And notwithstanding that her father was just a raisin and currant merchant—very wealthy though—he married her; a very well-matched couple, because Nina was also the most beautiful girl in the locality. Only that her beauty was entirely of a different type. She was a strapping woman, very robust and full of life; her cheeks looked as if they dripped blood, like those of a mountain country girl. It was the first time in the old aristocratic family where there was an influx of lower-class blood. And the degeneration, which would have otherwise made it [the family] extinct, might stop with Stefanos Kantis' descendants.[138]

The depiction of Rena, the chambermaid in *From the kitchen to the harem*, is a prime example of one of the many eugenic motifs employed by Xenopoulos:[139] "Rena was really beautiful; blonde, white, full, soft, with large arms and fantastic breasts. And if her rather misty and nearly colorless eyes did not resemble those of a lamb, and if her somewhat thick face and all her physiognomy did not have something animal-like, it would not have been impossible for someone to take her for a country girl. The other lines of her body were ever so *genteel*."[140]

Demos Spathis, the protagonist and first person narrator in *Honorable and dishonorable*, is aware of the association between the upper class and beauty: "As I grew older, I began to lose my looks. I must note this. ... With age my features would begin to lose their refinement, they would become thicker, and *they would betray my lower-class origins*. I would look like my father who, without being ugly, did not have anything very *genteel* about

138 Ibid., 12.

139 The novel was published in serial form in the newspaper *Ethnos* from December 4, 1923, to April 28, 1924. In this study I refer to the first publication in book form by the Vlassis Brothers in 1994.

140 Xenopoulos, *Απ' την κουζίνα στο χαρέμι*, 11 (my italics).

him. ... I don't care! I would become stronger in my body, my mind, my soul, than all of them. There is *aristocracy without ancestors: the aristocracy which starts from each person.*"[141] Demos hoped to be able to override his "inherited" trait using his own will to do so.

Books such as Samuel Smiles' *Self-Help* (which, coincidentally, was published in the same year as the *OS* and was very popular during the Victorian era) reflect the Lamarckian philosophy of volition as the catalyst to one's development. The book was translated into Greek and was entitled *Βοήθει σαυτόν*.[142] The book in Greek refers to "a genteel character" and "a true nobleman" (192). It is morally oriented and makes statements such as "genius is nothing without a heart" (193). While Lamarkism had lost its impact in the scientific world after Darwin's theory of evolution, it still featured in the social sciences and in literature until the 1920s; the issues dealt with were on self-help and self-improvement.

As discussed here, Xenopoulos' novels were part of a broad international literary discourse which was concerned with the perceived degeneration of society and which utilized a post-Darwinian approach. The eradication of degeneration was paramount to promote a regenerate society. Nitsa in *The Three-Sided Woman*, the degenerate version of the New Woman, would only survive in the story if she was remorseful for her emancipated ways. Xenopoulos' *The night of degeneration* has shown that he viewed hereditary illnesses as features of this degeneration, which culminated in his eugenic ideas on solving the problem. He would push for a view of hard inheritance through his narrators, which by the turn of the twentieth century was a dominant view among intellectuals in society. Such views by eugenicists were based on hereditarianism/neo-Darwinism, and from these concepts were drawn the "answers" to society's problems associated with issues such the salvation of humanity. While *The night of degeneration* was made available to the wider Greek readership when it was first published in the Athenian newspaper *Ethnos* in 1926, it was not readily accessible to them till 1994, when it was first published in book form by Vlassis Brothers. One can speculate why the novel only came out in serial form at that time: perhaps

141 Xenopoulos, *Τίμιοι και άτιμοι*, 48–49 (my italics).

142 See Smiles, *Βοήθει σαυτόν*. The Greek publication in book form referred to here is dated 1925. It was published earlier, including in 1873 by the publisher L. D. Villara in Athens. It was first published in the periodical *Pandora* (*Πανδώρα*) between 1871 to 1872.

thematically it did not fit with all the other entertaining novels that he wrote; it may not have been understood; or it is possible that it was not received very well by the public. A further possibility is that its true message of eugenics would have been very confronting to the general public.

EPILOGUE

> Literature does not merely hold up the mirror to nature and express more eloquently what is already well known and understood: it allows the exploration of what is perceived only dimly, if at all, the subversive anxieties that cannot be directly stated, because they challenge too vigorously the mores and taboos of society.
>
> ROSLYNN D. HAYNES[1]

This book has endeavored to contribute to the nascent scholarship on the influence of Darwinian, post-Darwinian, and other evolutionary ideas in the works of literary writers and other intellectuals in Greece in the period from 1880 to the 1930s. During this period, biological messages in much of the literature considered here were relevant only to those commentators attuned to the international scientific scene. While this literature was not given the significance it deserved, it is an integral component, along with the sociopolitical component, in the framing of the individual, the society, and the nation in Greece.

The exploration has also attempted to place the Greek works in the context of European literature. As early as the 1880s, Greek intellectuals were exploring and questioning Darwinism and its impact on literature and society by means of their publications in newspapers, periodicals, and creative literature. What this study has shown is that prominent literary figures and other intellectuals in Greece of the late nineteenth and early twentieth centuries were not just dealing with important issues from sociopolitical or philosophical perspectives but they were also contemplating avenues based on

1 Haynes, *From Faust to Strangelove*, 313.

Darwinian and associated post-Darwinian theories. At a time when Greece was aspiring to build a nation in relation to its politics, race, culture, national identity, religion, language, and gender equality, these internal issues were also being explored within an evolutionary framework of science and medicine; this was certainly not peculiar to Greece, as it was also seen in other countries in Europe and North America.

Darwinian ideas found expression in the fiction and journalism of Grigorios Xenopoulos. His ideas were not isolated and his works show themes common to other Greek writers and also other European and American writers. Xenopoulos was instrumental in bringing the ideas of Darwinism and its impact to the general public through his writings in *Children's Guidance*. Even though Kazantzakis had translated the *OS* into Greek in 1915, this was at a time when Darwin's theories were not taught in schools and only academics and intellectuals would have had knowledge of the translation and access to Greek science periodicals and other sources.

In Western countries there have been in the last decades numerous interdisciplinary studies which examine the literary responses to scientific discoveries and trends. Scientific advances, particularly in the biological sciences and genetics, many of them distinctly related to Darwinism, have philosophical and cultural implications and have aroused the literary mind. The complete mapping of the Human Genome in 2003 and its repercussions provide profound insights into the complex relationship between nature and nurture, and also into the "descent of man" in relation to other species. This has resulted in biologists asserting their role in the exploration of human behavior. Ongoing research on the human genome with its establishment of new genetic markers which localize human qualities will, and has already, influenced the literary world and intellectual commentary.

The debate still continues with the nature/nurture controversy. The new synthesis of epigenetic evolution (or "evo-devo"), whereby evolutionary changes or adaptations can occur outside the influence of genetics, has in the eyes of Lamarckians challenged the Darwinian contribution to the evolutionary synthesis. The "neuroplastic" property of our brain, whereby we can "will" changes to it anatomically, appears now to be heralded by many in the scientific and medical worlds. The question also is if we can biologically pass on these changes to the next generation. The hereditarian, epigenetic, and environmental perspectives oscillate in importance throughout the social sciences with the hope that environment—that is, "nurture"—will play the decisive role in improving individuals, society, and the nation. This has already provided fodder for the literary imagination.

What is needed now is not just the critical rereading of first-wave and second-wave Darwinism in nineteenth- and early twentieth-century literature, but also the reading of present-day literary works associated with the current evolutionary issues, which are surfacing. In the Greek sphere, further investigation of the literary responses of the early modern Greek writers to Darwinism will narrow the gap in critical literary scholarship, which in turn will provide a basis for comparative study with present-day interdisciplinary discourses in science and literature.

This book demonstrated the value of the arts, in their various forms, for communicating science by making it relevant, accessible, real, and often more palatable for the broader reading public. The arts provided a form of expression that anyone can understand, where science alone may not succeed in doing so. The importance of bringing to the fore the history of Darwinian application, and in particular eugenics, by literary writers and other intellectuals, is critical in the way we understand new scientific ideas, ideologies, and developments such as genetic engineering, stem-cell research, artificial selection in its various medical forms, and even antibiotic resistance. For instance, if one wishes to accept Palamas' application to *ideas* of Weismann's 1885 scientific theory of continuity, an application Palamas utilized in his writings, then we accept that *ideas* evolve but that with each generation, in essence, they can stay the same. This notion of an *idea* "in essence staying the same" can be applied to the *idea* of Darwinian and post-Darwinian theories. If a particular *idea* is misconstrued in its application, we could establish an ideology which is reminiscent of the horrors of past history. The point here is to be aware of the past history and of the fact that a previously dangerous *idea* in a newly evolved form is, in essence, a dangerous *idea*.

The 200th anniversary of Darwin's birth and the 150th anniversary of the publication of his *OS* in 2009 rekindled further enquiry and discourse on Darwinism. Highly relevant to the ideologies associated with shaping Greek society and literature is that in the last few years one can observe in the Greek press an increasing trend by academics writing about biological

Darwinism and attitudes towards it.[2] Certainly this has been a universal trend in Western countries over the last twenty years; however, the critical point here is that in terms of Greek Orthodoxy there is still no acknowledgment of, or official stance on, Darwin's theory of evolution. Individuals and groups within the Church have made statements in the past but they have all been unofficial.

It has been the intention of this book to open up an area of study that has not been developed in Greek literary history and historiography. I hope that it will shed new light on the way literary writers and other intellectuals have contributed to the shaping of the modern Greek intellectual community.

2 Zarimis, "Darwinism and Its Impact in the Recent Greek Press."

ADDENDUM

THE FEMALE SEX'S HANDBOOK

The following is an example of how Darwin's name was a selling point even when the content of the publication had little or nothing to do with his own work. A guide published in 1874 entitled *The female sex's handbook: A guide to the physical and moral upbringing of women* indicates that it was translated into Greek from German.[1] It actually appears to be originally derived from Erasmus Darwin's *A Plan for the Conduct of Female Education in Boarding Schools* (1797).[2]

The authors in the Greek handbook are referred to as Darwin and Hufeland, and in the foreword they are referred to as being among "men long famous in the European world and highly respected for their knowledge and their genius."[3] My investigation shows that Hufeland was not a coauthor of the English book.

The omission of the first names of the authors, in particular that of Darwin, implies to the Greek reader that it may have been written by Charles Darwin, whose work was most highly reputed.[4] To add to this, the Greek handbook does not disclose the date of the original English version, and so gave the impression that the information was current. The English version advised girls on their behavior and their role at that time in society. It highlighted that they should be encouraged to have an enquiring mind, which was quite progressive and more so than the Greek guide of seventy years later. Moreover, the Greek guide had added text targeted at Greek women

1 Darwin and Hufeland, *Εγκόλπιον του γυναικείου φύλου*. However, the foreword indicates that it was published in Greek by the newspaper *Ermoupolis* (*Ερμούπολις*), ca. 1867.

2 E. Darwin, *Plan for the Conduct of Female Education*, 16–20.

3 Erasmus Darwin (1731–1802), Charles Darwin's grandfather, a physician, naturalist, philosopher, poet, and deist who wrote *Zoonomia* (1797) and other works; after 1859 did not command anywhere near his grandson's powerful influence. Christopher Wilhelm Hufeland (1762–1836) was a German physician and wrote, among other works, *The Art of Prolonged Living* (1797), a guide on how to live longer.

4 Freeman's bibliography of Charles Darwin's work does not list this women's handbook, and it is highly improbable that he would have written it. See Freeman, *The Works of Charles Darwin*, 1–235.

and also contained highly sexist material, which is not present in the English book. This highlights the influence and prestige that the name Darwin had because of its association with Charles, not Erasmus. Had Erasmus' name been added to the Greek guide, it would not have had the same impact. It is clear that Charles Darwin's reputation is used to appropriate a discourse which promotes the woman's place in the home.

The Greek guide praised women for having superior moral and religious qualities and emphasized that a woman's priorities were as a wife and mother, placing her own needs last. It also provided guidance in issues such as her mental and physical conduct, and how to give selflessly to her husband and children. Her education needs were not considered as important as the male's, and she was not thought capable of the same level of learning. For instance, the guide stated that addition, subtraction, multiplication, and division should be taught to all children. However, "the more advanced parts of Arithmetic, Algebra, and Equations, as they are quite abstract, should be excluded from a woman's upbringing."[5] This is not present in the original English version. In contrast, it mentioned that "the higher parts of arithmetic, as algebra and fluxions, belong to the abstract sciences," which may be too hard for young minds if taught too early.[6]

5 Darwin and Hufeland, *Εγκόλπιον του γυναικείου φύλου*, 41.
6 E. Darwin, *Plan for the Conduct of Female Education*, 16–20.

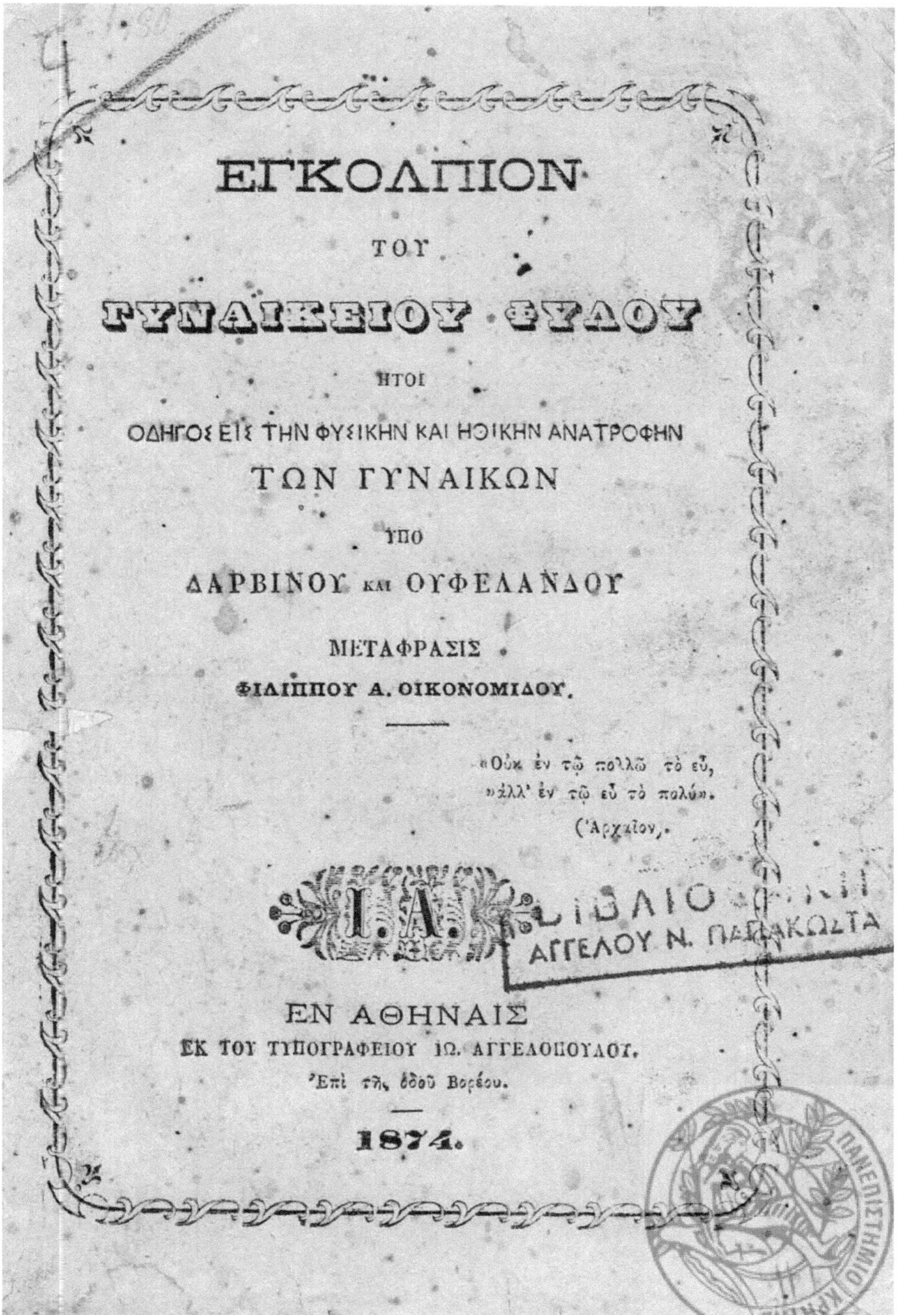
ΕΓΚΟΛΠΙΟΝ
ΤΟΥ
ΓΥΝΑΙΚΕΙΟΥ ΦΥΛΟΥ
ΗΤΟΙ
ΟΔΗΓΟΣ ΕΙΣ ΤΗΝ ΦΥΣΙΚΗΝ ΚΑΙ ΗΘΙΚΗΝ ΑΝΑΤΡΟΦΗΝ
ΤΩΝ ΓΥΝΑΙΚΩΝ
ΥΠΟ
ΔΑΡΒΙΝΟΥ και ΟΥΦΕΛΑΝΔΟΥ
ΜΕΤΑΦΡΑΣΙΣ
ΦΙΛΙΠΠΟΥ Α. ΟΙΚΟΝΟΜΙΔΟΥ.

«Οὐκ ἐν τῷ πολλῷ τὸ εὖ,
»ἀλλ' ἐν τῷ εὖ τὸ πολύ».
(Ἀρχαῖον).

Ι. Α.

ΕΝ ΑΘΗΝΑΙΣ
ΕΚ ΤΟΥ ΤΥΠΟΓΡΑΦΕΙΟΥ ΙΩ. ΑΓΓΕΛΟΠΟΥΛΟΥ.
Ἐπὶ τῆς ὁδοῦ Βορέου.
1874.

By Darwin and Hufeland. *The female sex's handbook: A guide to the physical and moral upbringing of women*. Reproduced with permission from Anemi Digital Library of Modern Greek Studies, University of Crete Library.

ΔΑΡΒΙΝΟΣ

Θέλω τὸ μέγα πνεῦμά σου, Δαρβῖνε, νὰ ὑμνήσω
Ὅμως ὁ νοῦς μου σταματᾷ στὸ τόσο μεγαλεῖο :
Ἀντὶ ἐμπρός, ἡ πένα μου ζητᾷ νὰ στρέψῃ ὀπίσω,
Σὰν ἄλογο ποῦ μυριστῇ στὸ δρόμο του θηρίο,
Ἢ νοιώσῃ ἄβυσσο ἐκεῖ π' ἀτάραχο βαδίζει,
Καὶ στέκεται, καὶ χλιμιντρᾷ, κι' ἀδιάκοπα ἀφρίζει !

—

Ν' ἀρχίσω πῶς ; ποιὸ ὄνομα ἁρμόζει νὰ σοῦ δώσω ;
Νὰ σ' ὀνομάσω ἄνθρωπο, θεό, ψυχή, σκουλῆκι ;
Στὸ γαλαζένιο οὐρανὸ τὸ νοῦ μου νὰ σηκώσω,
Ἢ μήπως κρύβεσαι βαθυὰ στὴ γῆ σὰν τὸ μερμῆκι ;
Νὰ σὲ ζητήσω σ' ἄγρια τῆς ἐρημιᾶς θηρία,
Ἢ στῶν πουλιῶν τὴ μαγική, Δαρβῖνε, μελῳδία ;

—

Μὴ βρίσκεσαι μέσ' τὴν ὑγρὴ τῆς θάλασσας ἀγκάλη,
Σὰν ἀστακὸς, σὰν φάλαινα, σὰ στρεῖδι, σὰ γαρίδα ;
Ἢ μήπως τὸ κεφάλι σου μεσ' τὰ φυτὰ προβάλλῃ
Σὰν μενεξές, σὰν ἀνανὰς, σὰ ῥόκα, σὰν τσουκνίδα ;
Ἄνοιξ' ἀκόμα μιὰ στιγμὴ τὸ στόμα σου καὶ κλεῖσε
Νὰ μᾶς εἰπῇς, εἶσαι καπνὸς, σκιά, πνοή, τί εἶσαι ;

Σὰ σκούληκας μέσα στῆς γῆς τὰ ἔγκατα τρυπώνεις,
Ἢ στὰ βαθυὰ τῆς θάλασσας σὰν βουτηχτὴς γυρίζεις.
Ἐκεῖ ἀπὸ τὰ σπλάχνα της ἕν' ἄτομο ξεχώνεις
Καὶ κόσμο νέο μὲ αὐτὸ καὶ σὺ, Δαρβῖνε, κτίζεις !
Ὡσὰν ταχυδακτυλουργὸς σπείρεις σὲ χίλιους τόπους
Ψάρια, πουλιὰ καὶ ἑρπετὰ, θηρία καὶ ἀνθρώπους !

—

Θυμοῦμαι, ὅταν μάθαινα, μικρὸ παιδὶ ἀκόμα
Πῶς ὁ Θεός μας ἔπλασε τὴ γῆ εἰς ἓξ ἡμέρας :
Πῶς εἶπε νὰ γενῇ τὸ φῶς τοῦ πλάστου μας τὸ στόμα
Κ' ὕστερα ἔπλασε μ' αὐτὸ τὸν ἥλιο, τοὺς ἀστέρας :
Ἔπλασε ζῶα καὶ φυτὰ—καὶ μὴν κανεὶς ξεχάσῃ
Πῶς τελευταῖο ἄφησε τὸν ἄνθρωπο νὰ πλάσῃ . . .

—

Καὶ βρίσκω τώρα πῶς πολὺ οἱ δύω κόσμοι 'μοιάζουν,
Πῶς ὁ πανάγαθος θεὸς καὶ ὁ καλὸς Δαρβῖνος
Τὸν κόσμο ἀπὸ κἄτι τι κ' οἱ δυὼ κατασκευάζουν,
Ὁ ἕνας ἀπὸ τὸ μηδέν, ἀπ' τὴ ζωὴ ἐκεῖνος . . . .
Αἴ, καὶ θεὸ ἂν μέσα σου, φιλόσοφε, δὲν κρύπτῃς,
Ὅμως αὐτὸν ἀληθινὰ ἐσὺ ἀποκαλύπτεις ! !

—Μὰ πῶς, ῥωτοῦν, ὁ τέλειος ἐπλάσθη τελευταῖος ;
Καὶ ἄλλοι—πῶς; ὁ ἄνθρωπος κατάγετ' ἀπὸ στρείδια ;
Καὶ νά ! θρησκευτικὸς ἀγὼν βγαίνει στὴ μέση νέος,
Καὶ στὸ Δαρβῖνο ῥίχνουνται οἱ εὐλαβεῖς σὰν φείδια!
Μὰ γιὰ σταθῆτε μιὰ στιγμή, θεοσεβεῖς, σταθῆτε,
Ἂν ἀγαπᾶτε τὸ θεὸ, τὸν Δάρβιν ἂν μισῆτε !

—

Ῥίχνω τὸ βλέμμα πίσω μου πενῆντα τόσο χρόνια
Καὶ βλέπω τὸν πλησίον μου μ' ἕνα ῥαβδὶ στὸ χέρι
Νὰ ἀνεβαίνῃ τὰ βουνὰ, νὰ περπατῇ στὰ χιόνια
Καὶ τόσος δρόμος σπίτι του τὸ δόλιο νὰ μὴν φέρῃ !
Σὰν τὸν περιπλανώμενο γυρίζει πάντα μόνος,
Ἡ ὥρα μέρα γίνεται, κ' ἡ 'μέρα μῆνας, χρόνος ! . .

—

Τώρα γοργὰ σὰν ἀστραπὴ βρίσκεται ὅπου θέλει,
Ξυπνάει στὴν ἀνατολή, βραδιάζει εἰς τὴ δύσι,
Τὸ λογισμό του κεραυνὸ, ὅπου θελήσῃ, στέλλει
Κι' ἀκόμα εἰς τοὺς οὐρανοὺς ζητᾷ νὰ φτερουγίσῃ !
Τέτοια κανένας στὰ παληὰ νὰ ἔλεγε ἂν τολμοῦσε
Σωστὰ θὰ τὸν ἐσούβλιζαν, τρελλὸς ἂν δὲν περνοῦσε.

—

Καὶ τώρα πῶς θαυμάζετε, ἂν ἀπὸ τὸ σκουλῆκι
Γεννήθηκε ὁ κωκωβιὸς, ἂν ἀπ' αὐτὸν χελώνια,
Ἀπ' τῆς χελώνες οἱ ἀετοὶ, ἀπ' τοὺς ἀητοὺς οἱ λύκοι,
Ἀπὸ τοὺς λύκους ἑρπετὰ, ὕστερα τὰ πεπόνια,

Στοὺς οὐρανούς μας τὰ πτηνὰ, τ' ἄνθη τῆς γῆς τὸ χῶμα,
Κι' ἀπ' τοὺς ἀνθοὺς καὶ τὰ πτηνὰ ὁ ἄνθρωπος ἀκόμα ;

—

Καὶ ζοῦνε ἄλλα, καὶ πολλὰ ἀπὸ αὐτὰ πεθαίνουν,
Καὶ ἄλλα μεταμόρφωσι παράδοξη λαμβάνουν,
Μένουν τὰ δυνατώτερα, τὰ ποιὸ ὡραῖα μένουν,
Καὶ τὰ παιδιὰ τὴ μάνα τους ἕνα καιρὸ ξεχάνουν !
Ὁμοιάζει ἐργοστάσιο ἀπέραντο ἡ φύσις
Ποῦ γίνεσαι γιὰ νὰ χαθῇς, πεθαίνεις γιὰ νὰ ζήσῃς...

—

Χιλιάδες ἔτυχε φοραῖς, νὰ ἰδῆτε, ποιὸ μεγάλη
Μία σταλαγματιὰ νεροῦ, τυριοῦ κανένα θρέμμα ;
Αἴ δὲν γνωρίσατε ἐκεῖ τοῦ βίου μας τὴν πάλη
Στοῦ κόσμου τοῦ ἀόρατου τὸ κάθε ἕνα βῆμα ;
Ἐκεῖ, ἐκεῖ θὰ νοιώσετε τὴν ἀλληλοσφαγία,
Ἐκεῖ τὴν μεταμόρφωσι καὶ τὴ δημιουργία . . .

—

Ὅλα ἀλλάζουν, τίποτα αἰώνιο δὲν μένει,
Ἡμέρα—νύχτα γίνεται δημιουργία νέα,
Βαδίζουν' ὅλα, καὶ κανὲν τὸ ἄλλο δὲν προσμένει
Κι' ὅλα κρατοῦν τὴν **πρόοδο** στὸ χέρι τους σημαία...
Ναὶ, ὅλοι τρέχετε ἐμπρὸς, πάντοτε προχωρῆτε,
Καὶ μὴ σταθῆτ' ἂν ὁ θεὸς δὲν σᾶς εἰπῇ—*Σταθῆτε* ! !

Dock.

Dock, "Darwin," *Don't Get Lost* [*Μη Χάνεσαι*], vol. 3, no. 289, May 1882, 4–5.
Courtesy of the Digital Collection Kosmopolis, University of Patras, Library and Information Center.

APPENDIX

DARWIN
(translation of poem "Δαρβίνος")

1

Darwin, I want to praise your great spirit
But my mind stops at such grandeur:
Instead of forward, my pen wants to turn back,
Like a horse which senses a beast on its course,
Or perceives an abyss there where it calmly treads,
And it stands, and neighs and froths incessantly!

2

How should I start? What name befits you?
Should I name you man, god, soul, worm?
Should I raise my mind to the azure sky,
Or maybe you are hiding deep in the earth like an ant?
Should I seek you among the wild beasts of the desert,
Or, Darwin, amongst the birds' magical melody?

3

Could you be found in the waters of the sea's open arms?
Like a lobster, like a whale, like an oyster, like a prawn?
Or maybe your head emerges in the plants
Like a violet, like a pineapple, like a rocket, like a bed of nettles?
Open your mouth again for a moment and close it,
Tell us, are you smoke, shade, breath, what are you?

4

Like a worm within the earth's depths you burrow,
Or in the depths of the sea like a diver you wander.
There from the sea's flesh you dig up an individual
And, Darwin, a new world with this you build!
Like a conjuror you sow in a thousand places
Fish, birds and reptiles, beasts and man!

5

I remember, when I was learning, still a small child,
That our God created the earth in six days:
That our creator's mouth said "Let there be light"
And with this he then formed the sun, the stars:
He created animals and plants—and don't anyone forget
That he left man to create last of all ...

6

And I find now that the two worlds are very much alike,
That both the most beneficent god and good Darwin
Construct the world from something,
The one from nothingness, from life the other ...
Ah philosopher, and if within, you don't conceal a god,
All the same you truly do reveal him!!

7

"But how," they ask, "was the perfect being created last?"
And others, "What? Does man originate from oysters?"
And see! A new religious struggle arises,
And the devout throw themselves like snakes at Darwin!
But wait a moment, pious ones, wait,
If you love god, if you hate Darwin!

8

I glance back fifty or so years
And I see my neighbor with a stick in hand
Ascending the mountains, walking in the snow
And all this walking doesn't bring the poor thing home!
Like a wanderer he travels around always on his own,
The hour becomes a day, and the day months, years! ...

9

Now quick as a flash he can be wherever he wants,
He wakes in the east, and night finds him in the west,
His thoughts, fast as a thunder bolt, wherever he wants he sends
And even in the skies he wants to fly!
If anyone dared to say such things in the past
They would truly have impaled him, if he wasn't considered mad.

10

And now how can you marvel, if from the worm
Was born the goby fish, if from it tortoises,
From tortoises the eagles, from eagles the wolves,
From wolves the reptiles, then rockmelons,
In our skies the birds, the blossoms from the earth's soil,
And from blossoms and birds even man was born?

11

And some things live, and many of these die,
And others take on a strange transformation,
The stronger ones remain, the finest ones remain,
And the children at some time will forget their mothers!
Nature resembles an endless factory
Where you are created so you can vanish, you die so you can live ...

12

You've happened a thousand times to see, magnified
A drop of water, a crumb of cheese?
Ah, didn't you recognize there our life's struggle?
In the world of the unseen every single move?
There, there you will understand the mutual destruction,
There also the transformation and creation ...

13

Everything changes, nothing remains forever,
New creation goes on night and day,
They all go on, and none waits for the other
And all hold **progress** in their hand like a flag ...
Yes, you all hurry forward, always keep advancing,
And don't stand still unless god tells you—*Stand* still!!

Dock[7]

7 Emphases in original.

ΤΟΠΟ ΑΛΛΑΖΕΙ;

Ὦ ἐξοχώτατε Ζωχιὲ, στὴ διδαχή σου
Ἵδρωσες αἷμα ἐπροχθὲς γιὰ νὰ μᾶς πείσῃς
Πῶς εἶν' ἀπὸ πιθήκους ἡ καταγωγή σου
Καθὼς καὶ ὅλης τσῆ ἀθρώπινης τσῆ φύσις.
Μεγάλο θάμα! μὴ ἀπὸ τσοὺ ἀπογόνους
Ποῦ βλέπεις γύρω σου, ἔκρινες καὶ τσοὺ προγόνους;

Ἔλεγες·—Ποιὸς μπορεῖ νὰ βάλῃ μὲ τὸ νοῦ του
Πῶς ἡ βαλανιδιὰ ποτὲ ἦταν βαλανίδι;
Καὶ ποιὸς μπορεῖ νὰ φαντασθῇ καὶ τὸν παποῦ του
Μαϊμοῦ μὲ ζαρωμένο κι' ἄσχημο μισίδι;
Μὰ ἐκ πείρας τὸ γνωρίζω ἀκροατήριό μου,
Ἔκαμα αὐτοψία εἰς τὸν ἑαυτό μου.—

Λοιπὸν, Δαρβῖνε, ὅπως ἡ βαλανιδία
Βγαίνει ὠχ' τὸ σπόρο μὲς στὴ γῆ ποῦ χώνουμε,
Κ' ἐμᾶς μαϊμοῦδες μᾶς ἐσπέρνουν μιὰ βραδύα
Καὶ τὴν αὐγὴ ἀθρῶποι ξεφυτρόνουμε;
Τί γίνετ' ἡ οὐρά μας; τόπο ἀλλάζει; πέ μου!
Ἢ μήκως καὶ μαραίνεται καὶ πέφτει 'γέ μου!...

ΒΟΛΤΑΙΡΟΣ.

Voltaire, "Does it change place?" *Rabagas* [Ραμπαγάς], year 2, no. 159, February 21, 1880, 2.
Reproduced with permission from the Hellenic Parliament Library, Athens.

DOES IT CHANGE PLACE?
(translation of poem "Τόπο αλλάζει;")

1

Oh your excellency Zochios, at your lecture
The day before you sweated blood so that you could convince us
That your descent is from apes
As is all humanity in nature
Great miracle! Possibly from the descendants
You see around you, did you also examine the ancestors?

2

You would say—Who would think
That the oak tree was ever an acorn?
And who can also imagine his grandfather
A monkey with a wrinkled and ugly mug?
But through experience, my audience, I know,
I did an inspection on myself.

3

Well, Darwin, like the oak tree
Which comes from the seed that we bury in the earth,
So we as monkeys they would sow us one night
And at dawn we would sprout as people?
What becomes of our tail? Does it change place? Tell me!
Or maybe even it withers and falls off, my son! ...

Voltaire

BIBLIOGRAPHY

Adamos, Takis. *Η λογοτεχνική κληρονομιά μας: από μια άλλη σκοπιά* [Our literary heritage: from another point of view]. Vol. 1. Athens: Kastaniotis, 1979.

Alden, Patricia. *Social Mobility in the English Bildungsroman: Gissing, Hardy, Bennett, and Lawrence*. Studies in Modern Literature 58. Ann Arbor, MI: UMI Research Press, 1986.

Aldrich, Michele. "United States." In *The Comparative Reception of Darwinism*, edited by Thomas F. Glick, 207–226. Austin: University of Texas Press, 1974.

Alec-Tweedie, Ethel. "Eugenics." *Fortnightly Review* 91, no. 545 (May 1912): 854–865.

Alter, Stephen G. *Darwinism and the Linguistic Image: Language, Race, and Natural Theology in the Nineteenth Century*. Baltimore: Johns Hopkins University Press, 1999.

Ambatzopoulou, Fragkiski. "Ο Κωστής Παλαμάς, και οι επιστήμες" [Kostis Palamas, and the sciences]. *Νέα Εστία*, no. 1771 (October 2004): 372–408.

———. "'Το ψιθύρισμα της Επιστήμης': ιατρικά θέματα στον Παλαμά" [The whispering of science: medical themes from Palamas]. *Αντί*, nos. 793–794 (July 2003): 40–43.

Amilitou, Eftychia. *Νικόλαος Σιγαλός* [Nicholas Sigalos]. Athens: Greek Literary and Historical Archives, 2002.

Anastasopoulou, Maria. "Feminist Discourse and Literary Representation in Turn-of-the-Century Greece: Kalliroë Siganou-Parren's 'The books of dawn.'" *Journal of Modern Greek Studies* 15 (1997): 1–28.

———. *Η συνετή απόστολος της γυναικείας χειραφεσίας. Καλλιρρόη Παρρέν: η ζωή και το έργο* [The prudent apostle of female emancipation. Kalliroi Parren: the life and work]. Athens: Heliodromion Society, 2003.

Angelopoulos, Athanasios. *Νέον λεξικόν της ελληνικής μυθολογίας* [New dictionary of Greek mythology]. Athens: Eleftheri Skepsi, 2000.

Aposkitou-Alexiou, Martha. "Τριάντα τέσσερα άγνωστα γράμματα του Νίκου Καζαντζάκη" [Thirty-four unknown letters by Nikos Kazantzakis]. *Αμάλθεια* 9, no. 35 (April–June 1978): 117–164.

Apostolides, Petros. "Επιστήμη και Τέχνη" [Science and arts]. *Ημερολόγιον Σκόκου* 10, no. 0 (1895): 25–35.

Apostolidou, Venetia. *Ο Κωστής Παλαμάς ιστορικός της νεοελληνικής λογοτεχνίας* [Kostis Palamas historian of modern Greek literature]. Athens: Themelio, 1992.

Argyropoulou, Roxani D., ed. *Η φιλοσοφική σκέψη στην Ελλάδα από το 1828 ώς το 1922* [Philosophical thought in Greece from 1828 to 1922]. Vol. 2, *Η φιλοσοφία μεταξύ επιστήμης και θρησκείας, 1876–1922* [The philosophy between science and religion, 1876–1922]. Athens: Gnosis, 1998.

Asher, David. "Schopenhauer and Darwinism." *Journal of Anthropology* 1, no. 3 (January 1871): 312–332.

Auerbach, Nina. "Magi and Maidens: The Romance of the Victorian Freud." In *Reading Fin de Siècle Fictions*, edited by Lynn Pykett, 22–38. London: Longman, 1996.

Babiniotis, Georgios. *Λεξικό της νέας ελληνικής γλώσσας* [Dictionary of the modern Greek language]. Athens: Kendro Lexikologias, 1998.

Bade, Patrick. *Femme Fatale: Images of Evil and Fascinating Women*. New York: Mayflower Books, 1979.

Bashford, Alison, and Philipa Levine, eds. *The Oxford Handbook of the History of Eugenics*. Oxford: Oxford University Press, 2010.

Bastias, Kostis. "Η τρίμορφη γυναίκα" [The three-sided woman]. *Δημοκρατία*, September 18, 1924, 2.

Bauer, Bernhard A. *Woman and Love*. Translated by Eden and Cedar Paul. London: Heinemann Medical, 1927.

Beaton, Roderick. *Introduction to Modern Greek Literature*. 2nd ed. Oxford: Clarendon Press, 1999.

Beer, Gillian. *Darwin's Plots: Evolutionary Narrative in Darwin, George Eliot and Nineteenth-Century Fiction*. 2nd ed. Cambridge, UK: Cambridge University Press, 2000.

———. "Darwin's Reading and the Fictions of Development." In *The Darwinian Heritage*, edited by David Kohn, 543–588. Princeton, NJ: Princeton University Press, 1985.

———. *Open Fields: Science in Cultural Encounter*. Oxford: Clarendon Press, 1996.

———. "Satire, Voice, and Nineteenth-Century Science." In *Η λογοτεχνία και οι προϋποθέσεις της: τιμητικό αφιέρωμα στην Τζίνα Πολίτη*, edited by Tina Krontiri and Katerina Kitsi Mytakou, 267–282. Thessaloniki: University Studio Press, 1999.

———. "Science in Literature." In *Companion to the History of Modern Science*, edited by R. C. Olby, G. N. Cantor, J. R. R. Christie, and M. J. S. Hodge, 783–797. London: Routledge, 1990.

Beizer, Janet. *Ventriloquized Bodies: Narrative of Hysteria in Nineteenth-Century France*. Ithaca, NY: Cornell University Press, 1994.

Bender, Bert. *The Descent of Love: Darwin and the Theory of Sexual Selection in American Fiction, 1871–1926*. Philadelphia: University of Pennsylvania Press, ca. 1996.

———. "Frank Norris on the Evolution and Repression of the Sexual Instinct." *Nineteenth-Century Literature* 54, no. 1 (June 1999): 73–103.

Bezas, Donatos. "Εμμανουήλ Ροΐδης" [Emmanuel Roidis]. In *Η παλαιότερη πεζογραφία*, edited by Nasos Vagenas, Giannis Dallas, and Kostas Stergiopoulos, 5, 8–89. Athens: Sokolis, 1997.

Bezirgan, Najma. "The Islamic World." In *The Comparative Reception of Darwinism*, edited by Thomas F. Glick, 375–387. Austin, TX: University of Texas Press, 1972.

Bien, Peter. *Kazantzakis: Politics of the Spirit*. 2 vols. Princeton, NJ: Princeton University Press, 1989–2007.

———. *Nikos Kazantzakis*. New York: Columbia University Press, 1972.

———, trans. and ed. *The Selected Letters of Nikos Kazantzakis*. By Nikos Kazantzakis. Princeton, NJ: Princeton University Press, 2012.

Birken, Lawrence. *Consuming Desire: Sexual Science and the Emergence of a Culture of Abundance, 1871–1914*. New York: Cornell University Press, 1988.

Blesios, Athanasios. *Κράτος, κοινωνία και έθνος στη νεοελληνική λογοτεχνία* [State, society, and nation in modern Greek literature]. Athens: Ekdoseis Sokoli, 2008.

Bont, Raf de. "'Foggy and Contradictory': Evolutionary Theory in Belgium, 1859–1945." In *The Reception of Charles Darwin in Europe*, edited by Eve-Marie Engels and Thomas Glick, 1:188–198. London: Continuum, 2008.

Bowler, Peter J. *Charles Darwin: The Man and the Influence*. Oxford: Basil Blackwell, 1990.

———. *The Eclipse of Darwinism: Anti-Darwinian Evolution Theories in the Decades around 1900*. Baltimore: Johns Hopkins University Press, 1983.

———. *Evolution: The History of an Idea*. Rev. ed. Berkeley: University of California Press, 1989.

———. *Evolution: The History of an Idea*. 3rd ed. Berkeley: University of California Press, 2009.

———. *The Non-Darwinian Revolution: Reinterpreting a Historical Myth*. Baltimore: Johns Hopkins University Press, 1988.

Bragg, Melvyn. *On Giants' Shoulders: Great Scientists and Their Discoveries; From Archimedes to DNA*. London: Hodder and Stoughton, 1998.

Bratchell, D. F. *The Impact of Darwinism: Texts and Commentary Illustrating Nineteenth Century Religious, Scientific, and Literary Attitudes*. Amersham: Avebury Publishing, 1981.

Browne, Porter Emerson. *A Fool There Was: A Story*. London, 1909.

Brunetière, Ferdinand. *L'évolution des genres dans l'histoire de la littérature* [The evolution of genres in the history of literature]. Vol. 1. Paris: Librairie Hachette, [1890] 1898.

Bullen, J. B. *The Expressive Eye: Fiction and Perception in the Work of Thomas Hardy*. Oxford: Clarendon Press, 1986.

Burkhardt, Frederick, et al., eds. *The Correspondence of Charles Darwin, 1821–1873*. 21 vols. Cambridge, UK: Cambridge University Press, 1985–2014.
Bury, J. B. *The Idea of Progress: An Inquiry into Its Origin and Growth*. London: Macmillan, 1920.
Buxhœveden, Daniel, and Gayle Woloschak, eds. *Science and the Eastern Orthodox Church*. England: Ashgate, 2011.
Cantor, Geoffrey, and Sally Shuttleworth, eds. *Science Serialized: Representations of the Sciences in Nineteenth-Century Periodicals*. Cambridge, MA: MIT Press, 2004.
Carroll, Joseph. *Evolution and Literary Theory*. Columbia: University of Missouri Press, 1995.
———. *Literary Darwinism: Evolution, Human Nature, and Literature*. New York: Routledge, 2004.
Chamberlain, J. Edward, and Sander L. Gilman, eds. *Degeneration: The Dark Side of Progress*. New York: Columbia University Press, 1985.
Charis, Petros. *Έλληνες πεζογράφοι* [Greek prose writers]. 2nd ed. Vol. 1. Athens: Kollaros, n.d.
———. *Έλληνες πεζογράφοι* [Greek prose writers]. Vol. 6. Athens: Kollaros, 1981.
———. *Έλληνες πεζογράφοι* [Greek prose writers]. Vol. 7. Athens: Kollaros, 1986.
———. "Πρώτοι απολογισμοί Α': ο πεζογράφος" [First reviews 1: the prose writer]. *Νέα Εστία* 50, no. 587 (1951): 128–135.
Childs, Donald J. *Modernism and Eugenics: Woolf, Eliot, Yeats, and the Culture of Degeneration*. Cambridge, UK: Cambridge University Press, 2006.
Christoulakis, Christos. "Ο σκεπτικισμός και η φυλή μας. Απάντηση του κ. Χρ. Χρηστουλάκη" [Skepticism and our race. Response by Mr. C. Christoulakis]. *Γράμματα* 3, no. 25 (1915): 81–93.
Conrad, Joseph. *Heart of Darkness*. Harmondsworth, UK: Penguin, [1902] 1983.
———. *Victory: An Island Tale*. Harmondsworth, UK: Penguin, [1915] 1984.
Cuddy, Lois A., and Claire M. Roche. *Evolution and Eugenics in American Literature and Culture, 1880–1940: Essays on Ideological Conflict and Complicity*. Lewisburg, PA: Bucknell University Press, 2003.
Culler, A. Dwight. "The Darwinian Revolution and Literary Form." In *The Art of Victorian Prose*, edited by George Levine and William Madden, 224–246. New York: Oxford University Press, 1968.
Dale, Peter Allan. *In Pursuit of a Scientific Culture: Science, Art, and Society in the Victorian Age*. Madison, WI: University of Wisconsin Press, 1989.
Darwin, Charles. *The Autobiography of Charles Darwin and Selected Letters*. Edited by Francis Darwin. New York: Dover Publications, [1892] 1958.
———. "A Biographical Sketch of an Infant." *Mind*, no. 7 (1877): 285–294.
———. *De l'origine des espèces par sélection naturelle ou des lois de transformation des êtres organisés* [The origin of species by natural selection or laws of transformation of organic beings]. 2nd ed. Translated by Clémence Royer. Paris: Guillaumin et Cie, 1866.
———. *The Descent of Man, and Selection in Relation to Sex*. 2 vols. London: John Murray, 1871.
———. *The Descent of Man, and Selection in Relation to Sex*. 2 vols. Princeton, NJ: Princeton University Press, 1981. Facsimile photoreproduction of the 1871 edition.
———. *The Descent of Man, and Selection in Relation to Sex*. 2nd ed. 2 vols. London: Penguin Books, [1879] 2004.
———. *The Expression of the Emotions in Man and Animals*. London: John Murray, 1872.
———. *The Expression of the Emotions in Man and Animals*. Chicago: University of Chicago Press, 1969.
———. *Η καταγωγή του ανθρώπου* [The descent of man]. Translated by Giannis Vistakis. Athens: Ekdoseis Anagnostidi, ca. 1976.
———. *Η καταγωγή του ανθρώπου* [The descent of man]. Translated by Vasos Vasiliou. Athens: Govostis, ca. 1977.
———. *Η καταγωγή των ειδών* [The origin of species]. Translated by Andreas Pangalos. Athens: Govostis, 1956. Second edition ca. 1974.
———. *Η καταγωγή των ειδών* [The origin of species]. Translated by the Teaching and Research Staff of the Biology Faculty at the University of Patras. Patras: Ekdoseis Panepistimiou Patron, 1997.

———. *On the Origin of Species by Means of Natural Selection, or the Preservation of Favoured Races in the Struggle for Life*. London: John Murray, 1859.
———. *The Origin of Species by Means of Natural Selection, or the Preservation of Favoured Races in the Struggle for Life*. Edited with an introduction by J. W. Burrow. Harmondsworth, UK: Penguin Books, 1968.
———. *Περί της γενέσεως των ειδών* [On the origin of species]. Translated by Nikos Kazantzakis. Athens: Fexis, 1915.
———. *The Variation of Animals and Plants under Domestication*. Vol. 2. London: John Murray, 1868.
Darwin, Erasmus. *A Plan for the Conduct of Female Education in Boarding Schools*. East Ardsley, UK: S. R. Publishers, [1797] 1968.
Darwin, and Hufeland. *Εγκόλπιον του γυναικείου φύλου ήτοι οδηγός εις την φυσικήν και ηθικήν ανατροφήν των γυναικών* [The female sex's handbook: A guide to the physical and moral upbringing of women]. Translated by Philippos Oikonomidis. Athens: J. Angelopoulos, 1874.
Dawkins, Richard. *The Blind Watchmaker*. London: Penguin Books, [1986] 1991.
———. *The Selfish Gene*. 2nd ed. Oxford: Oxford University Press, 1999.
Degler, Carl N. *In Search of Human Nature: The Decline and Revival of Darwinism in American Social Thought*. New York: Oxford University Press, 1991.
Delmouzos, Alexandros. *Το κρυφό σκολειό 1908–1911* [The secret school 1908–1911]. Thessaloniki: Aristotelian University of Thessaloniki, 2006.
Delon, Michel. *Ο Διαφωτισμός και η σημασία των διαβαθμίσεων. Les Lumières ou le sens des gradations* [The Enlightenment and the significance of gradations]. Translated from French by Anna Tambaki. Athens: Institute for Neohellenic Research, the National Hellenic Research Foundation, 2003. (Annual lecture K. Th. Dimaras, Bilingual edition.)
Delopoulos, Kyriakos. *Νεοελληνικά φιλολογικά ψευδώνημα (1800–2004)* [Modern Greek pseudonyms (1800–2004)]. 3rd ed. Athens: Kollaros, 2005.
Desmond, Adrian, and James Moore. *Darwin*. London: Michael Joseph, 1991.
Dijkstra, Bram. *Evil Sisters: The Threat of Female Sexuality and the Cult of Manhood*. New York: Alfred A. Knopf, 1996.
———. *Idols of Perversity: Fantasies of Feminine Evil in Fin-de-Siècle Culture*. New York: Oxford University Press, 1986.
Dimaras, K. Th. *Ιστορία της νεοελληνικής λογοτεχνίας: από τις πρώτες ρίζες ώς την εποχή μας* [History of modern Greek literature: from its first origins to our times]. 6th ed. Athens: Ikaros, 1975.
Dimiroulis, Dimitris. "Διαμαχή Ροΐδη-Βλάχου: ο ποιητής γεννιέται ή γίνεται;" [The clash between Roidis and Vlachos: is the poet born or made?]. *Το Δέντρο* 19, no. 129–130 (October–December 2003): 31–37.
———. *Εμμανουήλ Ροΐδης: η τέχνη του ύφους και της πολεμικής* [Emmanuel Roidis: the art and style of polemics]. Athens: Metaihmio, 2005.
Dock [pseud.]. "Δαρβίνος" [Darwin]. *Μη Χάνεσαι* 3, no. 289 (May 1882): 4–5.
Doulaveras, Vicky. "Τα δυο φύλα σε αστικά μυθιστορήματα του Γρηγορίου Ξενόπουλου" [The two sexes in the urban novels by Grigorios Xenopoulos]. PhD diss., University of Sydney, 1995.
———. "Κοινωνικές δομές και ιδεολογία στην τριλογία του Γρηγορίου Ξενόπουλου" [Social structures and ideology in the trilogy by Grigorios Xenopoulos]. Special issue, *Περίπλους*, July–December 1991, 131–141.
Dowling, Linda. "The Decadent and the New Woman in the 1890s." In *Reading Fin de Siècle Fictions*, edited by Lynn Pykett, 47–63. London: Longman, 1996.
Doyle, Arthur Conan. *The Hound of the Baskervilles*. Harmondsworth, UK: Penguin, 1981.
———. *Sherlock Holmes: The Complete Short Stories*. London: John Murray, 1961.
Dragoumis, Ion [Idas, pseud.]. *Ελληνικός πολιτισμός* [Greek civilization]. Athens: Neas Theseis, [1913] 1991.
———. "Ο σκεπτικισμός και η φυλή μας. Απάντηση Ίδα" [Skepticism and our race. Response by Ida]. *Γράμματα* 2, no. 21 (1914): 506–509.
Eckman, Paul. "Facial Expression and Emotion." *American Psychologist* 48 (1993): 384–392.

———. "Universals and Cultural Differences in Facial Expressions of Emotion." In *Nebraska Symposium on Motivation*, edited by J. Cole, 19:207–283. Lincoln: University of Nebraska Press, 1972.

Education Association [Εκπαιδευτικός Όμιλος]. "Ανάγκη ... ν'ανοιχτούνε στα κορίτσια κι' άλλοι δρόμοι" [Necessary ... that other options should be available to girls]."Υπόμνημα του Εκπαιδευτικού Ομίλου. Ανώτερα παρθεναγωγεία και διδασκαλία" [Memo of the Education Association. Upper girls' schools and teaching]. In *Εκπαίδευση και αγωγή των κοριτσιών: ελληνικοί προβληματισμοί (1830–1910). Ένα ανθολόγιο*, edited by Eleni Fournaraki, 591–598. Historical Archive of Greek Youth 11. Athens: Geniki Grammateia Neas Genias, 1987.

Ellegård, Alvar. *Darwin and the General Reader: The Reception of Darwin's Theory of Evolution in the British Periodical Press, 1859–1872*. Götenburg: Elanders Boktryckeri Aktiebolag, 1958.

Engels, Eve-Marie, and Thomas Glick, eds. *The Reception of Charles Darwin in Europe*. 2 vols. London: Continuum, 2008.

Farinou-Malamatari, G. "Γρηγόριος Ξενόπουλος" [Grigorios Xenopoulos]. In *Η παλαιότερη πεζογραφία μας*, edited by Nasos Vagenas, Giannis Dallas, and Kostas Stergiopoulos, 9:288–377. Athens: Sokolis, 1997.

———. "Το παρα-κείμενο της πεζογραφίας του Ξενόπουλου" [The paratext of the prose by Xenopoulos]. Special edition, *Νέα Εστία*, no. 1738 (October 2001): 372–402.

Faris, Robert E. L. "Evolution and American Sociology." In *Evolutionary Thought in America*, edited by Stow Persons, 160–181. Hamden, CO: Archon Books, [1950] 1968.

Ferri, Enrico. *Socialism and Positive Science: Darwin, Spencer, Marx*. 5th ed. London: Independent Labour Party, 1909.

Fielding, William J. *Woman: The Eternal Primitive*. Girard, KS: Halderman-Julius, 1927.

Fournaraki, Eleni, ed. *Εκπαίδευση και αγωγή των κοριτσιών: ελληνικοί προβληματισμοί (1830–1910). Ένα ανθολόγιο* [Girls' education and upbringing: Greek responses (1830–1910). An anthology]. Historical Archive of Greek Youth 11. Athens: Geniki Grammateia Neas Genias, 1987.

Fragkoglou, Giorgos. "Το λαϊκό αισθηματικό μυθιστόρημα του Γρηγορίου Ξενόπουλου (1912–1945)" [The popular love story by Grigorios Xenopoulos (1912–1945)]. Master's thesis, Aristotle University of Thessaloniki, 1985.

Freeman, R. B. *Charles Darwin: A Companion*. Folkestone, UK: Dawson, 1978.

———. *The Works of Charles Darwin: An Annotated Bibliographical Handlist*. 2nd ed. Folkestone, UK: Dawson, 1977.

Galton, Francis. *English Men of Science: Their Nature and Nurture*. 2nd ed. London: Frank Cass, [1874] 1970.

———. *Hereditary Genius: An Inquiry into Its Laws and Consequences*. London: Macmillan, 1869.

———. "Hereditary Talent and Character." *Macmillan's Magazine* 12 (June 1865): 165.

———. *Inquiries into Human Faculty and Its Development*. 2nd ed. London: J. M. Dent and Co., [1883] 1907.

———. "Kantsaywhere." In *The Life, Letters, and Labours of Francis Galton*, edited by Karl Pearson, 3a:411–425. London: Edward Arnold, 1897.

Gavriilidis, Vlasis [Caliban, pseud.]. "Δαρβίν-Λάτας" [Darwin-Latas]. *Μη Χάνεσαι* 1, no. 24 (March 1880): 1–8.

Georganta, Athina. "Από τον Αριστοτέλη στον Δαρβίνο" [From Aristotle to Darwin]. *Το Βήμα*, January 11, 2004, 43.

———. *Εμμανουήλ Ροΐδης: Η πορεία προς την Πάπισσα Ιωάννα (1860–1865)* [Emmanuel Roidis: the path towards Pope Joan (1860–1865)]. Athens: Istos, 1993.

Gianidis, Elisaios. *Γλώσσα και ζωή. Αναλυτική μελέτη του γλωσσικού ζητήματος* [Language and life. Analysis of the language question]. Athens, [1908] 1928. Excerpt of this work in *Η φιλοσοφική σκέψη στην Ελλάδα από το 1828 ώς το 1922. Η φιλοσοφία μεταξύ επιστήμης και θρησκείας, 1876–1922*, edited by Roxani D. Argyropoulou, 2:610–620. Athens: Gnostis, 1998.

Giannakidis, Kostas. "Ο πίθηκος έδωσε το μήλο στον Αδάμ και στην Εύα" [The ape gave the apple to Adam and Eve]. *Έψιλον*, Sunday supplement of *Ελευθεροτυπία*, no. 831, March 18, 2007, 38–47.

Giannios, Nikolaos. "Ο σκεπτικισμός και η φυλή μας. Απάντηση Ν. Γιαννιού" [Skepticism and our race. Response by N. Giannios]. *Γράμματα* 2, no. 21 (1914): 505–506.
Giannopoulos, Perikles. "Η σύγχρονος ζωγραφική" [Contemporary art]. In *Άπαντα*, edited by Vasilis Christou, 7–65. Athens: Eleftheri Skepsi, 1988.
Gibbons, Tom. *Rooms in the Darwin Hotel: Studies in English Literary Criticism and Ideas 1880–1920*. Nedlands: University of Western Australia Press, 1973.
Gissing, George. *Demos: A Story of English Socialism*. Brighton, UK: Harvester Press, [1886] 1972.
———. *The Nether World*. Brighton, UK: Harvester Press, [1889] 1974.
———. *The Unclassed*. Rev. ed. Brighton, UK: Harvester Press, [1895] 1976.
Glick, Thomas F., ed. *The Comparative Reception of Darwinism*. Austin: University of Texas Press, 1974.
Glinos, Dimitris. "Γιάννης Ψυχάρης: βιογραφικό και κριτικό σημείωμα" [Giannis Psycharis: biographical and commentary note]. Introduction to *Αγνή*, by Giannis Psycharis, 7–31. 2nd ed. Athens: Eleftheroudakis, 1930.
Gorky, Maxim. *The Artamonov Business*. Translated by Alec Brown. London: Hamish Hamilton, [1925] 1948.
Gottschall, Jonathan, and David Sloane Wilson, eds. *The Literary Animal: Evolution and the Nature of Narrative*. Illinois: Northwestern University Press, 2005.
Gourmont, Remy de. *The Natural Philosophy of Love*. Translated by Ezra Pound. London: The Casanova Society, 1926.
Grand, Sarah. "The New Aspect of the Woman Question." *North American Review*, no. 158 (1894): 270–276.
Greene, John C. *Darwin and the Modern World View*. Baton Rouge: Louisiana State University Press, 1961.
———. *Science, Ideology, and World View: Essays in the History of Evolutionary Ideas*. Berkeley: University of California Press, 1981.
Greenslade, William. *Degeneration, Culture and the Novel, 1880–1940*. Cambridge, UK: Cambridge University Press, 1994.
Gregory, Richard L., ed. *The Oxford Companion to the Mind*. 2nd ed. Oxford: Oxford University Press, 2004.
Grosse, Ernst. *The Beginnings of Art*. New York: D. Appleton, [1897] 1914.
Grosz, Elizabeth. *The Nick of Time: Politics, Evolution, and the Untimely*. Sydney: Allen and Unwin, 2004.
Hadziiossif, Christos. "Class Structure and Class Antagonism in Late Nineteenth-Century Greece." In *Greek Society in the Making, 1863–1913: Realities, Symbols and Visions*, edited by Philip Carabott, 3–95. Aldershot, UK: Ashgate Variorum, 1997.
Hamilton, Peter. "Policing the Face." In *The Beautiful and the Damned: The Creation of Identity in Nineteenth Century Photography*, edited by Peter Hamilton and Roger Hargreaves, 57–107. London: Lund Humphries in association with The National Portrait Gallery, 2001.
Hardy, Thomas. *A Pair of Blue Eyes*. New York: Macmillan, [1873] 1957.
Harker, L.A., and D. Keltner. "Expressions of Positive Emotion in Women's College Yearbook Pictures and Their Relationship to Personality and Life Outcomes across Adulthood." *Journal of Personality and Social Psychology* 80, no. 1 (2001): 112–124.
Haynes, Roslynn D. *From Faust to Strangelove: Representations of the Scientist in Western Literature*. Baltimore: Johns Hopkins University Press, 1994.
Heilmann, Ann. *New Woman Fiction: Women Writing First-Wave Feminism*. London: Macmillan, 2000.
Hemmings, F. W. J., and Giovanni Carsaniga. *The Age of Realism*. Brighton, UK: Harvester Press, 1978.
Henkin, Leo. *Darwinism in the English Novel: The Impact of Evolution on Victorian Fiction*. New York: Russell and Russell, 1963.
Hermann, Tomáš, and Michal Šimůnek. "Between Science and Ideology: The Reception of Darwin and Darwinism in the Czech Lands, 1859–1959." In *The Reception of Charles Darwin in Europe*, edited by Eve-Marie Engels and Thomas Glick, 1:199–216. London: Continuum, 2008.
Herzfeld, Michael. *Ours Once More: Folklore, Ideology, and the Making of Modern Greece*. New York: Pella, 1986.

Hopkins, Lisa. *Giants of the Past: Popular Fictions and the Idea of Evolution*. Lewisburg, PA: Bucknell University Press, 2004.

Hunt, James. "On the Doctrine of Continuity Applied to Anthropology." *Anthropological Review* 5, no 16 (January 1867): 110–120.

Huxley, Thomas. H. *Man's Place in Nature, and Other Anthropological Essays*. London: Macmillan, [1894] 1901. First published in essay form in 1863.

Huxley, Thomas H., and Julian Huxley. *Evolution and Ethics: 1893–1943*. London: The Pilot Press, 1947.

Hyman, Stanley Edgar. *The Tangled Bank: Darwin, Marx, Frazer and Freud as Imaginative Writers*. New York: Atheneum, 1974.

Ioannou, Giorgos, ed. *Το δημοτικό τραγούδι: παραλογές* [The folk song: ballads]. Athens: Ermis, 1975.

Irvine, William. "The Influence of Darwin on Literature." *Proceedings of the American Philosophical Society* 103, no. 5 (October 1959): 616–628. (Commemoration of the centennial of the publication of *OS*)

James, William. *The Principles of Psychology*. Cambridge, MA: Harvard University Press, 1983.

Jann, Rosemary. "Darwin and the Anthropologists: Sexual Selection and Its Discontent." *Victorian Studies* 37, no. 2 (1994): 286–306.

Jones, Greta. *Social Darwinism and English Thought: The Interaction between Biological and Social Theory*. Brighton, UK: Harvester Press, 1980.

Kairis, Michalis N. *Η ευγονία: διάλεξις γενομένη εν τη αιθούση της Σιναίας Ακαδημίας τη 24η Μαρτίου 1917* [Eugenics: lecture given in the hall of the Sinai Academy on March 24, 1917]. Athens: Adelfon Frantziskaki, 1917.

Kalopothakis, M. D. "Η θεωρία της εξελίξεως: υπό καθαρώς επιστημονικόν έλεγχον" [The theory of evolution: under purely scientific scrutiny]. *Παρνασσός* 13 (1890): 365–378.

Kambanis, Aristos. *Ιστορία της νέας ελληνικής λογοτεχνίας* [History of modern Greek literature]. 4th ed. Athens: Kollaros, n.d.

Karantonis, Andreas. *Νεοελληνική λογοτεχνία: φυσιογνωμίες* [Modern Greek literature: physiognomies]. 2 vols. Athens: D. N. Papadimas, 1977.

Karas, Giannis, ed. *Ιστορία των επιστημών: νεοελληνική βιβλιογραφία* [History of the sciences: modern Greek bibliography]. Athens: Kentro Neoellinikon Erevnon E.I.E., 1997.

Karkavitsas, Andreas. *Η λυγερή* [The fair maid]. Athens: Pella, [1890] n.d.

Kazantzakis, Nikos. *Αναφορά στον Γκρέκο: μυθιστόρημα* [Report to Greco: a novel]. 6th ed. Athens: Ekdoseis Eleni Kazantzaki, [1961] n.d.

———. *Ασκητική: Salvatores Dei* [Spiritual exercises: saviors of God]. 5th ed. Athens: Ekdoseis Eleni Kazantzaki, [1927] 1971.

——— [Akritas, pseud.]. "Επιστολαί από το Παρίσι. Οι Έλληνες σπουδασταί" [Letters from Paris. The Greek students]. *Νέα Εστία* 64 (September 1, 1958): 1284–1285. Reprinted from the newspaper *Νέον Άστυ*, March 8, 1908.

———. *Ο Καπετάν Μιχάλης: ελευτερία ή θάνατος* [Kapetan Mihalis: freedom or death]. 3rd ed. Athens: Ekdoseis Eleni Kazantzaki, [1955] 1974.

———. *The Last Temptation*. Translated by Peter A. Bien. Oxford: Bruno Cassirer, 1961.

——— [Petros Psiloritis, pseud.]. *Ο πρωτομάστορας* [The master builder]. Athens: Panathinaia, 1910. Published as a separate booklet by the periodical *Panathinaia*, which had printed the play in issue no. 20, June 1910, 131–144.

———. *Report to Greco*. Translated by Peter A. Bien. London: Bruno Cassirer, 1966.

———. *The Saviors of God: Spiritual Exercises*. Translated by Kimon Friar. New York: Simon and Schuster, 1960.

———. *Serpent and Lily: A Novella, with a Manifesto, "The Sickness of the Age."* Translated by Theodora Vasils. Berkeley: University of California Press, [1906] 1980.

Kitsi-Mitakou, Katerina. "Aquatic Spaces and Women's Places: A Comparative Reading of George Eliot's *The Mill on the Floss* and Alexandros Papadiamantis' *Η φόνισσα*." *Byzantine and Modern Greek Studies* 29, no. 2 (2005): 187–202.

Korasidou, Maria. *Οι άθλιοι των Αθηνών και οι θεραπευτές τους: φτώχεια και φιλανθρωπία στην ελληνική πρωτεύουσα τον 19ο αιώνα* [The Athens poor and their healers: nineteenth-century poverty and philanthropy in the Greek capital]. Historical Archive of Greek Youth 29. Athens: Geniki Grammateia Neas Genias, 1995.

Kordatos, Gianis (also Giannis). *Ιστορία της νεοελληνικής λογοτεχνίας από το 1453 ώς το 1961* [History of modern Greek literature from 1453 to 1961]. Vol. 1. Athens: Epikairotita, 1983.

———. *Μεγάλη ιστορία της Ελλάδας. 2500 π. Χ. πρόελληνες μέχρι 1924 ανακύρηξη της δημοκρατίας* [The great history of Greece. 2500 BC pre-Greeks to 1924 the proclamation of the republic]. Vol. 12, 1862–1900. Athens: Ekdoseis 20os Aionas, 1960.

Kostelenos, D. P. *Σύγχρονη ιστορία νεοελληνικής λογοτεχνίας: από την Άλωση ώς τις μέρες μας* [Contemporary history of modern Greek literature: from the fall to our times]. Athens: Pagoulatos Brothers, 1977.

Kouhtsoglou, Giannis. "Ο 'σπενσεριστής' Ξενόπουλος" [Xenopoulos the Spencerian]. *Νέα Εστία*, no. 112 (1982): 1015–1018.

Kourtidis, Aristotelis P. [Voltaire, pseud.]. "Τόπο αλλάζει;" [Does it change place?]. *Ραμπαγάς* 2, no. 159 (February 21, 1880): 2.

Kouvelas, Ilias. "Η θεωρία που προκαλεί ακόμη και σήμερα" [The theory that challenges even today]. *Βήμα Ιδεών*, no. 21 (January 2009): 19–22. http://www.vimaideon.gr//Article.aspx?d=20090101&nid=11084655&sn=%CE%9A%CE%A5%CE%A1%CE%99%CE%9F%20%CE%A4%CE%95%CE%A5%CE%A7%CE%9F%CE%A3&spid=1478. Last accessed January 2014.

Krimbas, Costas. *Δαρβινικά* [Darwinian ideas]. Athens: Ermis, 1986.

———. *Δαρβινισμός και η ιστορία του έως τις μέρες μας* [Darwinism and its history until our present]. Athens: Okeanida, 2009.

———. "Ο δαρβινισμός στην Ελλάδα. Τα πρώτα βήματα: η αλληλογραφία Χελδράιχ-Δαρβίνου, Μηλιαράκης, Νικολαΐδης, Ζωχιός, Σουγκράς" [Darwinism in Greece. The first steps: the correspondence of Heldreich-Darwin, Miliarakis, Nicolaidis, Zochios, Sougras]. In *Θραύσματα κατόπτρου*, 81–108. Athens: Themelio, 1993. Also published in *Τα Ιστορικά*, no. 2 (1984): 335–348; both revised and extended versions of *Materia Medica Greca* 10, no. 5 (1982): 465–471.

———. *Εκτείνοντας τον δαρβινισμό και άλλα δοκίμια* [Extending Darwinism and other essays]. Athens: Nefeli, 1998.

———. "Η έρευνα στην Ελλάδα: 1977" [Research in Greece: 1977]. In *Θραύσματα κατόπτρου*, 225–239. Athens: Themelio, 1993.

Kritikos, Theodore. "Science and Religion in Greece at the End of the Nineteenth Century." *Historein* 1 (1999): 35–50.

Lamni, Zaharoula M. "Περί της μητρικής στοργής. Φυσικός και ηθικός νόμος της φύσεως" [On maternal affection. The natural and moral law of nature]. In *Αττικόν Ημερολόγιον του Δίσεκτου Έτους 1896*, edited by Irinaios Asopios, 470–473. Athens: Korinnis, 1896.

Lankester, Edwin Ray. *Degeneration: A Chapter in Darwinism*. London: Macmillan, 1880.

Leahey, Thomas Hardy. *A History of Psychology: Main Currents in Psychological Thought*. 5th ed. Upper Saddle River, NJ: Prentice Hall, 2000.

Leatherdale, William. "The Influences of Darwinism on English Literature and Literary Ideas." In *The Wider Domain of Evolutionary Thought*, edited by David Oldroyd and Ian Langham, 1–26. Dordrecht, Holland: D. Reidel, 1983.

Leeuwenburgh, Bart, and Jannecke van der Heide. "Darwin on Dutch Soil: The Early Reception of His Ideas in the Netherlands." In *The Reception of Charles Darwin*, edited by Eve-Marie Engels and Thomas Glick, 1:175–187. London: Continuum, 2008.

Lemprière, John. *Bibliotheca Classica: Or, a Classical Dictionary*. 3rd ed. London, [1788] 1838.

Leon, George B. *The Greek Socialist Movement and the First World War: The Road to Unity*. East European Monographs 18. Boulder, CO: Eastern European Quarterly, 1976.

Levine, George. *Darwin and the Novelists: Patterns of Science in Victorian Fiction*. Cambridge, MA: Harvard University Press, 1988.

Lombroso, Cesare, and William Ferrero. *The Female Offender*. London: Peter Owen Limited, [1895] 1959.

Lombroso-Ferrero, Gina. *Criminal Man: According to the Classification of Cesare Lombroso*. Montclair, NJ: Patterson Smith, [1911] 1972.
Lyell, Charles. *The Geological Evidence of the Antiquity of Man with Remarks on Theories of the Species by Variation*. London: John Murray, 1863.
Mackridge, Peter. *Language and National Identity in Greece, 1766–1976*. Oxford: Oxford University Press, 2010.
Makridis, Vasilios. "Εξελικτική και Ορθοδοξία στην Ελλάδα: απόπειρες εναρμόνισης σε κριτική θεώρηση" [Evolution and orthodoxy in Greece: a critical review on the attempts to harmonize]. *Νεύσις*, no. 7 (Winter 1998): 173–220.
Malafantis, Konstantinos D. *Οι "Αθηναϊκαί Επιστολαί" του Γρηγορίου Ξενόπουλου στη "Διάπλασιν των Παίδων" (1896–1947)* [The Athenian letters of Grigorios Xenopoulos in the Children's Guidance (1896–1947)]. Athens: Astir (A. and E. Papadimitriou), 1995.
———. "Ο Γρηγόριος Ξενόπουλος και το 'διαπλαστικό έργο' του" [Grigorios Xenopoulos and his formative work]. Special issue, *Νέα Εστία* 150, no. 1738 (October 2001): 494–501.
Malanos, Timos. *Δειγματολόγιο* [Sample book]. Athens: Fexis, 1962.
Manier, Edward. *The Young Darwin and His Cultural Circle: A Study of Influences which Helped Shape the Language and Logic of the First Drafts of the Theory of Natural Selection*. Dordrecht, Holland: D. Reidel, 1978.
Marías, Julián. *History of Philosophy, 1914–2005*. New York: Dover Publications, 1967.
Marx, Karl. *A Contribution to the Critique of Political Economy*. Translated by S.W. Ryazanskaya. London: Lawrence and Wishart, [1859] 1971. http://www.marxists.org/archive/marx/works/1859/critique-pol-economy/index.htm. Last accessed January 2014.
Marx, Karl, and Frederick Engels. *Selected Works in One Volume*. London: Lawrence and Wishart, 1973.
Mastrodimitris P. D. *Εισαγωγή στη νεοελληνική φιλολογία* [Introduction to modern Greek literature]. 7th ed. Athens: Domos, 2005.
Mavroeidi-Papadaki, Sofia. "Η γυναίκα στο έργο του Ξενόπουλου" [Woman in Xenopoulos' work]. *Ιόνιος Ανθολογία*, no. 126 (1939): 77–81.
Mayr, Ernst. "Darwin's Influence on Modern Thought." *Scientific American* 283, no. 1 (July 2000): 79–83.
———. *The Growth of Biological Thought: Diversity, Evolution, and Inheritance*. Cambridge, MA: Belknap Press, 1982.
———. *What Evolution Is*. New York: Basic Books, 2001.
M. D. S. [pseud.]. "Πέντε διαφορετικές παιδικές φυσιογνωμίες" [Five different physiognomies of children]. *Η Διάπλασις των Παίδων*, no. 27 (June 7, 1941): 209–210; no. 28 (June 14, 1941): 220; no. 29 (June 21, 1941): 224.
Menardos, Simos. *Εμμανουήλ Ροΐδης* [Emmanuel Roidis]. Athens: Kollaros, 1918.
Middleton, D. J. N., and P. Bien. *God's Struggler: Religion in the Writings of Nikos Kazantzakis*. Macon, GA: Mercer University Press, 1996.
Milful, Alison. "Songs of the Siren: Women Writers and the Femme Fatale." PhD diss., University of New South Wales, 2000.
Miliarakis, Spyridon. "Ο Δάρβιν και η θεωρία του: μια εκατονταετηρίς" [Darwin and his theory: a centenary]. *Εστία*, April 9–15, 1909, 5
———. "Ο Δάρβιν και η θεωρία του: ο δαρβινισμός εν Ελλάδι" [Darwin and his theory: Darwinism in Greece]. *Εστία*, April 16–17, 1909, 5.
———. "Κάρολος Δάρβιν: βιογραφικόν σχεδίασμα μικρού τινος παιδίου" [Charles Darwin: the biographical sketch of an infant]. *Εστία*, no. 104 (1877): 817–824.
———. "Κάρολος Δάρβιν" [Charles Darwin]. *Εστία* 8, no. 186 (1879): 451–456.
———. *Αι ψυχικαί ιδιότητες των ζώων* [The mental traits of animals]. Athens: P. D. Sakellarios, 1926.
Miller, Geoffrey. "Arts of Seduction." In *Evolution, Literature, and Film: A Reader*, edited by Brian Boyd, Joseph Carroll, and Jonathan Gottschall, 156–173. New York: Columbia University Press, 2010.

Montagu, Jennifer, *The Expressions of the Passions: The Origin and Influence of Charles Le Brun's "Conférence sur l'expression générale et particulière."* New Haven: Yale University Press, 1994.

Morton, Peter. *The Vital Science: Biology and the Literary Imagination, 1860–1900.* Boston: George Allen and Unwin, 1984.

Mousmoutis, Dionysis. "Δυο επιστολές του Γρηγορίου Ξενόπουλου στον Τίμο Μαλάνο" [Two letters by Grigorios Xenopoulos to Timos Malanos]. *Νέα Εστία* 150, no. 1738 (2001): 515–531.

Moysidos, M. M. "Ευγονική αποστείρωσις: αρχαί, μέθοδοι, εφαρμογή" [Eugenic sterilization: beginnings, methods, implementation]. *Έρευνα*, September–October 1934, 1–68.

Murphy, Gardner. *Historical Introduction to Modern Psychology.* Rev. ed. London: Routledge and Kegan Paul, 1960.

Murray, John Middleton, ed. *Poems of John Keats.* London: Peter Nevill, 1948.

Mylogiannis, Giorgos [G. Peristatikos, pseud.]. *Πορτραίτα* [Portraits]. Athens: Karanasis, 1987.

Nicolaidis, Efthymios. "Ιστοριογραφία των επιστημών" [Historiography of the sciences]. *Historiography of Modern and Contemporary Greece 1833–2002. Proceedings of the 4th International Conference of the Institute for Neohellenic Research and the National Hellenic Research Foundation* 1 (2004): 527–538.

———. *Science and Eastern Orthodoxy: From the Greek Fathers to the Age of Globalization.* Translated by Susan Emanuel. Baltimore: Johns Hopkins University Press, 2011.

Nies, Betsy L. *Eugenic Fantasies: Racial Ideology in the Literature and Popular Culture of the 1920s.* London: Routledge, 2002.

Nordau, Max. *Degeneration.* 2nd ed. Translator not mentioned. New York: Howard Fertig, [1895] 1968.

———. *Εκφυλισμός* [Degeneration]. Translated by Angelos Vlachos. Athens: Fexis, 1911.

Numbers, Ronald L., and John Stenhouse, eds. *Disseminating Darwinism: The Role of Place, Race, Religion, and Gender.* Cambridge, UK: Cambridge University Press, 1999.

Oldroyd, David R. *Darwinian Impacts: An Introduction to the Darwinian Revolution.* Atlantic Highlands, NJ: Humanities Press, 1980.

Oldroyd David, and Ian Langham, eds. *The Wider Domain of Evolutionary Thought.* Dordrecht, Holland: D. Reidel, 1983.

O'Neill, Klara L. "The Unfinished Ape: Mediation and Modern Greek Identity in Iakovos Pitsipios' *Ο πίθηκος Ξουθ.*" *Journal of Modern Greek Studies* 21, no. 1 (2003): 67–110.

Ouida [Marie Louise de la Ramée]. "The New Woman." *The North American Review* 158, no. 450 (1894): 610–619.

Palamas, Kostis. "Αντιρρωμαντικό στάσιμο του ποιητικού χορού" [The anti-romantic song of the poetic chorus] (1934). In *Άπαντα*, edited by the Kostis Palamas Institute, 8:503–518. Athens: Govostis, 1960.

———. *Άπαντα* [Collected works]. Edited by the Kostis Palamas Institute. 2nd ed. 16 vols. Athens: Govostis, 1960. Vol. 17, "Τα ευρετήρια" [Index volume], edited by G. P. Savvides, G. Kehagioglou, and the Kostis Palamas Institute. Athens: Biris, [1962] 1984.

———. "Από την αφορμή ενός λόγου για το εθνικό έπος των νεωτέρων Ελλήνων" [With the opportunity of a word on the national epic of the newer Greeks] (1907). In *Άπαντα*, edited by the Kostis Palamas Institute, 6:487–511. Athens: Govostis, 1960.

———. "Βιβλία για τα παιδιά" [Books for the children] (1907). In *Άπαντα*, edited by the Kostis Palamas Institute 6: 542–559. Athens: Govostis, 1960.

———. "Δυο παλιοτράγουδα" [Two old songs] (ca. 1913). In *Άπαντα*, edited by the Kostis Palamas Institute, 4:449–452. Athens: Govostis, 1960.

———. "Ο Δωδεκάλογος του Γύφτου" [The twelve words of the gypsy] (1907). In *Άπαντα*, edited by the Kostis Palamas Institute, 3:285–479. Athens: Govostis, 1960.

———. "Η εθνική γλώσσα" [The national language] (1905). In *Άπαντα*, edited by the Kostis Palamas Institute, 6:245–261. Athens: Govostis, 1960.

———. "Εισηγητική έκθεσις δια την απονομήν του επάθλου 'Βικέλα'" [Introductory essay on the bestowal of the Vikelas award] (1928). In *Άπαντα*, edited by the Kostis Palamas Institute, 16:447–451. Athens: Govostis, 1960.

———. "Η Λάμια" [The lamia] (ca. 1913). In *Άπαντα*, edited by the Kostis Palamas Institute, 5:394–395. Athens: Govostis, 1960.

———. "Ο λυρισμός του εμείς" [The lyricism of us] (1933). In *Άπαντα*, edited by the Kostis Palamas Institute, 10:498–507. Athens: Govostis, 1960.

———. "Ένας μεγάλος φίλος των Ελλήνων" [A great friend of the Greeks] (1917). In *Άπαντα*, edited by the Kostis Palamas Institute, 8:440–445. Athens: Govostis, 1960.

———. "Μετά την παράστασιν του Άμλετ" [After the show on Hamlet] (1899). In *Άπαντα*, edited by the Kostis Palamas Institute, 16:143–148. Athens: Govostis, 1960.

———. "Το νέο βιβλίο της Κυρίας Δέλτα" [The new book by Mrs. Delta] (1911). In *Άπαντα*, edited by the Kostis Palamas Institute, 8:81–87. Athens: Govostis, 1960.

———. "Ένας πατριώτης" [A patriot] (1899). In *Άπαντα*, edited by the Kostis Palamas Institute, 16:159–163. Athens: Govostis, 1960.

———. "Η ποίηση στον περασμένο τον αιώνα" [Poetry in the last century] (1901). In *Άπαντα*, edited by the Kostis Palamas Institute, 6:191–204. Athens: Govostis, 1960.

———. "Ο ποιητής René Arcos" [The poet René Arcos] (1909). In *Άπαντα*, edited by the Kostis Palamas Institute, 8:88–93. Athens: Govostis, 1960.

———. "Πώς τραγουδούμε το θάνατο της κόρης" [How we sing the death of the daughter] (1917). In *Άπαντα*, edited by the Kostis Palamas Institute, 8:361–391. Athens: Govostis, 1960.

———. "Σημείωμα για τον Παύλο Νιρβάνα" [Note for Pavlos Nirvanas] (1907). In *Άπαντα*, edited by the Kostis Palamas Institute, 16:536–537. Athens: Govostis, 1960.

———. "Σημειώματα στο περιθώριο" [Notes in the margin] (1908). In *Άπαντα*, edited by the Kostis Palamas Institute, 10:63–134. Athens: Govostis, 1960.

———. *The Twelve Words of the Gypsy*. Translated by Theodore Ph. Stephanidis and George C. Katsimbalis. Tennessee: Memphis State University Press, 1975.

Pancaldi, Giuliano. *Darwin in Italy: Science across Cultural Frontiers*. Bloomington: Indiana University Press, 1991.

Papa, Katina. "Η αλληλογραφία μου με τον Γρηγόριο Ξενόπουλο: 15 Δεκ. 1921" [My correspondence with Grigorios Xenopoulos: 15 Dec. 1921]. *Νέα Εστία*, 1952, 163–169.

Papadiamantis, Alexandros. *Η φόνισσα* [The murderess]. Athens: Ellinika Grammata, [1903] 2006.

Papandreou, George. "Ο σκεπτικισμός και η φυλή μου. Απάντηση του κ. Γ. Παπανδρέου" [Skepticism and our race. Response by Mr. G. Papandreou]. *Γράμματα* 3, no. 25 (1915): 93–95.

Papanoutsos, Evangelos. *Α. Δελμούζος: η ζωή του, επιλογή από το έργο του* [A. Delmouzos: his life and selected works]. Athens: Morfotiko Idryma Ethnikis Trapezis, 1978.

Parren, Kalliroi, ed. "Αι αδικούσαι και αι αδικούμεναι" [The injustices and the female underprivileged]. *Εφημερίς των Κυριών*, no. 552 (November 22, 1898): 1–3.

———, ed. "Η Βασίλισσα περιοδεύει" [The Queen tours]. *Εφημερίς των Κυριών*, no. 9 (May 3, 1887): 3–4.

———. *Τα βιβλία της αυγής: η μάγισσα* [The books of dawn: the witch]. Athens: Paraskevas Leonis, 1901.

———, ed. "Το κάλλος και η εργασία" [Beauty and work]. *Εφημερίς των Κυριών*, no. 595 (November 14, 1899): 1–2.

———, ed. "Μητέρες παιδοκτόνοι" [Maternal infanticide]. *Εφημερίς των Κυριών*, no. 12 (May 24, 1887): 4.

———. *Η χειραφετημένη* [The emancipated woman]. Athens: Ekati, 1999.

Paul, Harry W. "Religion and Darwinism: Varieties of Catholic Reaction." In *The Comparative Reception of Darwinism*, edited by Thomas F. Glick, 403–436. Austin: University of Texas, 1974.

Peacock, Thomas Love. *The Works of Thomas Love Peacock*. London: Constable, [1817] 1824.

Pearson, Karl. *The Life, Letters, and Labours of Francis Galton*.Vol 3a. Cambridge: Cambridge University Press, 1930.

Peckham, Morse. "Darwinism and Darwinisticism." *Victorian Studies* 3, no. 1 (1959): 19–40.

Petit, C., W. Savage, and E. Renoir. *Dictionnaire français-anglais*. Paris: Librairie Hachette, 1946.

Pick, Daniel. *Faces of Degeneration: A European Disorder, ca. 1848–ca. 1918*. Cambridge, UK: University Press, 1989.

Politi, Jina. "Δαρβινικό κείμενο και *Η φόνισσα* του Παπαδιαμάντη" [Darwinian text and Papadiamantis' *The murderess*]. In *Συνομιλώντας με τα κείμενα*, 155–181. Agra, Athens, 1996.
———. "Η μυθιστορηματική κατεργασία της ιδεολογίας: ανάλυση της *Λυγερής* του Ανδρέα Καρκαβίτσα." [A novelistic treatment of ideology: analysis of Andreas Karkavitsas' *The fair maid*]. In *Συνομιλώντας με τα κείμενα*, 63–127. Agra, Athens, 1996.
Politis, Linos. *A History of Modern Greek Literature*. Oxford: Clarendon Press, [1973] 1975.
Politis, Nikolaos G. *Νεοελληνική μυθολογία* [Modern Greek mythology]. 2 vols. Athens: Karl Wilberg and N. A. Nakis, 1871-1874.
———. *Παραδόσεις* [Traditions]. Vol. 1. Athens: Akadimia Athinon, 1904.
Pouliopoulos, Nikos D. *Ο Νίκος Καζαντζάκης και τα παγκόσμια ιδεολογικά ρεύματα* [Nikos Kazantzakis and the global ideological currents]. Vol. 1. Athens, 1972.
Praz, Mario. *The Romantic Agony*. Translated by Angus Davidson. 2nd ed. London: Oxford University Press, [1930] 1970.
Prevelakis, Pandelis, ed. *Τετρακόσια γράμματα του Καζαντζάκη στον Πρεβελάκη* [Four hundred letters by Kazantzakis to Prevelakis]. 2nd ed. Athens: Ekdoseis Eleni Kazantzaki, 1984.
Preveziotou, Kornelia. "*Η γυναίκα (η κακή)*" [The (evil) woman]. *Βοσπορίς*, no. 31 (1903): 357–358. Cited in Angelika Psarra, ed. "Το μυθιστόρημα της χειραφέτησης ή η 'συνετή' ουτοπία της Καλλιρρόης Παρρέν." In *Η χειραφετημένη*, by Kalliroi Parren, 480. Athens: Ekati, 1999.
Prinou, Lucia, Lia Halkia, and Costas Skordoulis. "Θεωρία της εξέλιξης: η αναγκαιότητα της διδασκαλίας της και η περιπέτειά της στο ελληνικό σχολείο" [Theory of evolution: the necessity of its teaching and its path in the Greek school]. In *Πρακτικά 4ου Πανελλήνιου Συνεδρίου του Εθνικού και Καποδιστριακού Πανεπιστημίου, "Για τη διδακτική των φυσικών επιστημών και τις νέες τεχνολογίες στην εκπαίδευση"*, edited by Vassilis Tselfes, Petros Kariotoglou, and M. Patsadakis, 4:260–266. Athens: University of Athens, 2004.
Promitzer, Christian, Sevasti Trubeta, and Marius Turda, eds. *Health, Hygiene and Eugenics in South-Eastern Europe to 1945*. Budapest: Central European University Press, 2011.
Psarra, Angelika, ed. "Το μυθιστόρημα της χειραφέτησης ή η 'συνετή' ουτοπία της Καλλιρρόης Παρρέν" [The novel of emancipation or the prudent utopia of Kalliroi Parren]. Afterword to *Η χειραφετημένη*, by Kalliroi Parren, 407–486. Athens: Ekati, 1999.
Pykett, Lynn. *Engendering Fictions: The English Novel in the Early Twentieth Century*. London: Edward Arnold, 1995.
———. *The "Improper" Feminine: The Women's Sensation Novel and the New Woman Writing*. London: Routledge, 1992.
———, ed. *Reading Fin de Siècle Fictions*. London: Longman, 1996.
Ragkavis, Kleon. *Ιουλιάνος ο Παραβάτης: ποίημα δραματικόν εις μέρη πέντε* [Julian the apostate: dramatic poem in five parts]. Athens: Ek tou Typografeiou Petrou Perri, 1877.
Rekas, Jan, ed. *Echoes of Old Athens*. Kensington: New South Wales University Press, 1988.
Richards, Evelleen. "Darwin and the Descent of Woman." In *The Wider Domain of Evolutionary Thought*, edited by David Oldroyd and Ian Langham, 57–111. Dordrecht, Holland: D. Reidel, 1983.
Richards, Robert J. *Darwin and the Emergence of Evolutionary Theories of Mind and Behavior*. Chicago: The University of Chicago Press, 1987.
Richardson, Angelique. *Love and Eugenics in the Late Nineteenth Century: Rational Reproduction and the New Woman*. Oxford: Oxford University Press, 2008.
Rivers, Christopher. *Face Value: Physiognomical Thought and the Legible Body in Marivaux, Lavater, Balzac, Gautier, and Zola*. Madison, WI: University of Wisconsin Press, 1994.
Roberts, Mary Louise. *Civilization without Sexes: Reconstructing Gender in Postwar France, 1917–1927*. Chicago: University of Chicago Press, 1994.
———. *Disruptive Acts: The New Woman in Fin-de-Siècle France*. Chicago: University of Chicago Press, 2002.
Roidis, Emmanuel. "Αγγέλου Βλάχου κωμωδίαι" [Angelos Vlachos' comedies] (1871). In *Άπαντα (1868–1879)*, edited by Alkis Angelou, 2:28–52. Athens: Ermis, 1978.
———. *Άπαντα* [Collected works]. Edited by Alkis Angelou. 5 vols. Athens: Ermis, 1978.

———. *Αφηγηματικά κείμενα* [Narrative texts]. Edited by Apostolos Sahinis and Dimitris Dimiroulis. Athens: Institute Kosta and Eleni Ourani, 1995.
———. "Η εορτή του Όνου κατά τον μεσαίωνα" [The anniversary of Onos in the middle ages]. In *Άπαντα*, edited by Alkis Angelou. Vol. 2, 1868–1879, 9–18. Athens: Ermis, 1978.
———. "Ιστορία ενός πιθήκου" [Story of an ape]. In *Άπαντα*, edited by Alkis Angelou. Vol. 5, 1894–1904, 347–353. Athens: Ermis, 1978.
———. "Ιστορία ορνιθώνος" [Story of a hen-coop]. In *Άπαντα*, edited by Alkis Angelou. Vol. 5, 1894–1904, 218–286. Athens: Ermis, 1978.
———. "Σκαλάθυρμα" [Short treatise]. In *Άπαντα*, edited by Alkis Angelou. Vol. 3, 1880–1890, 188–190. Athens: Ermis, 1978.
Roppen, Georg. *Evolution and Poetic Belief: A Study in Some Victorian and Modern Writers.* Oslo: Oslo University Press, 1956.
Royer, Clémence, trans. "Préface de la première édition" [Preface of the first edition]. In *De l'origine des espèces par sélection naturelle ou des lois de transformation des êtres organisés*, by Charles Darwin, xv–xxxviii. 2nd ed. Paris: Guillaumin et Cie, 1866. (Page numbers based on Palamas' actual copy of the book.)
Rozetti, Dora [pseud.]. *Η ερωμένη της* [Her female lover]. Athens: Metaichmio, [1929] 2005.
Russell, Bertrand. *Our Knowledge of the External World: As a Field for Scientific Method in Philosophy.* London: George Allen and Unwin, [1914] 1961.
Russett, Cynthia Eagle. *Sexual Science: The Victorian Construction of Womanhood.* Cambridge, MA: Harvard University Press, 1989.
Ryazanskaya, S. W., ed. *Marx, Engels: Selected Correspondence.* 3rd ed. Translated by I. Lasker. Moscow: Progress Publishers, 1975.
Sahinis, Apostolos. *Το νεοελληνικό μυθιστόρημα* [The modern Greek novel]. 5th ed. Athens: Kollaros, 1980.
Sanderson, Stephen K. *Social Evolutionism: A Critical History.* Cambridge, MA: Basil Blackwell, 1990.
Sarrigelis, Mihalis. "Η θεωρία της εξέλιξης στην εκπαίδευση." [The theory of evolution in education]. TVXS-TV Χωρίς Σύνορα, January 2014. http://tvxs.gr/news/paideia/i-theoria-tis-ekseliksis-stin-ekpaideysi. Last accessed January 2014.
Siegel, Sandra. "Literature and Degeneration: The Representation of 'Decadence.'" In *Degeneration: The Dark Side of Progress*, edited by J. Edward Chamberlain and Sander L. Gilman, 199–219. New York: Columbia University Press, 1985.
Skliros, George. "Διαλεκτικός Υλισμός" [Dialectic materialism]. *Ο Νουμάς*, no. 280 (1908): 2–8.
———. "Ο σκεπτικισμός και η φυλή μας" [Skepticism and our race]. *Γράμματα* 2, no. 17 (1913): 289–291.
———. "Ο σκεπτικισμός και η φυλή μας. Απάντηση του κ. Γ. Σκληρού" [Skepticism and our race. Response by Mr. G. Skliros]. *Γράμματα* 3, no. 25 (1915): 95–104.
———. *Τα σύγχρονα προβλήματα του ελληνισμού* [The contemporary problems of Greece]. Alexandria: Grammata, 1919.
Small, H., and T. Tate, eds. *Literature, Science, Psychoanalysis, 1830–1970: Essays in Honour of Gillian Beer.* Oxford: Oxford University Press, 2003.
Smiles, Samuel. *Βοήθει σαυτόν: εκ του αγγλικού* [Self-help: from the English]. Translator not mentioned. Athens: Kollaros, 1925.
Sotiriadou, Anthi. "Η εμφάνιση της θεωρίας της εξέλιξης των ειδών, δεδομένα από τον ελληνικό χώρο" [The emergence of the theory of evolution of the species: data from the Greek domain]. PhD diss., Aristotle University of Thessaloniki, 1990.
Sougras, Spyridon. *Η νεοτάτη του υλισμού φάσις: ήτοι ο δαρουϊνισμός και το ανυπόστατον αυτού* [The most recent phase of materialism: Darwinism and its unfoundedness]. Athens: Typois Efimeridos Ton Syzitiseon, 1876.
Souvestre, Emile. *Riche et pauvre* [Rich and poor]. Paris: La Librairie Nouvelle, [1836] 1870.
Spencer, Herbert. *First Principles.* The Thinker's Library 62. 6th ed. London: Watts, 1946.
Spengler, Joseph J. "Evolutionism in American Economics, 1880–1946." In *Evolutionary Thought in America*, edited by Stow Persons, 202–267. Hamden, CO: Archon Books, [1950] 1968.

Stamatakos, Ioannis. *Λεξικόν της νέας ελληνικής γλώσσης* [Dictionary of the modern Greek language]. Vol. 3. Athens: Vivliopromitheftiki, 1971.

Starr, Jordan David. *The Blood of the Nation: A Study of the Decay of Races through the Survival of the Unfit*. Boston: American Unitarian Association, 1906.

Stavridi-Patrikiou, Rena, ed. *Δημοτικισμός και κοινωνικό πρόβλημα* [Demoticism and the social problem]. Athens: Ermis, 1976.

Stavropoulou, Eri. "Ανδρέας Καρκαβίτσας" [Andreas Karkavitsas]. In *Η παλαιότερη πεζογραφία μας: από τις αρχές της ώς τον Πρώτο Παγκόσμιο Πόλεμο*, edited by Nasos Vagenas, Giannis Dallas, and Kostas Stergiopoulos, 8:174–251. Athens: Sokolis, 1997.

Stepan, Nancy. "Biological Degeneration: Races and Proper Places." In *Degeneration: The Dark Side of Progress*, edited by J. Edward Chamberlain and Sander L. Gilman, 97–121. New York: Columbia University Press, 1985.

Stevenson, Lionel. *Darwin among the Poets*. Russell and Russell: New York, [1932] 1963.

Stevenson, Robert Louis. *The Strange Case of Dr Jekyll and Mr Hyde*. Harmondsworth, UK: Penguin, [1885] 1979.

Stewart, Charles. *Demons and the Devil: Moral Imagination in Modern Greek Culture*. Princeton, NJ: Princeton University Press, 1991.

Stocking, George W. *Race, Culture, and Evolution: Essays in the History of Anthropology*. New York: The Free Press, 1968.

Stoker, Bram. *Dracula*. London: Arrow Books, [1897] 1974.

Strindberg, August. *Miss Julie*. Drama Classics. Translated by Kenneth McLeish. London: Nick Hern Books, [1888] 1995.

Stylianopoulou, Fotini. "Δαρβίνος και συγκινήσεις" [Darwin and emotions]. *Ελευθεροτυπία*, science supplement no. 23 (March 2, 2006): 26–30.

Surprenant, Céline. "Darwin and Proust." In *The Literary and Cultural Reception of Charles Darwin in Europe*, edited by Thomas F. Glick and Elinor Shaffer, 4:431–457. London: Bloomsbury Academic, 2014.

Taine, Hippolyte A. *History of English Literature*. Translated by H. Van Lauen. 4 vols. London, [1864] 1890.

Theodorou, Vassiliki, and Despina Karakatsani. "Eugenics and Puericulture. Medical Attempts to Improve the Biological Capital in Interwar Greece." In *Health, Hygiene and Eugenics in South-Eastern Europe to 1945*, edited by Christian Promitzer, Sevasti Trubeta, and Marius Turda, 299–323. Budapest: Central European University Press, 2011.

Thrylos, Alkis [Eleni Ouranis]. "Ο πεζογράφος Ξενόπουλος" [Xenopoulos the prose writer]. *Νέα Εστία* 69 (1961): 103–109.

———. "*Η τρίμορφη γυναίκα* κινηματόδραμα σε 3 μέρη και 12 εικόνες" [The three-sided woman, a drama in 3 parts and 12 pictures]. *Δημοκρατία*, September 20, 1924, 2.

Trencsényi, Balázs. *The Politics of "National Character": A Study in Interwar East European Thought*. Abingdon, Oxon; New York: Routledge, 2012.

Trotter, David. *The English Novel in History, 1895–1920*. London: Routledge, 1993.

Trubeta, Sevasti. "Eugenic Birth Control and Prenuptial Health Certification in Interwar Greece." In *Health, Hygiene and Eugenics in South-Eastern Europe to 1945*, edited by Christian Promitzer, Sevasti Trubeta, and Marius Turda, 271–298. Budapest: Central European University Press, 2011.

———. *Physical Anthropology, Race, and Eugenics in Greece (1880s–1970s)*. Balkan Studies Library 11. Leiden: Brill, 2013.

Turda, Marius. *Modernism and Eugenics*. New York: Palgrave Macmillan, 2010.

Turda, Marius, and Aaron Gillette. *Latin Eugenics in Comparative Perspective*. London: Bloomsbury, 2014.

Typaldos, Zisimos. *Η δίκη των πιθήκων* [The trial of the apes]. Ermoupolis, Syros: Typois Patridos, 1881.

Tytler, Graeme. *Physiognomy in the European Novel: Faces and Fortunes*. Princeton: Princeton University Press, 1982.

Tziovas, Dimitris. *The Other Self: Selfhood and Society in Modern Greek Fiction*. Lanham, MD: Lexington Books, 2003.
University of Missouri–Kansas City. School of Law 2002. "Tennessee vs John Scopes: the Monkey Trial, 1925." http://www.law.umkc.edu/faculty/projects/ftrials/scopes/scopes.htm. Last accessed January 2014.
Unnamed [editorial management of *Grammata*, including Dimitrios Zachariadis]. "Ο σκεπτικισμός και η φυλή μας. Η γνώμη των *Γραμμάτων*" [Skepticism and our race. The opinion of the *Grammata* management]. *Γράμματα* 3, no. 25 (1915): 107–110.
Valetas, G. *Η γενιά του '80: ο νεοελληνικός νατουραλισμός και οι αρχές της ηθογραφίας* [The generation of '80: modern Greek naturalism and the beginnings of ethography]. Athens: Ekdoseis Grammatologikou Kentrou, 1981.
Varika, Eleni. *Η εξέγερση των κυριών: η γένεση μιας φεμινιστικής συνείδησης στην Ελλάδα 1833–1907* [The uprising of women: the birth of a feminist conscience in Greece 1833–1907]. Athens: Idryma Erevnas kai Paideias tis Emporikis Trapezas tis Ellados, 1987.
Vassis Konstantinos. "Γρηγόριος Ξενόπουλος, ένα φαινόμενο πολυγραφίας" [Grigorios Xenopoulos, a prolific phenomenon]. In *Επικύρωση του Γρηγορίου Ξενόπουλου: πενήντα χρόνια από την εκδημία του*, edited by Tasos Athanasiadis, 22–46. Athens: Tetradia Efthynis, 2001.
Vikelas, Dimitrios. *Η Γη του Πυρός (εκ της περιηγήσεως του Δαρβίνου 1832–1834. Περιήγησις των γάλλων Βίλλεμς και Ρουσσώ 1890)* [Tierra del Fuego (from the voyage of Darwin 1832–1834. Voyage of the Frenchmen Willems and Rousseau 1890)]. Athens: Syllogos pros Diadosin Ofelimon Vivlion, 1900.
Vitti, Mario. *Ιστορία της νεοελληνικής λογοτεχνίας* [History of modern Greek literature]. Athens: Odysseas, 1987. First published in Italian in 1971.
Vlahakis, George. "Science in Society in 19th Century Greece: The Journals." In *Science, Technology and the 19th Century State. Conference Proceedings Syros, 9–10 July, 1999*, edited by Efthymios Nicolaidis and Konstantinis Chatzis, 117–123. Athens: Institute for Neohellenic Research, National Hellenic Research Foundation, 2000.
Vlastos, Petros. "Πραγματισμός" [Pragmatism]. *Ο Νουμάς*, no. 288 (March 23, 1908): 11.
———. "Ο σκεπτικισμός και η φυλή μας. Απάντηση στον κ. Σκληρό" [Skepticism and our race. Response to Mr. Skliros]. *Γράμματα* 2, no. 21 (1914): 509–510.
Vlisidis, Thrasyvoulos S. "Η κατά Δαρβίνον θεωρία της εξελίξεως των ενόργανων όντων" [The Darwin theory of evolution of organic beings]. *Η Φύσις*, no. 23 (April 1904): 385–389.
Voutieridis, Ilias. "Μελέτες" [Studies]. *Αναγέννηση*, no. 6 (1927): 375–378.
———. *Σύντομη ιστορία της νεοελληνικής λογοτεχνίας (1000–1930)* [Short history of modern Greek literature (1000–1930)]. 2nd ed. Athens: Dimitrios N. Papadimas, 1966.
Weininger, Otto. *Sex and Character*. Translator from German not mentioned. 6th ed. London: William Heinman, [1903] 1912.
Weismann, August. *Essays upon Heredity and Kindred Biological Problems*. Edited and translated by E. B. Poulton, S. Schönland, and A. E. Shipley. Vol 1. Oxford: Clarendon Press, 1889.
Wellek, René. *Concepts of Criticism*. New Haven: Yale University Press, 1963.
Wells, H. G. *The Island of Dr Moreau*. London: Benn, London, [1896] 1927.
Werckmeister, O. K. "Marx on Ideology and Art." *New Literary Journal* 4, no. 3 (1973): 501–519.
Wolf, Eva Nagel. "Elenore Abbott, Illustrator." *The International Studio* 67, no. 266 (March 1919): xxvii.
Woloschak, Gayle. "The Compatibility of the Principle of Biological Evolution with Eastern Orthodoxy." *St. Vladimir's Theological Quarterly* 2, no. 55 (2011): 209–231.
Woolf, Virginia. *A Room of One's Own*. London: Hogarth Press, 1929.

Publications by Grigorios Xenopoulos
From the Collected works:

Xenopoulos, Grigorios. *Άπαντα* [Collected works]. Edited by Stavros Biris. 2nd ed. 12 vols. Athens: Biris, 1972.
———. "Το βραχιόλι" [The bracelet]. 10:326–332.

———. "Οι Βρυκόλακες" [Ghosts]. 11:359–361.
———. "Η ζωή μου σαν μυθιστόρημα: αυτοβιογραφία" [My life as a novel: autobiography]. 1:57–363.
———. "Ο κατήφορος" [The downfall]. 5:9–354.
———. "Πλούσιοι και φτωχοί" [Rich and poor]. 2:9–318.
———. "Στέλλα Βιολάντη: έρως εσταυρωμένος" [Stella Violanti: love crucified]. 1:501–545.
———. "Στέφανος Γρανίτσας" [Stefanos Granitsas]. 11:150–151.
———. "Τερέζα Βάρμα-Δακόστα: ένας σύγχρονος Μεσαίωνας" [Tereza Varma-Dacosta: the Middle Ages today]. 9:12–147.
———. "Η τιμή του αδελφού" [The brother's honor]. 6:5–310.
———. "Τίμιοι και άτιμοι" [Honorable and dishonorable]. 3:11–283.
———. "Η τρίμορφη γυναίκα" [The three-sided woman]. 8:213–436.
———. "Τυχεροί και άτυχοι" [Fortunate and unfortunate]. 4:9–238.

"Athenian Letters" [Αθηναϊκές επιστολές] in the *Children's Guidance* magazine [*Η Διάπλασις των Παίδων*]:
Xenopoulos, Grigorios. "Αγόρια και κορίτσια" [Boys and girls]. *Η Διάπλασις των Παίδων*, no. 8, January 23, 1916, 65.
———. "Οι αθεϊσταί" [The atheists]. *Η Διάπλασις των Παίδων*, no. 21, April 26, 1914, 167.
———. "Αισιοδοξία" [Optimism]. *Η Διάπλασις των Παίδων*, no. 48, November 2, 1929, 572.
———. "Εις απάντησιν" [A response]. *Η Διάπλασις των Παίδων*, no. 32, July 11, 1909, 263
———. "Αποκάλυψις" [Revelation]. *Η Διάπλασις των Παίδων*, no. 13, February 28, 1915, 105.
———. "Η δίκη του ... Δαρβίνου" [The trial of ... Darwin]. *Η Διάπλασις των Παίδων*, no. 37, August 15, 1925, 292.
———. "Μια εξέλιξη" [An evolution]. *Η Διάπλασις των Παίδων*, no. 28, June 23, 1923, 220.
———. "Η επιστήμη" [Science]. *Η Διάπλασις των Παίδων*, no. 37, September 5, 1898, 291.
———. "Η επιστήμη κι ο Θεός" [Science and God]. *Η Διάπλασις των Παίδων*, no. 16, March 31, 1923, 124.
———. "Θεόδωρος Χελδράιχ" [Theodor Heldreich]. *Η Διάπλασις των Παίδων*, no. 36, September 7, 1902, 282–283.
———. "Θεολογία αργόσχολων" [Theology of the idle]. *Η Διάπλασις των Παίδων*, no. 33, July 18, 1942, 235.
———. "Η θεωρία του Δαρβίνου" [Darwin's theory]. *Η Διάπλασις των Παίδων*, no. 16, March 18, 1939, 139.
———. "Θρησκεία" [Religion]. *Η Διάπλασις των Παίδων*, no. 10, February 7, 1915, 79.
———. "Θρησκεία και επιστήμη" [Religion and science]. *Η Διάπλασις των Παίδων*, no. 1, December 5, 1931, 8.
———. "Και τα φυτά είναι ... ζώα" [And the plants are ... animals]. *Η Διάπλασις των Παίδων*, no. 30, June 22, 1940, 263.
———. "Η καλλιτέρα εποχή" [A better period]. *Η Διάπλασις των Παίδων*, no. 37, September 14, 1896, 291.
———. "Κάμιλλος Φλαμμαριόν" [Camille Flammarion]. *Η Διάπλασις των Παίδων*, no. 30, June 27, 1925, 236.
———. "Λίγο νερό στο κρασί" [A little water in the wine]. *Η Διάπλασις των Παίδων*, no. 28, June 12, 1921, 220.
———. "Μια νέα κοσμοθεωρία" [A new cosmic theory]. *Η Διάπλασις των Παίδων*, no. 16, March 19, 1927, 140.
———. "Σοβαρά τα πράγματα" [Things are serious]. *Η Διάπλασις των Παίδων*, no. 15, March 24, 1923, 116.
———. "Ο Σοσιαλιστής" [The socialist]. *Η Διάπλασις των Παίδων*, no. 22, April 30, 1916, 175.
———. "Τα σοφίσματα της αθεΐας" [The sophisms of atheism]. *Η Διάπλασις των Παίδων*, no. 11, December 11, 1917, 87.
———. "Φεμινισμός" [Feminism]. *Η Διάπλασις των Παίδων*, no. 18, April 3, 1921, 140.

Xenopoulos, Grigorios, ed. *Η Διάπλασις των Παίδων*. Complete digitized collection of the magazine, 1879–1948. Digitized in 2001 by the Hellenic Literary and Historical Archive (E. L. I. A.), Athens.

First publications in book form by Vlassis Brothers, Athens:
Xenopoulos, Grigorios. *Απάνεμα βράδια* [Leeward nights]. Athens: Vlassis Brothers, 1984.
———. *Απ' την κουζίνα στο χαρέμι* [From the kitchen to the harem]. Athens: Vlassis Brothers, 1984.
———. *Ο αρραβωνιαστικός μου* [My fiancé]. Athens: Vlassis Brothers, 1984.
———. *Ο γιος μου κι η κόρη μου* [My son and my daughter]. Athens: Vlassis Brothers, 1984.
———. *Ο Κοσμάκης Α': το πρωτοξύπνημα* [Kosmakis 1: the first awakening]. Athens: Vlassis Brothers, 1984.
———. *Ο Κοσμάκης Β': το κέντρον* [Kosmakis 2: the center]. Athens: Vlassis Brothers, 1984.
———. *Ο Κοσμάκης Γ': τελευταία όνειρα* [Kosmakis 3: last dreams]. Athens: Vlassis Brothers, 1984.
———. *Ο Κοσμάκης Δ': ο γυρισμός* [Kosmakis 4: the return]. Athens: Vlassis Brothers, 1984.
———. *Η νύχτα του εκφυλισμού* [The night of degeneration]. Athens: Vlassis Brothers, 1994.
———. *The Three-Sided Woman: A Novel*. Translated by Barbara Kent. Athens: Vlassis Brothers, 1992.

Other publications, including articles and essays:
Xenopoulos, Grigorios. "Ο Αύγουστος Στρίντμπεργ και 'Η Δεσποινίς Τζούλια'" [August Strindberg and Miss Julie]. In *Γρηγόριος Ξενόπουλος: επιλογή κριτικών κειμένων* [Grigorios Xenopoulos: selected critical texts], edited by G. Farinou-Malamatari, 166–173. Athens: Vlassis Brothers, 2002.
———. "Γράμματα, τέχνη, επιστήμη: φιλολογία" [Letters, art, science: literature]. *Παναθήναια* 2 (March 15, 1902): 364.
———. "Η διασκεδαστική τέχνη" [The art of entertaining]. *Έρευνα*, 1939, 3–20.
———. "Το ένστικτον του θανάτου" [The instinct of death]. *Παναθήναια*, October 15, 1904, 8–11; and October 31, 1904, 41–46.
———. *Το μυστικό της Κοντέσσας Βαλέραινας* [The secret of Contessa Valerena]. Athens: Kollaros, 1928.
———. "Ντόρα Ρωζέττη: *Η ερωμένη της*, ρομάντσο" [Dora Rozetti: her female lover, a romance novel]. *Νέα Εστία*, no. 71 (December 1, 1929): 1035–1037.
———. "Σύντομη αυτοβιογραφία" [Short autobiography]. Special edition, *Νέα Εστία* 50, no. 587 (December 1951): 5–7.
———. "Η *Τρίμορφη γυναίκα*, ο φεμινισμός και ο ανθρωπισμός. Απάντησις του κ. Ξενόπουλου στη κριτική του Άλκη Θρύλου" [The three-sided woman, feminism, and humanism. Mr. Xenopoulos' response to Alkis Thrylos' critique]. *Δημοκρατία*, September 28, 1924, 3.

Zachariadis, Dimitrios. "Σκεπτικισμός, αισθητισμός και οι θεολόγοι της κοινωνιολογίας. Απάντηση στον κ. Σκληρό" [Skepticism, aestheticism, and the theologians of sociology. Response to Mr. Skliros]. *Γράμματα* 2, no. 21 (1914): 510–520.
———. "Ο σκεπτικισμός και η φυλή μας. Απάντηση του κ. Δ. Ζαχαριάδη" [Skepticism and our race. Response by Mr. D. Zachariadis]. *Γράμματα* 3, no. 25 (1915): 104–107.
Zarimis, Maria. "Darwinian Reception and Selected Responses of Two Prominent Greek Literary Writers." In *The Literary and Cultural Reception of Charles Darwin in Europe*, edited by Thomas F. Glick and Elinor Shaffer, 4:649–666. London: Bloomsbury Academic, 2014.
———. "Darwinian Thought in Grigorios Xenopoulos' 'Athenian Letters.'" In *Greek Research in Australia: Proceedings of the Sixth Biennial International Conference of Greek Studies 2005*, edited by Elizabeth Close, Michael Tsianikas, and George Couvalis, 245–260. Adelaide: Flinders University, 2007.
———. "Darwinism and Its Impact in the Recent Greek Press." Special edition, *Modern Greek Studies (Australia and New Zealand): A Journal for Greek Letters* 16 (2012): 155–171.
———. "Darwinism in Modern Greek Literature and a Rereading of Grigorios Xenopoulos' *Πλούσιοι και φτωχοί*" [Darwinism in Modern Greek Literature and a Rereading of Grigorios

Xenopoulos' Rich and poor]. In *Ο ελληνικός κόσμος ανάμεσα στην εποχή του Διαφωτισμού και στον εικοστό αιώνα: Πρακτικά του Γ' Ευρωπαϊκού Συνεδρίου Νεοελληνικών Σπουδών (ΕΕΝΣ) Βουκουρέστι, 2–4 Ιουνίου 2006*, edited by Konstantinos A. Dimadis, 3:393–402. Athens: Ellinika Grammata, 2007.

———. "Dock and the Darwin Debate in Greece." In *Greek Research in Australia: Proceedings of the Seventh Biennial International Conference of Greek Studies in Australia 2007*, edited by Elizabeth Close, George Couvalis, George Frazis, Maria Palaktsoglou, and Michael Tsianikas, 369–384. Adelaide: Flinders University, 2009.

———. "The Influence of Darwinian Ideas on Greek Literary Writers of the Late Nineteenth and Early Twentieth Centuries: The Case of Emmanuel Roidis." *The Historical Review* 4 (2007): 171–190.

Zola, Emile. "The Experimental Novel." In *Documents of Modern Literary Realism*, edited by George J. Becker, 162–196. Princeton, NJ: Princeton University Press, 1963.

———. *Thérèse Raquin*. Translated by Robin Buss. London: Penguin Books, [1867] 2004.

INDEX

For Product Safety Concerns and Information please contact our EU representative GPSR@taylorandfrancis.com Taylor & Francis Verlag GmbH, Kaufingerstraße 24, 80331 München, Germany

Batch number: 10703952

Printed by Printforce, the Netherlands